Finite Markov Processes and Their Applications

Marius Iosifescu

The Center for Mathematical Statistics

Romanian Academy of Science

Dover Publications, Inc.

Mineola, New York

Copyright

Bibliographical Note

This Dover edition, first published in 2007, is a corrected republication of the work originally published in the Wiley Series in Probability and Mathematical Statistics by John Wiley & Sons, New York, in 1980. For the present edition, several typographical errors have been corrected within the text, and an Errata list now appears on pages 9–12.

Library of Congress Cataloging-in-Publication Data

Iosifescu, Marius.
Finite Markov processes and their applications / Marius Iosifescu. — Dover ed.
p. cm.
Originally published: Chichester ; New York : Wiley, c1980, in series: Wiley series in probability and mathematical statistics.
Includes bibliographical references and index.
ISBN 0-486-45869-5
1. Markov processes. I. Title.

QA274.7.I6713 2007
519.2'3—dc22
2006102957

Manufactured in the United States of America
Dover Publications, Inc., 31 East 2nd Street, Mineola, N.Y. 11501

This book is dedicated to the memory of my father
Victor Iosifescu (1905—1973)

Preface

This book is devoted to the study of finite Markov chains and processes. (In accordance with the usage in the Romanian school of probability since the 1940's we stick to using "chain" when the time parameter is discrete and "process" when it is continuous. We think that a good terminology should stress the nature of the set of time parameters.) Finite Markov chains constitute the simplest but at the same time the most fundamental example of a sequence of dependent random variables. Nowadays, finite Markov chains and processes are still investigated by mathematicians and finite Markovian models are extensively used in applications.

The present work is a revised and expanded edition of a book published in Romanian in 1977 which treated only the discrete parameter case. It should be said that scarcely a page has escaped some emendation, deletion or addition.

There are 8 chapters. To make the book self-contained the first chapter is a review of those notions and results from probability theory and linear algebra that are basic for the theory of Markov chains and processes. The experienced reader may start with the second chapter, which treats the fundamental concepts of homogeneous finite Markov chain theory, and at the same time illustrates with some examples the applicability of Markovian models.

Chapters 3 and 4 are devoted to the study of the two basic types of homogeneous finite Markov chains: absorbing and ergodic chains. The results obtained are gradually applied to the examples considered in Chapter 2.

Chapter 5 contains a complete study of the general properties of homogeneous finite Markov chains regardless of their type. Various extensions of the concept of a homogeneous finite Markov chain are also indicated.

Chapter 6 aims at emphasizing the fundamental part played by homogeneous finite Markov chains in mathematical modelling in psychology and genetics.

Chapter 7 treats the basic facts of nonhomogeneous finite Markov chain theory, a topic of current interest and in rapid expansion.

Chapter 8 (which did not exist in the Romanian edition) takes up the study of Markovian dependence in continuous time. It should also be considered as an elementary introduction to the study of continuous parameter stochastic processes. By taking the constructive approach we have avoided the use of the concept of separability and the intricacies involved therein and have added (we hope) a substantial number of potential readers.

We wish to draw the reader's attention to the bibliography. It greatly exceeds the number of works quoted in the course of the book and should be consulted with the purpose of discovering historical sources, parallel research and starting points for new investigations.

Acknowledgements

Practically all of the book describes the work of other mathematicians, some of which is nowadays part of the folklore. Sometimes they get explicit credit for their contributions and ideas (this has been especially pursued as to Romanian mathematicians the more so as a certain tendency of neglect and oblivion is apparent).

The author wishes to acknowledge the technical help he has received from Mrs. Adriana Grădinaru, Dr. Şerban Grigorescu, Dr. Ştefan Niculescu, Mr. Mihai-Florin Popescu, and Dr. Aurel Spătaru of the Centre of Mathematical Statistics.

Special thanks are due to an anonymous reviewer, who tried "to make the book read as if it was written by an Englishman, albeit with some archaic turns of phrase"

Finally, it is a pleasure to thank John Wiley & Sons for co-editing this book.

MARIUS IOSIFESCU

Bucharest, October 1979

Contents

Errata

Page 22	Line –3	For: is no Read: is no nontrivial
Page 45	Line 12	For: *null* Read: *zero*
Page 47	Line 15	For: null Read: zero
Page 47	Line –9	For: null Read: zero
Page 48	Line –15	For: nonnull Read: nonzero
Page 48	Line –14	For: nonnull Read: nonzero
Page 59	Line 9	For: $p_2 = (\mathbf{p}_2\,(i))$ Read: $\mathbf{p}_2 = (\mathbf{p}_2\,(i))$
Page 64	Line 9	For: $B = \{(X(n),\ X(n+1), -.. \in$ Read: $B = \{(X(n),\ X(n+1), \ldots)$
Page 65	Lines 16 & 17	For: $\sum\limits_{i,\ldots,i_{n-1}\in S}$ Read: $\sum\limits_{i_1,\ldots,i_{n-1}\in S}$
Page 87	Line –11	Add: We assume all the states dealt with are return states. It is easy to see that the results obtained hold for nonreturn states, too.
Page 87	Line –2	For: $\{X(\tau_i) = i\}$ is the sure event Ω, Read: the event $\{\tau_i = k\}$ implies the event $\{X(\tau_i) = i\}$,

Page 90 Line 4 For: (since $\{X(\tau_j)=j\}$ is the sure event Ω)
Read: (since the event $\{\tau_j = m\}$ implies the event $\{X(\tau_j) = j\}$)

Page 103 Line 2 For: gte Read: get

Page 103 Line –6 This line should read:

$$= \delta(i,j)\left(\mathsf{P}_i\left(X\left(1\right) \notin T\right) + \sum_{k\in T} p\left(i,k\right)\mathsf{P}_k\left(\nu\left(j\right) = 0\right)\right) +$$

$$\sum_{n\geq\delta(i,j)+1}\ \sum_{k\in T} n^s \mathsf{P}_i\left(\nu\left(j\right) = n \,|\, X\left(1\right) = k\right)\mathsf{P}_i\left(X\left(1\right) = k\right)$$

[since $\mathsf{P}_i\left(\nu\left(i\right) = 1\right) = \mathsf{P}_i\left(X\left(1\right) \notin T\right) + \sum_{k\in T} p\left(i,k\right)\mathsf{P}_k\left(\nu\left(i\right) = 0\right)$]

Page 105 Lines – 8, –7, –6 These lines should read:

Clearly,

$$\nu' = 1 + \sum_{X(0)\neq j\in T} u(j)$$

and since whatever $i \in T$

Page 119 Line –11 For: $\sum_{r=0}^{s}$ Read: $(-1)^s \sum_{r=0}^{s}$

Page 122 Line –14 For: regardless Read: regardless of

Page 123 Line 11 For: nonnull Read: nonzero

Page 123 Line 18 For: *nonnull* Read: *nonzero*

Page 123 Line –8 For: nonnull Read: nonzero

Page 124 Line –7 For: $> \sum_{k\in S} \delta_k$ Read: $\geq \sum_{k\in S} \delta_k$

Page 124	Line –1	For: [248].	Read: [52], pp. 121–122.
Page 126	Line 17	For: *nonnull*	Read: *nonzero*
Page 126	Line –15	For: nonnull	Read: nonzero
Page 126	Line –10	For: nonnull	Read: nonzero
Page 145	Line –6	For: nonnull	Read: nonzero
Page 145	Line –4	For: null	Read: zero
Page 151	Line 7	For: $(-1)^k\,(k!a_s$	Read: $(-1)^s\,(s!a_s$
Page 152	Line –1	For: null	Read: zero
Page 159	Line 19	For: null	Read: zero
Page 164	Line 8	For: $I_k\,(\sqrt{uv})$	Read: $I_{\lvert k\rvert}\,(\sqrt{uv})$
Page 168	Line 3	For: nonnull	Read: nonzero
Page 168	Line 5	For: nonnull	Read: nonzero
Page 169	Line –14	For: $\mathbf{T}_2$	Read: $\widehat{\mathbf{T}}^2$
Page 169	Line –13	For: $\mathbf{T}^n$	Read: $\widehat{\mathbf{T}}^n$
Page 202	Line 3	For: $a_{l\cdot}(j)_r$	Read: $a_{lr}(j)_r$
Page 202	Line –7	For: $\boldsymbol{\mu}'_i(n+1) = \mathbf{C}\,\boldsymbol{\mu}'_i(n)$, Read: $\boldsymbol{\mu}_i(n+1) = \mathbf{C}\,\boldsymbol{\mu}_i(n)$,	
Page 219	Line 12	For: that is the chain is weakly ergodic. Read: and, e.g., $\Pi(m,n) = \Pi(m)$ for $1 \le n < m_1$, that is, the chain is weakly ergodic.	

Page 219	Line –15	For: diverges if and only if Read: diverges if and only if whatever $t \geq 1$
Page 219	Line –14	For: $\prod_{s\geq 1}$ Read: $\prod_{s\geq t}$
Page 219	Line –9	Delete: (7.4) holds, thus implying that
Page 220	Line 14	For: cannot be obtained due to the fact that μ is not proper. Read: can be also obtained using the fact that $\mu(\mathbf{\Pi}) \geq 1/r$ for any stable $r \times r$ matrix $\mathbf{\Pi}$ and reasoning as in the proof of Theorem 7.3.
Page 226	Line –10	For: A more general result is Read: A related result is
Page 226	Line –1	For: $m \geq 0$ Read: $m \geq 0$ and do not depend on m and $i \in S$.
Page 233	Line –7	For: $n \geq 1$ Read: $m \geq 0$, $n \geq 1$
Page 243	Line –5	For: $0 \leq l \leq n$ Read: $0 \leq l \leq n+1$
Page 255	Line 7	For: $0 < h\,\tau_{h,j} - \tau_j \leq h$. Read: $0 \leq h\tau_{h,j} - \tau_j \leq h$ if h is sufficiently small.
Page 267	Line 2	For: nonnull Read: nonzero

Introduction

Let us consider a system (that might be a physical, a biological, or a social one), which, at discrete time moments $n = 0, 1, 2, \ldots$, is in one of the "states" $1, \ldots, r$, and denote by $X(n)$ the state of the system at time n. Suppose that the evolution of the system is probabilistic (equivalently: stochastic or random). This means that instead of knowing the successive states $X(0), X(1), \ldots$ we only know the probabilities

$$\mathsf{P}(X(0) = i_0, \ldots, X(n) = i_n)$$

of all the finite sequences of states $i_0, \ldots, i_n$. The set of these probabilities is known as the *law of evolution* of the system. Clearly, some consistency conditions are to be satisfied. Indeed, the random events $\{X(0) = i_0, \ldots, X(n) = i_n, X(n+1) = i_{n+1}\}$, $1 \leqslant i_{n+1} \leqslant r$, are pairwise disjoint and their union is the random event $\{X(0) = i_0, \ldots, X(n) = i_n\}$*). Therefore

$$\mathsf{P}(X(0) = i_0, \ldots, X(n) = i_n)$$

$$= \sum_{i_{n+1}=1}^{r} \mathsf{P}(X(0) = i_0, \ldots, X(n) = i_n, X(n+1) = i_{n+1}),$$

whatever states $i_0, \ldots, i_n$ and the nonnegative integer n. Notice that from the law of evolution we can derive the so called *conditional* law of evolution of the system. This consists of the conditional probabilities

$$\mathsf{P}(X(n+1) = i_{n+1} | X(n) = i_n, \ldots, X(0) = i_0)$$

of the various states of the system at time $n+1$ given the evolution of the system until time n. Indeed, the above conditional probability equals

$$\frac{\mathsf{P}(X(0) = i_0, \ldots, X(n) = i_n, X(n+1) = i_{n+1})}{\mathsf{P}(X(0) = i_0, \ldots, X(n) = i_n)}$$

assuming the denominator to be nonnull. Conversely, knowing the conditional temporary law and the initial probabilities $\mathsf{P}(X(0) = i_0)$,

*) In other words, the sequence of states $i_0, \ldots, i_n$ may be followed by one of the states $1, \ldots, r$, these possibilities being mutually exclusive.

$1 \leqslant i_0 \leqslant r$, allows us to derive the law of evolution of the system. We have indeed

$$\begin{aligned} &\mathsf{P}(X(0) = i_0, \ldots, X(n) = i_n) \\ &= \mathsf{P}(X(n) = i_n | X(n-1) = i_{n-1}, \ldots, X(0) = i_0) \times \\ &\times \mathsf{P}(X(n-1) = i_{n-1} | X(n-2) = i_{n-2}, \ldots, X(0) = i_0) \times \ldots \\ &\ldots \times \mathsf{P}(X(1) = i_1 | X(0) = i_0)\, \mathsf{P}(X(0) = i_0). \end{aligned}$$

The simplest case is obviously that in which the conditional probability $\mathsf{P}(X(n+1) = i_{n+1} | X(n) = i_n, \ldots, X(0) = i_0)$ does not depend on $i_0, \ldots, i_n$, thus being equal to $\mathsf{P}(X(n+1) = i_{n+1})$, i.e., the one in which knowledge of the past does not give any information about the future. In this case the above equality yields

$$\mathsf{P}(X(0) = i_0, \ldots, X(n) = i_n) = \mathsf{P}(X(0) = i_0) \ldots \mathsf{P}(X(n) = i_n),$$

an equation which exhibits the (stochastic) independence of the successive states of the system. The case of independence has been thoroughly investigated in classical probability theory.

The subject matter of this book is the study of the next most general case. Called *Markovian*, this case is characterized by the fact that the conditional probability $\mathsf{P}(X(n+1) = i_{n+1} | X(n) = i_n, \ldots, X(0) = i_0)$ equals $\mathsf{P}(X(n+1) = i_{n+1} | X(n) = i_n)$. In other words, the impact of the past on the future evolution of the system is concentrated in the state present at the last moment at which the system was observed. This is also expressed by saying that the system is *memoryless*: once in a certain state at a given time, the way in which the system reached that state does not at all affect its future evolution. *)

Of course, we may also imagine a continuous time evolution in which changes of state may occur at any time and not only at multiples of a given time unit.

In this setting the state of the system at time $t \geqslant 0$ is denoted by $X(t)$ and the Markovian case defined by requiring that

$$\begin{aligned} &\mathsf{P}(X(t_{n+1}) = i_{n+1} | X(t_n) = i_n, \ldots, X(t_0) = i_0) = \\ &= \mathsf{P}(X(t_{n+1}) = i_{n+1} | X(t_n) = i_n) \end{aligned}$$

whatever the increasing sequence $0 = t_0 < t_1 < \ldots < t_n < t_{n+1}$ of time points and the natural integer n.

Very few mathematical concepts enjoy potentialities comparable with those of the concept of Markovian dependence. The contents of the present book support, we hope, this assertion. Moreover, the reader should be warned that the scope of the book has not allowed us to approach the concept in its full generality.

*) Notice that this is not the *complete* absence of memory that characterizes the independent case. A Markovian system retains its *most recent* recollection of the past.

The theoretical potentialities of the concept of Markovian dependence can be compared only with its wealth of applications. Besides the applied fields exemplified in the text there are many others. Among these we mention the following: statistical quality control [221 and 402], reliability ([12], Ch. 5, and [27], Ch. 12), biology and medicine [166], demography [96 and 97], geography [62], geophysics ([324], §78), social mobility [14, 186, and 312], education [116, 185, 254 and 369], economics [195], § 7.7 and [366]), pollution [291], marketing [88, 112, 145 and 237], finances [66 and 317], mobility of labour [14 and 36], circulation of banknotes [16], metric theory of numbers [245, 246 and 322].

Last but not least, the theory of Markovian dependence and of its generalizations is a domain in which Romanian mathematicians have made contributions of lasting value. Details are given throughout the book and in the concluding historical notes.

CHAPTER 1

Elements of Probability Theory and Linear Algebra

This is an introductory chapter intended to review notions and results from probability theory and linear algebra which are essential for setting up the theory of finite Markov chains and processes. In the author's intention it should offer a sound basis for the understanding of the other chapters. The reader wanting to deepen the topics treated here can profitably consult the books [8,18, 53, and 359].

1.1 RANDOM EVENTS

1.1.1. Probability theory originated in physical observations associated with simple gambling games. It was found, for example, that if a "fair" coin was tossed a large number of times, the relative frequency of heads, i.e., the ratio of the number of heads to the total number of tosses was very near to 1/2. Nowadays, after three centuries of evolution, probability theory is able to explain such phenomena on the basis of a logically constructed body of concepts and results.

1.1.2. The first basic concept of probability theory is that of sample space. Let us consider a random experiment, i.e., any physical operation which can result in one of many possible outcomes and the particular outcome of which cannot be predicted in advance with certainty. As simple examples of random experiments we can indicate the toss of a coin, the roll of a die, the selection of a card from a standard deck of 52, etc. The *sample space* of a random experiment is the set of all possible outcomes for the experiment. The elements of the sample space are named *sample points*.

The traditional notation for a sample space is the capital Greek letter Ω. For example, if a coin is tossed once we may take $\Omega = \{h, t\}$, where h means the occurrence of a head and t the occurrence of a tail. If a coin is tossed twice, this is a different random experiment and the corresponding sample space may be $\Omega = \{hh, ht, th, tt\}$. Similarly, if a die is rolled once we may take $\Omega = \{1, 2, 3, 4, 5, 6\}$, and if it is rolled twice then Ω might consist of the 36 pairs of numbers (i, j), $1 \leqslant i, j \leqslant 6$. It should be noticed the sample space of a random experiment need not be unique, i.e., there could be more than one reasonable way of specifying all the possible outcomes of the experiment. For example, if we roll a die once we could use as sample space the one mentioned above, but another possible sample space consists of just two points corresponding to the outcomes "even" and "odd". In fact, the sample space to be used will be dictated by the nature of the particular problem under consideration.

We should not belive that every sample space is a finite set. For example, the sample space for the experiment consisting of the toss of a coin until we get a head may be $\Omega = \{1, 2, 3, ...\}$ which is a countably infinite set. In general, any set Ω can be viewed as the sample space of a suitable random experiment: the only requirement on Ω is that a one-to-one correspondence can be established between the outcomes of the experiment and the points of Ω.

1.1.3. Let us now define the concept of a random event. A *random event* associated with a random experiment, therefore with a sample space, is the truth set of a statement about the outcome of the experiment, i e., the subset of the sample space consisting of those sample points for which the statement is true. For example, in the case of the experiment consisting of tossing twice a coin when $\Omega = \{$hh, ht, th, tt$\}$ the random event "the two tosses have the same outcome" is $\{$hh, tt$\}$. Thus, a random event is nothing but a subset of the sample space. (This assertion will be made precise later on — see 1.2.4.)

Random events will be denoted by capital letters $A, B, C, ...$

Any sample point belonging to a random event A is said to be *favourable* to A. The random event A is said to *occur* if and only if the sample point (i.e., the outcome of the experiment) is favourable to A. The whole sample space Ω is said to be the *sure* (or *certain*) event and the empty set $\emptyset$ is said to be the *impossible* event. This nomenclature is easily justified: the sure event must occur on any given performance of the experiment while the impossible event can never occur in a given performance of the experiment.

A random event A is said to *imply* a random event B if and only if $A \subset B$, i.e., if A is a subset of B. Again, the terminology is easily justified. Two random events A and B are said to be *equivalent* if and only if either of them implies the other, i.e., if and only if $A = B$ in the set theoretical meaning.

1.1.4. Let us now introduce some operations by which new random events can be formed from old ones. They correspond to the construction of compound statements starting from given ones by means of the logical operations of disjunction, conjunction and negation.

Let A and B be random events in the same sample space. Define the *union* of A and B denoted by $A \cup B$ as the random event consisting of those sample points belonging to either A and B or both.

Next, define the *intersection* of A and B denoted by $A \cap B$ as the random event consisting of those sample points belonging to both A and B. Clearly, if A implies B then $A \cup B = B$ and $A \cap B = A$. Finally, define the *complement* of A denoted by A^c as the random event consisting of those sample points that do not belong to A.

For example, considering the random experiment that consists of the roll of a die when $\Omega = \{i\colon 1 \leqslant i \leqslant 6\}$, if $A = \{i \text{ is odd}\}$, $B = \{i \geqslant 4\}$, then

$$A \cup B = \{i \text{ is odd or } i \geqslant 4\} = \{1, 3, 4, 5, 6\},$$

$$A \cap B = \{i \text{ is odd and } i \geqslant 4\} = \{5\},$$

$$A^c = \{i \text{ is not odd}\} = \{2, 4, 6\},$$

$$B^c = \{i < 4\} = \{1, 2, 3\}.$$

More generally, the union of n random events $A_1, \ldots, A_n$ denoted by $A_1 \cup \ldots \cup A_n$ or $\bigcup_{i=1}^{n} A_i$ is the random event consisting of those sample points that belong to at least one of the A_i, $1 \leqslant i \leqslant n$. Similarly, the union of countably many random events $A_1, A_2, \ldots$. denoted by $\bigcup_{i \geqslant 1} A_i$ is the random event consisting of those sample points that belong to at least one of the A_i, $i \geqslant 1$.

Also, the intersection of n random events $A_1, \ldots, A_n$ denoted by $A_1 \cap \ldots \cap A_n$ or $\bigcap_{i=1}^{n} A_i$ is the random event consisting of those sample points that belong to all the A_i, $1 \leqslant i \leqslant n$. Similarly, the intersection of countably many random events $A_1, A_2, \ldots$ denoted by $\bigcap_{i \geqslant 1} A_i$ is the random event consisting of those sample points that belong to all the A_i, $i \geqslant 1$.

Two random events A and B in the same sample space are said to be *mutually exclusive* or *disjoint* if and only if they have no sample points in common, i.e., if and only if it is impossible that both A and B occur during the same performance of the experiment. In symbols, A and B are mutually exclusive if and only if $A \cap B = \emptyset$. More generally, n random events $A_1, \ldots, A_n$ are said to be mutually exclusive or disjoint if and only if $A_i \cap A_j = \emptyset$, $i \neq j$, $1 \leqslant i, j \leqslant n$. This means that no more than one of the events can occur during the same performance of the experiment. Similarly, countably many random events $A_1, A_2, \ldots$ are said to be mutually exclusive or disjoint if and only if $A_i \cap A_j = \emptyset$ for $i \neq j$, $i, j \geqslant 1$.

A finite or countably infinite family of mutually exclusive random events is said to be *exhaustive* (or to be a *partition* of the sample space) if and only if their union is Ω.

The most important properties of the operations with random events are listed below:

$$A \cup A = A, \ A \cap A = A, \ \ A \cap \Omega = A, \ \ A \cup \varnothing = A,$$

$$A \cup \Omega = \Omega, \ A \cup A^c = \Omega, \ A \cap A^c = \varnothing, \ A \cap \varnothing = \varnothing,$$

$$A \cup B = B \cup A, \ A \cup (B \cup C) = (A \cup B) \cup C,$$

$$A \cap B = B \cap A, \ A \cap (B \cap C) = (A \cap B) \cap C,$$

$$A \cap \Big(\bigcup_{i \geqslant 1} B_i\Big) = \bigcup_{i \geqslant 1} (A \cap B_i),$$

$$\Big(\bigcup_{i \geqslant 1} A_i\Big)^c = \bigcap_{i \geqslant 1} A_i^c, \ \ \Big(\bigcap_{i \geqslant 1} A_i\Big)^c = \bigcup_{i \geqslant 1} A_i^c$$

(the last two equations are known as *De Morgan's laws*). It follows from the last three equations by taking respectively $B_i = \varnothing$, $A_i = \varnothing$ and $A_i = \Omega$ for $i > n$ that

$$A \cap \Big(\bigcup_{i=1}^{n} B_i\Big) = \bigcup_{i=1}^{n} (A \cap B_i),$$

$$\Big(\bigcup_{i=1}^{n} A_i\Big)^c = \bigcap_{i=1}^{n} A_i^c, \ \Big(\bigcap_{i=1}^{n} A_i\Big)^c = \bigcup_{i=1}^{n} A_i^c.$$

1.2 PROBABILITY

1.2.1. Let us now consider the concept of probability. Common sense demands it should give a measure of what is likely to happen, i.e., a measure of possibility. Intuitively this is associated with the observable frequency of occurrence. More precisely, for a random event A, its probability $\mathsf{P}(A)$ should reflect the long-run relative frequency of A in a large number of independent repetitions of the random experiment with which A is associated (see 1.1.1). Thus $\mathsf{P}(A)$ should be a number between 0 and 1. In particular, $\mathsf{P}(\varnothing)$ should equal 0 and $\mathsf{P}(\Omega)$ should equal 1. Next, if $A_1, \ldots, A_n$ are disjoint random events, then the number of occurrences of $A_1 \cup \ldots \cup A_n$ in any given number of performances of the experiment obtains by adding the numbers of occurrences of the A_i, $1 \leqslant i \leqslant n$, hence the probability of $A_1 \cup \ldots \cup A_n$ should equal the sum of the probabilities of the A_i, $1 \leqslant i \leqslant n$. Consequently, the probability should possess an additivity property for disjoint random events.

1.2.2. The above nonmathematical considerations are reflected in the definition of the *mathematical* concept of probability. Let us first consider the case where the sample space Ω is either finite or countably infinite. In this case *all* the subsets of Ω will be qualified as random events. Let us denote the class of these subsets by $\mathcal{P}(\Omega)$. Notice that $\mathcal{P}(\Omega)$ satisfies (trivially) the following properties:

$$\Omega \in \mathcal{P}(\Omega); \tag{1.1}$$

$$A_1, A_2, \ldots \in \mathcal{P}(\Omega) \text{ implies } \bigcup_{i \geq 1} A_i \in \mathcal{P}(\Omega); \tag{1.2}$$

$$A \in \mathcal{P}(\Omega) \text{ implies } A^c \in \mathcal{P}(\Omega). \tag{1.3}$$

By *probability measure* (*probability* for short) on $\mathcal{P}(\Omega)$ we shall mean a function P which assigns a nonnegative number $\mathsf{P}(A)$ to each random event $A \in \mathcal{P}(\Omega)$, such that $\mathsf{P}(\Omega) = 1$, and which is *countably additive*, that is

$$\mathsf{P}\Big(\bigcup_{i \geq 1} A_i\Big) = \sum_{i \geq 1} \mathsf{P}(A_i),$$

whenever the random events $A_1, A_2, \ldots$ are disjoint.

It follows from the above equation by taking $A_i = \varnothing$, $i \geq 1$, that $\mathsf{P}(\varnothing) = 0$ and then by taking $A_i = \varnothing$, $i > n$, that

$$\mathsf{P}\Big(\bigcup_{i=1}^{n} A_i\Big) = \sum_{i=1}^{n} \mathsf{P}(A_i),$$

whatever the disjoint random events $A_1, \ldots, A_n$. The last equation shows us the probability P is also *finitely additive*. Notice that countable additivity is *not* implied by finite additivity. The (somewhat unnatural) requirement of countable rather than simply finite additivity for a probability leads to a much richer mathematical theory.

Since $A \cup A^c = \Omega$, $A \cap A^c = \varnothing$, finite additivity implies

$$\mathsf{P}(A^c) = 1 - \mathsf{P}(A)$$

and as $\mathsf{P}(A^c) \geq 0$ we find that $\mathsf{P}(A) \leq 1$ whatever $A \in \mathcal{P}(\Omega)$. More generally if $A \subset B$ then $\mathsf{P}(A) \leq \mathsf{P}(B)$. This follows from the fact that the events A and $B \cap A^c$ are disjoint and $B = A \cup (B \cap A^c)$.

A useful consequence of countable additivity is that for any descending sequence $A_1 \supset A_2 \supset \ldots$ of random events

$$\mathsf{P}\Big(\bigcap_{i \geq 1} A_i\Big) = \lim_{n \to \infty} \mathsf{P}(A_n). \tag{1.4}$$

Equation (1.4) implies that for any ascending sequence $A_1 \subset A_2 \subset \ldots$ of random events

$$\mathsf{P}\Big(\bigcup_{i \geq 1} A_i\Big) = \lim_{n \to \infty} \mathsf{P}(A_n). \tag{1.4'}$$

[Apply (1.4) to the descending sequence $A_1^c \supset A_2^c \supset \ldots$.]

We can construct a probability on $\mathcal{P}(\Omega)$ as follows. Let us associate with each sample point $\omega_i \in \Omega$ a nonnegative number $p(i)$ such that the sum of all the $p(i)$ equals 1. Define $\mathsf{P}(\omega_i) = p(i)$ and for any given random event $A \in \mathcal{P}(\Omega)$ put

$$\mathsf{P}(A) = \sum_{(i\,:\,\omega_i \in A)} p(i).$$

It is easily verified that P so defined is indeed a probability on $\mathcal{P}(\Omega)$. Conversely, *any* probability P on $\mathcal{P}(\Omega)$ can be constructed in this manner by taking $p(i) = \mathsf{P}(\omega_i)$.

It is clear from the above that the assignment of probabilities to random events is not unique. As far as mathematics is concerned the use of any probability measure (leading to a corresponding assignment of probabilities to random events) is allowed. In practical problems the actual specification of the values of the probabilities of the random events must come from the statistical analysis of the results provided by repeated performances of the experiment under consideration.

1.2.3. In the special case where $\Omega = \{\omega_1, \ldots, \omega_m\}$ and $p(i) = 1/m$, $1 \leqslant i \leqslant m$, we have

$$\mathsf{P}(A) = \frac{\text{number of sample points in } A}{\text{total number of sample points in } \Omega}$$

$$= \frac{\text{number of favourable outcomes}}{\text{number of possible outcomes}}.$$

This is the so-called *classical* definition of probability. Clearly, its use needs quite restrictive conditions be met.

The special case above can be realized by an urn containing m balls differing only in colour. It is then natural to assume that the drawings of the balls are equally likely, that is that the probability of drawing any one of the balls equals $1/m$. If the urn contains n_i balls of colour i, $1 \leqslant i \leqslant r$, $\sum_{i=1}^{r} n_i = m$, then the probability of drawing a ball of colour i will equal n_i/m. Notice that both the experiment consisting of the toss of a (fair) coin and the experiment consisting of the roll of a (fair) die can be considered as special cases of the experiment consisting of the drawing of a ball from the urn we have just described. In the first case $m = 2$, $n_1 = n_2 = 1$, and colours 1 and 2 correspond to head and tail, respectively. In the second case $m = 6$, $n_1 = n_2 = n_3 = n_4 = n_5 = n_6 = 1$, and colours 1, 2, 3, 4, 5, 6 correspond to the six faces of the die.

1.2.4. In the case where the sample space Ω is uncountably infinite it is not always possible to allow all its subsets to qualify as random events. The trouble is that it may happen that there is no probability measure on $\mathcal{P}(\Omega)$. A typical case is Ω = the set of real numbers. This difficulty can be overcome by only allowing certain subsets of Ω to count as random events. We shall consider as random events *just* the elements of a class $\mathcal{K}$ of subsets of Ω that satisfies conditions (1.1), (1.2) and (1.3) where $\mathcal{P}(\Omega)$ is replaced by $\mathcal{K}$ (obviously, these conditions are no longer trivially satisfied). Such a class of subsets is said to be a *σ-algebra* of subsets of Ω. By analogy with the case previously considered, a probability measure on $\mathcal{K}$ shall be a function P which assigns a nonnegative number $\mathsf{P}(A)$ to each random event $A \in \mathcal{K}$, such that $\mathsf{P}(\Omega) = 1$, and which is countably additive that is

$$\mathsf{P}\Big(\bigcup_{i\geqslant 1} A_i\Big) = \sum_{i\geqslant 1} \mathsf{P}(A_i),$$

whatever the disjoint random events $A_1, A_2, \ldots$. All the properties of probability emphasized in the case $\mathcal{K} = \mathcal{P}(\Omega)$ extend to the case $\mathcal{K} \neq \mathcal{P}(\Omega)$. The choice of the σ-algebra $\mathcal{K}$ is a problem whose solution should take into account the specific features of the random experiment considered and the possibility of defining a probability on $\mathcal{K}$. Thus, in the case where $\Omega =$ the set of real numbers, one can take $\mathcal{K} =$ the smallest σ-algebra of subsets of Ω that contains all the intervals of the real line. The elements of $\mathcal{K}$ are called one-dimensional Borel sets*). It can be shown that it is possible to define many different probability measures on $\mathcal{K}$.

Let us note that $\Omega^c = \emptyset \in \mathcal{K}$ and if $A_1, A_2, \ldots \in \mathcal{K}$ then $A_1^c, A_2^c, \ldots \in \mathcal{K}$, hence $\bigcup_{i\geqslant 1} A_i^c \in \mathcal{K}$. Then one of De Morgan's laws implies that $\bigcap_{i\geqslant 1} A_i = \Big[\bigcup_{i\geqslant 1} A_i^c\Big]^c \in \mathcal{K}$. Clearly, these remarks would have been trivial in the case $\mathcal{K} = \mathcal{P}(\Omega)$. It can also be shown that, in the general case $\mathcal{K} \neq \mathcal{P}(\Omega)$, countable additivity of the probability P implies equation (1.4).

The triple $(\Omega, \mathcal{K}, \mathsf{P})$ is said to be a *probability space*. It represents the ground on which probability theory is built.

1.2.5. Let $(A_n)_{n\geqslant 1}$ be a sequence of random events (elements of the σ-algebra $\mathcal{K}$). On account of the above considerations $\bigcup_{n\geqslant 1}\bigcap_{m\geqslant n} A_m$ and $\bigcap_{n\geqslant 1}\bigcup_{m\geqslant n} A_m$ are also random events. They are called the *limit inferior* (or *lower limit*) and the *limit superior* (or *upper limit*) of the sequence considered, and are denoted by $\liminf_{n\to\infty} A_n$ and $\limsup_{n\to\infty} A_n$, respectively. The sequence $(A_n)_{n\geqslant 1}$ is said to be *convergent* with limit A (notation: $\lim_{n\to\infty} A_n = A$) if and only if $\liminf_{n\to\infty} A_n = \limsup_{n\to\infty} A_n = A$. It is not difficult to see that the limit inferior consists of the sample points which belong to all but any finite number of the A_n. Similarly, the limit superior consists of the sample points which belong to infinitely many A_n.

---

*) Named after the great French mathematician Émile Borel (1871–1956).

Equations (1.4) and (1.4′) allow us to assert that

$$\mathsf{P}(\liminf_{n\to\infty} A_n) = \lim_{n\to\infty} \mathsf{P}\Big(\bigcap_{m\geq n} A_m\Big), \quad \mathsf{P}(\limsup_{n\to\infty} A_n) = \lim_{n\to\infty} \mathsf{P}\Big(\bigcup_{m\geq n} A_m\Big).$$

The notions of limit inferior and limit superior of a sequence of random events can be connected with the more familiar notions of limit inferior and limit superior of a sequence of functions. If for a random event A we define its *indicator* χ_A as the function defined by

$$\chi_A(\omega) = \begin{cases} 1, & \text{if } \omega \in A, \\ 0, & \text{if } \omega \notin A, \end{cases}$$

then the indicator of $\liminf_{n\to\infty} A_n$ is the function $\liminf_{n\to\infty} \chi_{A_n}$ and the indicator of $\limsup_{n\to\infty} A_n$ is the function $\limsup_{n\to\infty} \chi_{A_n}$. The sequence $(A_n)_{n\geq 1}$ is convergent if and only if the sequence of functions $(\chi_{A_n})_{n\geq 1}$ is convergent. More precisely, $\lim_{n\to\infty} A_n = A$ if and only if $\lim_{n\to\infty} \chi_{A_n} = \chi_A$. Hence, in particular, it follows that if $\lim_{n\to\infty} A_n^i = A^i$, $1 \leq i \leq r$, then

$$\lim_{n\to\infty} A_n^1 \cup \ldots \cup A_n^r = A^1 \cup \ldots \cup A^r,$$

$$\lim_{n\to\infty} A_n^1 \cap \ldots \cap A_n^r = A^1 \cap \ldots \cap A^r.$$

1.3 DEPENDENCE AND INDEPENDENCE

1.3.1. Consider a probability space $(\Omega, \mathcal{K}, \mathsf{P})$. Two random events $A, B \in \mathcal{K}$ are said to be (stochastically) independent if and only if $\mathsf{P}(A \cap B) = \mathsf{P}(A)\ \mathsf{P}(B)$. Empirical support for the above concept comes from the fact that with random events A and B arising from two *physically* independent experiments, i.e., such that a particular outcome of one of them has no effect on the outcome observed in the other*), the frequencies of occurrence of $A \cap B$, A and B in any number of performances of the combined experiment verify the above factorization.

Similarly, three random events $A, B, C \in \mathcal{K}$ are said to be (stochastically) independent if and only if $\mathsf{P}(A \cap B) = \mathsf{P}(A)\ \mathsf{P}(B)$, $\mathsf{P}(B \cap C) = \mathsf{P}(B)\mathsf{P}(C)$, $\mathsf{P}(C \cap A) = \mathsf{P}(C)\ \mathsf{P}(A)$, $\mathsf{P}(A \cap B \cap C) = \mathsf{P}(A)\ \mathsf{P}(B)\ \mathsf{P}(C)$. In general, if $(A_i)_{i\in I}$ is an arbitrary family of random events then the A_i are said to be *(stochastically) independent* if and only if

$$\mathsf{P}(A_{i_1} \cap \ldots \cap A_{i_k}) = \mathsf{P}(A_{i_1}) \ldots \mathsf{P}(A_{i_k}).$$

whatever the finite set of distinct indices $i_1, \ldots, i_k \in I$.

*) As examples one can quote two tosses of the same or different coins, two drawings of one card with replacement from the same deck of playing cards, etc.

It is easily seen that if the A_i, $i \in I$, are independent then the same is true of the B_i, $i \in I$, where each B_i may be either A_i or A_i^c.

Notice that the concept of (stochastic) independence is relative to the given probability P unlike, e.g., the concept of disjointness.

1.3.2. We illustrate the concept of independence by considering Bernoulli trials. A *Bernoulli trial* is a random experiment with two possible outcomes, α ("success") and $\bar{\alpha}$ ("failure"), having probabilities p and q respectively, $p + q = 1$. Tossing of a coin is the best known example of a Bernoulli trial.

Consider the random experiment consisting of n independent Bernoulli trials defined as follows. Its sample space Ω_n, is the set of all the 2^n ordered sequences of length n with components α and $\bar{\alpha}$. Assign to any sample point $\omega \in \Omega_n$ with k components α and $n-k$ components $\bar{\alpha}$ probability $p^k q^{n-k}$. The number of such points is clearly $\binom{n}{k}$ and since on account of the binomial theorem

$$\sum_{k=0}^{n} \binom{n}{k} p^k q^{n-k} = (p + q)^n = 1,$$

the assignment of probabilities is legitimate. It is not difficult to check that defining the random events

$$\begin{aligned} A_i &= \{\omega \in \Omega_n\colon \text{ the } i\text{th component of } \omega \text{ is } \alpha\} \\ &= \{\text{the outcome of the } i\text{th Bernoulli trial is } \alpha\}, \end{aligned}$$

our assignment of probabilities implies that $\mathsf{P}(A_i) = p$, $\mathsf{P}(A_i^c) = 1 - p = q$, $1 \leqslant i \leqslant n$, and hence makes independent the events B_i, $1 \leqslant i \leqslant n$, where each B_i may be either A_i or A_i^c.

1.3.3. Two random events $A, B \in \mathcal{K}$ which are not independent are said to be *dependent*. For two dependent random events A and B the equation $\mathsf{P}(A \cap B) = \mathsf{P}(A)\,\mathsf{P}(B)$ is no longer true. It is replaced by the equation $\mathsf{P}(A \cap B) = \mathsf{P}(A)\,\mathsf{P}(B \mid A)$, where

$$\mathsf{P}(B \mid A) = \frac{\mathsf{P}(A \cap B)}{\mathsf{P}(A)}$$

is called the *conditional* probability of B given A. Obviously, if $\mathsf{P}(A) = 0$ the ratio defining $\mathsf{P}(B \mid A)$ makes no sense (it becomes the "indeterminate" 0/0). Whenever we write a conditional probability $\mathsf{P}(B \mid A)$ we shall impose the proviso that $\mathsf{P}(A) > 0$ even if not explicitely mentioned. For any fixed random event $A \in \mathcal{K}$ the function denoted P_A which assigns to each $B \in \mathcal{K}$ the number $\mathsf{P}(B \mid A)$ is, as is easily seen, a probability measure on $\mathcal{K}$.

Notice that in the case where A and B are independent we have

$$\mathsf{P}(B \mid A) = \mathsf{P}(B), \ \mathsf{P}(A \mid B) = \mathsf{P}(A)$$

in accordance with the intuitive feeling that the occurrence of one of the events does not change the probability of occurrence of the other.

We shall repeatedly use the following formulae involving conditional probabilities.

Proposition 1.1 (*General multiplicative formula*). *For arbitrary events* $A_1, \ldots, A_n \in \mathcal{K}$, *we have*

$$\mathsf{P}(A_1 \cap A_2 \cap \ldots \cap A_n)$$
$$= \mathsf{P}(A_1)\ \mathsf{P}(A_2 \mid A_1)\ \mathsf{P}(A_3 \mid A_1 \cap A_2) \ldots \mathsf{P}(A_n \mid A_1 \cap \ldots \cap A_{n-1}), \qquad (1.5)$$

whenever $\mathsf{P}(A_1 \cap \ldots \cap A_{n-1}) > 0$.

Proposition 1.2 (*Formula of total probability*). *Let* $A_1, A_2, \ldots \in \mathcal{K}$ *be either a finite or a countably infinite partition of* Ω. *For any random event* $B \in \mathcal{K}$ *we have*

$$\mathsf{P}(B) = \Sigma \mathsf{P}(B \cap A_i), \qquad (1.6)$$

Also

$$\mathsf{P}(B) = \Sigma' \mathsf{P}(B \mid A_i)\, \mathsf{P}(A_i), \qquad (1.7)$$

where the sum is taken over those i *for which* $\mathsf{P}(A_i) > 0$.

The *proof* of the first proposition requires just the use of the very definition of conditional probability. The second proposition follows at once if we remark that the random events $B \cap A_i$, $1 \leqslant i \leqslant n$, are disjoint and their union is B.

1.4 RANDOM VARIABLES. MEAN VALUES

1.4.1. Intuitively, by random variable on a probability space $(\Omega, \mathcal{K}, \mathsf{P})$ we should mean a quantity that is measured in connection with a random experiment for which Ω is the sample space. Assume that for an outcome ω of the experiment, a measuring process yields a number $X(\omega)$. The function X that assigns to any ω the real number $X(\omega)$ is said to be a ***real valued random variable*** if and only if the set $\{\omega : X(\omega) < x\}$ of the sample points ω for which $X(\omega) < x$ is a random event ($\in \mathcal{K}$) whatever the real number x. The above restriction is introduced to enable us to speak of the probability that X takes on a given value (in the case where the set of the values of X is finite or countably infinite) or of the probability that X takes on a value in any specified interval (in the case where the set of the values of X is uncountably infinite)*).

*) On the other hand it is interesting to remark that the definition of a random variable does not involve the probability P but just Ω and $\mathcal{K}$.

Notice that in the case where Ω is finite or countably infinite, when any subset of Ω is a random event, the concept of a real valued random variable does coincide with that of a real valued function.

It may be shown that if X is a real valued random variable, then the sets $\{\omega: X(\omega) \in B\}$ are also random events whatever the one dimensional Borel set B. This fact suggests the following definition of the concept of a random variable with values in an arbitrary set M, endowed with a σ-algebra $\mathcal{M}$ of subsets: an *M-valued random variable* on a probability space $(\Omega, \mathcal{K}, \mathsf{P})$ is an M-valued function X defined on Ω such that $\{\omega: X(\omega) \in B\} \in \mathcal{K}$ whatever $B \in \mathcal{M}$. In particular, a random variable with values in r-dimensional Euclidean space is said to be an *r-dimensional random vector.*

In what follows, to simplify the notation, we shall drop the sample point ω, i.e., we shall abbreviate $\{\omega: X(\omega) < x\}$ and $\{\omega: X(\omega) \in B\}$ to $\{X < x\}$ and $\{X \in B\}$, respectively.

1.4.2. A random variable X is said to be *discrete* if and only if the set M of its values is finite or countably infinite. If $M = (x_1, x_2, \ldots)$, then the $\{X = x_i\}$ are random events and the knowledge of the probabilities $\mathsf{P}(X = x_i) = p_X(x_i)$, $i \geq 1$, is sufficient for computing the probability $\mathsf{P}(X \in B)$, where B is an arbitrary subset of M. Clearly, we have

$$\mathsf{P}(X \in B) = \sum_{x_i \in B} p_X(x_i),$$

and, in particular,

$$1 = \mathsf{P}(X \in M) = \sum_{i \geq 1} p_X(x_i).$$

A nonnegative function defined on a finite or countably infinite set such that the sum of all its values equals 1 is called a (discrete) *probability distribution* over the set considered. Consequently, the function p_X will be called the (probability) distribution of the discrete random variable X.

The simplest example of a discrete random variable is that of a "variable" assuming just one value (therefore a constant). Here are a few important examples of discrete random variables.

The observed number of successes in n independent Bernoulli trials (see 1.3.2) is a discrete random variable called *binomial with parameters n and p*. Its values are $0, 1, \ldots, n$, and its probability distribution (also called *binomial with parameters n and p*) is given by

$$p(k) = \binom{n}{k} p^k q^{n-k}, \quad 0 \leq k \leq n.$$

A celebrated limiting case of the binomial distribution with parameters n and p is the *Poisson distribution with parameter* λ that obtains taking $p = p(n) \to 0$ as $n \to \infty$ in such a way that $np \to \lambda > 0$. For this distribution

$$p(k) = e^{-\lambda} \frac{\lambda^k}{k!}, \; k = 0, 1, \ldots .$$

A discrete random variable with the above distribution is also called *Poisson with parameter* λ.

Another remarkable example of a discrete random variable is the number of independent Bernoulli trials necessary to be performed in order to get the first success, or, equivalently, the *waiting time* for success to occur. This discrete random variable is called *geometric with parameter* p. Its values are the natural integers 1, 2, ... and its probability distribution (also called *geometric with parameter* p) is given by $p(k) = q^{k-1}p, \; k = 1, 2, \ldots$.

The last example we consider is that of a random variable taking m arbitrary distinct values $x_1, \ldots, x_m$ each with probability $1/m$*), i.e.,

$$p(x_i) = 1/m, \; 1 \leqslant i \leqslant m.$$

This is called the *uniform* distribution over the set $(x_1, \ldots, x_m)$. (Compare with 1.2.3.)

1.4.3. A real valued random variable X is called (*absolutely*) *continuous* if and only if there exists a nonnegative function p_X defined on the real line such that

$$\int_{-\infty}^{\infty} p_X(x)\, dx = 1$$

and

$$\mathsf{P}(a < X < b) = \int_a^b p_X(x)\, dx,$$

whatever the real numbers $a < b$. The function p_X is called the (*probability*) *density function* of the random variable X.

Consequently, for a continuous random variable X we know the probability that X takes on a value in any specified interval. It is easily seen that all the random events $\{X = x\}$ have probability 0 (although they are *not* impossible).

It can be shown that

$$\mathsf{P}(X \in B) = \int_B p_X(x)\, dx$$

whatever the one dimensional Borel set B (the integral here is to be understood as a Lebesgue integral).

---

*) The values $x_1, \ldots, x_m$ are said to occur "at random".

We shall consider two important examples of density functions. For the first one

$$p(x) = \begin{cases} \lambda e^{-\lambda x}, & \text{if } x \geqslant 0, \\ 0, & \text{if } x < 0. \end{cases}$$

We have $p(x) \geqslant 0$ and

$$\int_{-\infty}^{\infty} p(x)\,dx = \lambda \int_{0}^{\infty} e^{-\lambda x}\,dx = -e^{-\lambda x}\Big|_{0}^{\infty} = 1,$$

hence p is a density function. It is called *exponential with parameter* λ. A continuous variable with exponential density function is said to be an *exponential* random variable.

Proposition 1.3. *An exponential random variable* X *has no memory, that is*

$$\mathsf{P}(X > x + y \mid X > x) = \mathsf{P}(X > y),$$

whatever the nonnegative numbers x *and* y.

Proof. We have

$$\mathsf{P}(X > x + y \mid X > x) = \frac{\mathsf{P}(\{X > x + y\} \cap \{X > x\})}{\mathsf{P}(X > x)}$$

$$= \frac{\mathsf{P}(X > x + y)}{\mathsf{P}(X > x)} \quad [\text{since } \{X > x + y\} \subset \{X > x\}]$$

$$= \frac{\lambda \int_{x+y}^{\infty} e^{-\lambda u}\,du}{\lambda \int_{x}^{\infty} e^{-\lambda u}\,du} = \frac{-e^{-\lambda u}\Big|_{x+y}^{\infty}}{-e^{-\lambda u}\Big|_{x}^{\infty}} = \frac{e^{-\lambda(x+y)}}{e^{-\lambda x}}$$

$$= e^{-\lambda y} = \int_{y}^{\infty} e^{-\lambda u}\,du = \mathsf{P}(X > y).$$

The lack of memory of the exponential density is fundamental to the theory of Markov processes (see Chapter 8). The exponential density is also a suitable model for various types of "waiting time" problems concerning service times, break-downs in certain industrial equipment, telephone calls, failures of electron tubes, occurrence of accidents to individuals, etc. In many such cases it is reasonable to assume that the time until a certain random event occurs is an exponential variable. In this context Proposition 1.3 can be interpreted as follows: if we have already spent some time in waiting for a certain random event to occur, then the distribution of further waiting time is identical to that of the original waiting time.

We should note that the "memoryless" property characterizes the exponential density, i.e., if a nonnegative random variable satisfies the condition

$$\mathsf{P}(X > x + y \mid X > x) = \mathsf{P}(X > y)$$

for all $x, y \geqslant 0$, then X is an exponential random variable*).

The second important example we discuss is that of the *normal* density function with parameters a and σ defined as

$$p(x) = \frac{1}{\sigma\sqrt{2\pi}} e^{-\frac{(x-a)^2}{2\sigma^2}}, \quad -\infty < x < \infty,$$

where $-\infty < a < \infty$ and $\sigma > 0$ are given constants. The fact that

$$\int_{-\infty}^{\infty} p(x)\, dx = 1$$

can be proved by the following instructive and well-known trick. Setting

$$\alpha = \int_{-\infty}^{\infty} p(x)\, dx,$$

we have (using polar coordinates $x = a + \rho \cos\theta$, $y = a + \rho \sin\theta$)

$$\alpha^2 = \left(\int_{-\infty}^{\infty} p(x)\, dx\right)\left(\int_{-\infty}^{\infty} p(y)\, dy\right) = \frac{1}{2\pi\sigma^2}\int_{-\infty}^{\infty}\int_{-\infty}^{\infty} e^{-\frac{(x-a)^2+(y-a)^2}{2\sigma^2}}\, dx\, dy$$

$$= \frac{1}{2\pi\sigma^2}\int_0^{2\pi} d\theta \int_0^{\infty} \rho e^{-\frac{\rho^2}{2\sigma^2}}\, d\rho = \frac{1}{2\pi\sigma^2}\, 2\pi \left(-\sigma^2 e^{-\frac{\rho^2}{2\sigma^2}}\right)\Bigg|_0^{\infty} = 1,$$

whence $\alpha = 1$ (since $\alpha > 0$).

The normal density was introduced by C.F. Gauss (1777—1855) as the probability density that should govern the errors of observations in order to make the arithmetical mean of a set of observations the best estimate (in a certain sense involving least squares) of the "true" value of the quantity measured. Actually, the errors of observations are at best only approximately normally distributed. The time when everybody believed in the normal law of errors — experimenters conceiving of it as a mathematical theorem and mathematicians conceiving of it as an empirical law—has gone long ago. However, without minimizing the practical importance of the normal density function, we should stress its paramount theoretical importance as a central concept in the asymptotic problems of probability theory.

---

*) Notice that if X is a geometric random variable with parameter p, then $\mathsf{P}(X > n) = q^n$ whatever the nonnegative integer n. Hence $\mathsf{P}(X > m + n \mid X > n) = \mathsf{P}(X > m + n)/\mathsf{P}(X > n) = q^{m+n}/q^n = \mathsf{P}(X > m)$ whatever the nonnegative integers m and n. But the above equation is *not* true for *all* nonnegative real numbers m and n (e.g., for $m = n = 1/2$).

1.4.4. Another way of characterizing real valued random variables is by means of distribution functions. The *distribution function* of a real valued random variable X is defined as

$$F_X(x) = \mathsf{P}(X < x), -\infty < x < \infty.$$

It can be shown that F_X is nondecreasing (that is, $x < y$ implies $F_X(x) \leqslant F_X(y)$), that $\lim_{x \to -\infty} F_X(x) = 0$, that $\lim_{x \to \infty} F_X(x) = 1$, and that F_X is continuous from the left, i.e.,

$$\lim_{\substack{y \to x \\ y < x}} F_X(y) = F_X(x).$$

Notice also that $\mathsf{P}(a \leqslant X < b) = F_X(b) - F_X(a)$, whatever the real numbers $a < b$.

For a discrete real valued random variable X with values $x_1, x_2, \ldots$ the distribution function F_X has a jump of magnitude $p_X(x_i) = \mathsf{P}(X = x_i)$ at $x = x_i$, $i = 1, 2, \ldots$ and remains constant between jumps. (In other words, F_X is a step function.) Therefore, given the probability distribution of X, we can construct the distribution function F_X. Conversely, given F_X we can determine p_X:

$$p_X(x_i) = \lim_{0 < h \to 0} F_X(x_i + h) - F_X(x_i).$$

For a continuous random variable X with density p_X we have

$$F_X(x) = \int_{-\infty}^{x} p_X(u)\, du, -\infty < x < \infty.$$

Thus, F_X is an absolutely continuous function and the density function p_X of X turns out to be the derivative of F_X.

For example, in the case of an exponential random variable with parameter λ

$$F(x) = \begin{cases} 0, \text{ if } x < 0 \\ \int_0^x \lambda e^{-\lambda u}\, du = 1 - e^{-\lambda x}, \text{ if } x \geqslant 0. \end{cases}$$

1.4.5. Starting with a given collection of random variables, we can construct new ones by operating on them in various ways. For example, it is easily proved that if X and Y are real valued random variables on the same probability space, then $X+Y$, XY and X/Y $(Y \neq 0)$—therefore, in particular, $X + a$ and aX, where a is an arbitrary real number—are also random variables. Clearly, $X+Y$ is the random variable taking the value $X(\omega) + Y(\omega)$ at ω, etc.

More generally, if f is a measurable function (e.g., continuous) of n real variables and $X_1, \ldots, X_n$ are real valued random variables, then $f(X_1, \ldots, X_n)$, which is defined as the function taking the value $f(X_1(\omega), \ldots \ldots, X_n(\omega))$ at ω, is also a random variable. This may be summarized as follows: functions of random variables are still random variables. The extension to the case of random variables with an arbitrary set of values is easily made.

1.4.6. Let X be a discrete random variable with distribution $p_X(x_i) = p_i$, $i = 1, 2, \ldots$, and f a real or complex valued function defined on the set of values of X. Define the *mean value* (also called *expectation, expected value, average value,* or *mean*) of $f(X)$, denoted by $\mathsf{E}(f(X))$, to be the sum of the series $\Sigma f(x_i)\, p_i$ (assuming it to be absolutely convergent*), that is $\Sigma |f(x_i)|\, p_i < \infty$).

Similarly, if X is a continuous random variable with density function p_X and f a real or complex measurable function of a real variable then we define the *mean value* of $f(X)$ denoted by $\mathsf{E}(f(X))$ as the integral $\int_{-\infty}^{\infty} f(x)\, p_X(x)\, dx$ $\left(\text{assuming it to be absolutely convergent**), that is } \int_{-\infty}^{\infty} |f(x)|\, p_X(x)\, dx < \infty\right)$.

In particular, for $f(x) = x^m$, m real,

$$\mathsf{E}(X^m) = \Sigma x_i^m p_i \text{ or } \mathsf{E}(X^m) = \int_{-\infty}^{\infty} x^m p_X(x)\, dx$$

is called the m th *moment* of the real valued random variable X.

For $m = 1$ we get the *mean value* or *mean* of X denoted by $\mathsf{E}(X)$.

The mth moment of $X - \mathsf{E}(X)$ is called the mth *central* moment of X. In particular, the second central moment of X is called the *variance* of X and denoted by var (X) or $\sigma^2(X)$. It is easily proved that $\sigma^2(X) = \mathsf{E}(X^2) - (\mathsf{E}(X))^2$. The square root $\sigma(X)$ of $\sigma^2(X)$ is called the *standard deviation* of X. The celebrated Bienaymé—Čebyšev inequality

$$\mathsf{P}(|X - \mathsf{E}(X)| > a) \leqslant \sigma^2(X)/a^2, \; a > 0,$$

shows that the variance of X may be interpreted as a measure of dispersion of the values of X.

It can be shown, from the very definitions of mean and variance, that:

1) If $X = c =$ constant, then $\mathsf{E}(X) = c$;
2) If $X \geqslant Y$, then $\mathsf{E}(X) \geqslant \mathsf{E}(Y)$;
3) $\mathsf{E}(aX) = a\mathsf{E}(X)$ whatever the real number a;
4) $\mathsf{E}(X + Y) = \mathsf{E}(X) + \mathsf{E}(Y)$;
5) $\sigma^2(aX) = a^2\sigma^2(X)$ whatever the real number a;
6) $\sigma^2(X + Y) = \sigma^2(X) + \sigma^2(Y) + 2\mathsf{E}((X - \mathsf{E}(X))(Y - \mathsf{E}(Y)))$.

The mean value of $(X - \mathsf{E}(X))\,(Y - E(Y))$ is called the *covariance* of X and Y and denoted by cov(X, Y). Obviously, cov$(X, X) =$ var(X).

---

*) This condition is trivially satisfied in the case where the set of values of X is finite.

**) This condition is trivially satisfied in the case where f is a bounded function.

For a binomial random variable X with parameters n and p we have

$$\mathsf{E}(X) = \sum_{k=0}^{n} k\binom{n}{k} p^k q^{n-k} = np$$

$\left(\text{this follows by setting } x = p/q \text{ in the identity } n(1+x)^{n-1} = \sum_{k=0}^{n} k\binom{n}{k} x^{k-1}\right.$ that comes from differentiating the identity $(1+x)^n = \sum_{k=0}^{n} \binom{n}{k} x^k$ with $\left.\text{respect to } x\right)$,

$$\mathsf{E}(X^2) = \sum_{k=0}^{n} k^2 \binom{n}{k} p^k q^{n-k} = npq + n^2p^2$$

$\left(\text{this follows by making use of the identity } n(n-1)(1+x)^{n-2} = \right.$ $= \sum_{k=0}^{n} k(k-1)\binom{n}{k} x^{k-2}$ that comes from differentiating the identity $\left. n(1+x)^{n-1} = \sum_{k=0}^{n} k\binom{n}{k} x^{k-1} \text{ with respect to } x\right)$, and

$$\operatorname{var}(X) = \mathsf{E}(X^2) - (\mathsf{E}(X))^2 = npq.$$

The mean and variance of a Poisson random variable with parameter λ are both equal to λ. The mean and variance of a geometric random variable with parameter p are $1/p$ and $(1-p)/p^2$, respectively. The proof of these assertions is left to the reader.

For an exponential random variable with parameter λ we have

$$\mathsf{E}(X) = \lambda \int_0^\infty x e^{-\lambda x}\, dx = \frac{1}{\lambda}\int_0^\infty t e^{-t}\, dt = \frac{1}{\lambda}\left(-te^{-t} - e^{-t}\right)\Big|_0^\infty = \frac{1}{\lambda},$$

$$\mathsf{E}(X^2) = \lambda \int_0^\infty x^2 e^{-\lambda x}\, dx = \frac{1}{\lambda^2}\int_0^\infty t^2 e^{-t}\, dt = \frac{1}{\lambda^2}\left(-t^2e^{-t} - 2te^{-t} - 2e^{-t}\right)\Big|_0^\infty = \frac{2}{\lambda^2},$$

$$\operatorname{var}(X) = \frac{1}{\lambda^2}.$$

The mean and variance of a random variable with normal density function with parameters a and σ are a and σ^2, respectively. (Try to prove it!) Thus we have a probabilistic interpretation of the parameters of the normal density function.

Another important concept that comes from considering a special function f is that of the characteristic function of a real valued random variable. Let us take $f(x) = \exp(itx) = e^{itx}$, t any real number, $i = \sqrt{-1}$ the imaginary unit. Given a real valued random variable X, by *characteristic function* (or *Fourier transform*) of X, denoted by φ_X, we mean the mapping $t \to \mathsf{E}(\exp itX)$. Notice that since $|\exp(itx)| = 1$ the characteristic function of a real valued random variable always exists.

It can be shown that the existence of the mth moment of X (m any natural number) implies the existence of the mth derivative of the characteristic function φ_X at $t = 0$ and

$$\left.\frac{d^m \varphi_X(t)}{dt^m}\right|_{t=0} = \mathrm{i}^m \mathsf{E}(X^m).$$

(The converse statement is true only for the even values of m.)

For discrete random variables taking only nonnegative integer values it is more convenient to use the (probability) generating function instead of the characteristic function. This concept corresponds to the choice $f(x) = z^x$, z any complex number of modulus $\leqslant 1$. Given a nonnegative integer valued random variable X, by *(probability) generating function* of X, denoted by g_X, we mean the mapping $z \to \mathsf{E}(z^X)$. Clearly, if $\mathsf{P}(X = k) = p_k$, $k = 0, 1, 2, \ldots$, then

$$g_X(z) = \sum_{k \geqslant 0} p_k z^k,$$

the series being absolutely convergent for $|z| \leqslant 1$*). Thus $\varphi_X(t) = g_X(e^{it})$ and it is easy to show that

$$\mathsf{E}(X) = g'_X(1), \ \mathsf{E}(X^2) = g'_X(1) + g''_X(1).$$

More generally, for random variables taking only nonnegative values it is convenient to use the Laplace transform, a concept corresponding to the choice $f(x) = e^{-\lambda x}$, λ any complex number with nonnegative real part. Given a nonnegative valued random variable X, by *Laplace transform* of X, denoted by L_X, we mean the mapping $\lambda \to \mathsf{E}(e^{-\lambda X})$. Notice that since $|e^{-\lambda x}| \leqslant 1$ for $\mathrm{Re}\,\lambda \geqslant 0$, $x \geqslant 0$, the Laplace transform of a nonnegative random variable always exists. Clearly, $\varphi_X(t) = L_X(-it)$ and for a nonnegative integer valued random variable X we have $L_X(\lambda) = g_X(e^{-\lambda})$. Next, if the nonnegative random variable X has density function p_X then

$$L_X(\lambda) = \int_0^\infty e^{-\lambda u} p_X(u)\, du \text{ **)}.$$

---

*) More generally, given a sequence $(a_k)_{k \geqslant 0}$ of real or complex numbers, one may consider its generating function $g(z) = \sum_{k \geqslant 0} a_k z^k$, provided the power series has a non-vanishing radius of convergence.

**) More generally, given a nonnegative function h defined on $[0, \infty)$, one may consider its Laplace transform $L_h(\lambda) = \int_0^\infty e^{-\lambda u} h(u)\, du$. Laplace transforms are widely used in differential equations, operational calculus, and engineering applications.

It should be noted that any one of the characteristic function, the (probability) generating function, and the Laplace transform of a random variable uniquely determine its distribution function.

1.4.7. We have previously defined the notion of independence for random events. This can be used to define independence of random variables. Let $X_1, \ldots, X_n$ be random variables on the same probability space taking values in an arbitrary set M endowed with a σ-algebra $\mathcal{M}$ of subsets. They are said to be *independent* if and only if*)

$$\mathsf{P}(X_1 \in B_1, \ldots, X_n \in B) = \mathsf{P}(X_1 \in B_1) \ldots \mathsf{P}(X_n \in B_n),$$

for all $B_1, \ldots, B_n \in \mathcal{M}$. In particular, discrete random variables $X_1, \ldots, X_n$ are independent if and only if the equations

$$\mathsf{P}(X_1 = x_1, \ldots, X_n = x_n) = \mathsf{P}(X_1 = x_1) \ldots \mathsf{P}(X_n = x_n)$$

are satisfied for all choices of $x_1, \ldots, x_n$. In other words, the random variables $X_1, \ldots, X_n$ are independent if and only if, A_i being a random event involving X_i alone, that is of the form $A_i = \{X_i \in B_i\}$, $1 \leqslant i \leqslant n$, the A_i are independent.

It can be shown that if X and Y are independent real valued random variables then $\mathsf{E}(XY) = \mathsf{E}(X)\ \mathsf{E}(Y)$, whence $\operatorname{cov}(X, Y) = 0$, which in turn implies Bienaymé's celebrated equality $\operatorname{var}(X + Y) = \operatorname{var}(X) + \operatorname{var}(Y)$. (The converse statement is not generally true.)

A very important property is that the characteristic function of a sum of independent real valued random variables is the product of the characteristic functions of the summands. The above statement remains true when the characteristic function is replaced by the generating function (for nonnegative integer valued random variables) or the Laplace transform (for nonnegative random variables).

In classical probability theory the study of (infinite) sequences of independent real valued random variables is a central theme. Such a sequence $(X_i)_{i\geqslant 1}$ is characterized by the fact that for any natural number n the random variables $X_1, \ldots, X_n$ are independent. By way of illustration we shall state, under somewhat restrictive conditions, three fundamental theorems. All of them refer to a sequence $(X_i)_{i\geqslant 1}$ of independent real valued random variables with common distribution function.

Theorem 1.4 (*Law of large numbers*). *If the mean value* $\mathsf{E}(X_1)$ $(= \mathsf{E}(X_2) = \ldots)$ *exists, then*

$$\lim_{n\to\infty} \frac{X_1(\omega) + \ldots + X_n(\omega)}{n} = \mathsf{E}(X_1)$$

for all sample points ω *except possibly on a random event of probability* 0.

---

*) To simplify the notation the comma will replace the intersection sign $\cap$. Thus $\{X_1 \in B_1, \ldots, X_n \in B_n\}$ will abbreviate $\{X_1 \in B_1\} \cap \ldots \cap \{X_n \in B_n\}$.

Theorem 1.5 (*Central limit theorem*). *If* $\mathsf{E}(X_1)$ $(= \mathsf{E}(X_2) = \dots) = m$ *and* $\operatorname{var}(X_1)$ $(= \operatorname{var}(X_2) = \dots) = \sigma^2 > 0$, *then*

$$\lim_{n\to\infty} \mathsf{P}\left(\frac{X_1 + \dots + X_n - nm}{\sigma\sqrt{n}} < x\right) = \frac{1}{\sqrt{2\pi}}\int_{-\infty}^{x} e^{-u^2/2}\, du$$

whatever the real number x.

Theorem 1.6 (*Law of iterated logarithm*). *Under the assumptions of the previous theorem*

$$\limsup_{n\to\infty} \frac{X_1(\omega) + \dots + X_n(\omega) - nm}{\sigma\sqrt{2n \ln\ln n}} = 1,$$

$$\liminf_{n\to\infty} \frac{X_1(\omega) + \dots + X_n(\omega) - nm}{\sigma\sqrt{2n \ln\ln n}} = -1$$

for all sample points ω *except possibly on a random event of probability* 0.

1.5 RANDOM PROCESSES

1.5.1. In the preceding paragraph we spoke about (infinite) sequences of independent random variables. However, the existence of such mathematical objects should be demonstrated.

To fix our ideas, we shall assume that all the random variables involved are discrete: this will in fact be sufficiently general for our subsequent needs.

Notice first that the existence of n random variables $X_1, \dots, X_n$ with given *joint probability distribution* $p_{1\dots n}$ defined as

$$p_{1\dots n}(x_1, \dots, x_n) = \mathsf{P}(X_1 = x_1, \dots, X_n = x_n)$$

can be proved without any difficulty. Indeed, it is sufficient to take as sample space Ω the set of all n-tuples $(x_1, \dots, x_n)$ of possible values of the variables considered, assign probability $p_{1\dots n}(x_1, \dots, x_n)$ to the sample point $(x_1, \dots, x_n)$ and define $X_i(x_1, \dots, x_n) = x_i$, $1 \leqslant i \leqslant n$.

The special case of n independent random variables amounts to $p_{1\dots n}(x_1, \dots, x_n) = p^{(1)}(x_1) \dots p^{(n)}(x_n)$, where $p^{(1)}, \dots, p^{(n)}$ are the probability distributions of the variables considered.

The problem whose solution will implicitly solve the problem of the existence of an infinite sequence of independent random variables is to find conditions under which, for a given sequence of joint probability distributions $p_{1\dots n}$, $n \geqslant 1$, there exists an infinite sequence of random variables such that, for all $n \geqslant 1$, the joint probability distribution of the first n variables be identical to $p_{1\dots n}$.

Observe first that $p_{1\dots n}$ should necessarily verify certain natural compatibility conditions implied by their very definition. Indeed, we have

$$\begin{aligned} p_{1\dots n}(x_1, \dots, x_n) &= \mathsf{P}(X_1 = x_1, \dots, X_n = x_n) \\ &= \mathsf{P}(X_1 = x_1, \dots, X_n = x_n, X_{n+1} = \text{arbitrary}) \\ &= \sum_{x_{n+1}} \mathsf{P}(X_1 = x_1, \dots, X_n = x_n, X_{n+1} = x_{n+1}) \qquad (1.8) \\ &= \sum_{x_{n+1}} p_{1\dots n+1}(x_1, \dots, x_n, x_{n+1}). \end{aligned}$$

It follows from a general theorem (P.J. Daniell, A.N. Kolmogorov) that, conversely, for *consistent* joint probability distributions $p_{1\dots n}$, $n \geqslant 1$, i.e., such that (1.8) is verified for all possible values $x_1, \dots, x_n$, and $n \geqslant 1$, an infinite sequence with the properties indicated above does always exist. More precisely, taking as sample space Ω the set of all infinite sequences $(x_1, x_2, \dots)$, as $\mathcal{K}$ the smallest σ-algebra that contains all the sets of the form $(x_1, \dots, x_n, x_{n+1} = \text{arbitrary}, x_{n+2} = \text{arbitrary}, \dots)$ for all possible values $x_1, \dots, x_n$ and $n \geqslant 1$ and defining $X_i(x_1, x_2, \dots) = x_i$, $i \geqslant 1$, there exists a probability measure P on $\mathcal{K}$ such that

$$\mathsf{P}(X_1 = x_1, \dots, X_n = x_n) = p_{1\dots n}(x_1, \dots, x_n)$$

for all possible values $x_1, \dots, x_n$ and $n \geqslant 1$.

In the special case of independent random variables where $p_{1\dots n}(x_1, \dots, x_n) = p^{(1)}(x_1) \dots p^{(n)}(x_n)$, the consistency conditions (1.8) are easily verified as $\sum_{x_n} p^{(n)}(x_n) = 1$ for all $n \geqslant 1$. Consequently, considering an infinite sequence of independent random variables with given probability distributions is quite legitimate from a purely mathematical point of view.

1.5.2. Let us now examine a more complicated situation. Assume that, for any point t of the real nonnegative semiaxis $[0, \infty)$, we are given a discrete random variable taking values in a finite or countably infinite set S. The family of random variables $(X(t))_{t \geqslant 0}$ is said to be a *random* (or *stochastic*) *process*. The *finite dimensional joint probability distributions* of such a process are defined as the joint probability distributions of the random variables $X(t_1), \dots, X(t_n)$ for arbitrary values $t_1 < \dots < t_n$, $n \geqslant 1$, that is

$$p_{t_1\dots t_n}(x_1, \dots, x_n) = \mathsf{P}(X(t_1) = x_1, \dots, X(t_n) = x_n).$$

We propose to take up the problem corresponding to that considered for infinite sequences of random variables: to find conditions under which, for a given family of finite dimensional joint probability distributions $p_{t_1\dots t_n}$, $t_1 < \dots < t_n$, $n \geqslant 1$, there exists a random process whose finite dimensional joint probability distributions coincide with the given ones.

In the present case the compatibility conditions the $p_{t_1 \dots t_n}$ should verify are as follows:

$$
\begin{aligned}
&\sum_{x_1} p_{t_1 \dots t_n}(x_1, \dots, x_n) = p_{t_2 \dots t_n}(x_2, \dots, x_n), \\
&\sum_{x_n} p_{t_1 \dots t_n}(x_1, \dots, x_n) = p_{t_1 \dots t_{n-1}}(x_1, \dots, x_{n-1}), \\
&\sum_{x_k} p_{t_1 \dots t_{k-1} t_k t_{k+1} \dots t_n}(x_1, \dots, x_{k-1}, x_k, x_{k+1}, \dots, x_n) \\
&\qquad = p_{t_1 \dots t_{k-1} t_{k+1} \dots t_n}(x_1, \dots, x_{k-1}, x_{k+1}, \dots, x_n), \quad 1 < k < n.
\end{aligned}
\tag{1.9}
$$

It follows from the same Daniell-Kolmogorov theorem that, for *consistent* finite dimensional joint probability distributions $p_{t_1 \dots t_n}$, $t_1 < \dots < t_n$, $n \geqslant 1$, i.e., such that equations (1.9) are verified, a random process with the properties indicated above does always exist. More precisely, taking as sample space Ω the set of all functions $x:[0, \infty) \to S$ (i.e., defined on $[0, \infty)$ with values in S), as $\mathcal{K}$ the smallest σ-algebra that contains all the sets of the form $(x(t_i) = x_i,\ 1 \leqslant i \leqslant n,\ x(t) \in S,\ t \neq t_i,\ 1 \leqslant i \leqslant n)$ for all possible values $x_i \in S$, $1 \leqslant i \leqslant n$, and $n \geqslant 1$ and defining $X(t, x(\cdot)) = x(t)$, $t \geqslant 0$, there exists a probability measure P on $\mathcal{K}$ such that

$$\mathsf{P}(X(t_1) = x_1, \dots, X(t_n) = x_n) = p_{t_1 \dots t_n}(x_1, \dots, x_n)$$

for all possible values $x_1, \dots, x_n \in S$, $t_1 < \dots < t_n$, $n \geqslant 1$.

1.6 MATRICES

1.6.1. Let r and s be positive integers. A rectangular array of rs real (or complex) numbers written in the form

$$
\begin{pmatrix}
a(1,1) & a(1,2) & \cdots & a(1,s) \\
a(2,1) & a(2,2) & \cdots & a(2,s) \\
\vdots & \vdots & & \vdots \\
a(r,1) & a(r,2) & \cdots & a(r,s)
\end{pmatrix}
$$

is called an $r \times s$ *matrix* with real (or complex) *entries* (equivalently: *elements* or *components*).

The quantity $a(i,j)$ is called the ijth entry of the matrix, $1 \leqslant i \leqslant r$, $1 \leqslant j \leqslant s$.

The quantities $a(i,1)$, $a(i,2)$, ..., $a(i,s)$ constitute the ith *row* of the matrix, and the quantities $a(1,j)$, $a(2,j)$, ..., $a(r,j)$ constitute the jth *column*, $1 \leqslant i \leqslant r$, $1 \leqslant j \leqslant s$. Therefore an $r \times s$ matrix has r rows and s columns.

Three kinds of matrices have special importance. A matrix with the same number of rows and columns, that is an $r \times r$ matrix, is called a (*square*) matrix of *order* r. A matrix with one single row, that is an

$1 \times s$ matrix, is called an *s-dimensional row vector*. A matrix with one single column, that is an $r \times 1$ matrix, is called an *r-dimensional column vector*.

1.6.2. Matrices (vectors) will be denoted by upper-case (lower-case) boldface letters of the Latin or Greek alphabets. If the *ij*th entry of the matrix $\mathbf{A}$ is $a(i, j)$, $1 \leqslant i \leqslant r$, $1 \leqslant j \leqslant s$, then we shall use the shorthand notation $\mathbf{A} = (a(i, j))$ or $\mathbf{A} = (a(i, j))_{1 \leqslant i \leqslant r,\, 1 \leqslant j \leqslant s}$. Similarly, if the *i*th component of the column vector $\mathbf{a}$ is $a(i)$, $1 \leqslant i \leqslant r$, we shall write $\mathbf{a} = (a(i))$ or $\mathbf{a} = (a(i))_{1 \leqslant i \leqslant r}$. The *transpose* $\mathbf{A}'$ of the matrix $\mathbf{A} = (a(i, j))$ is defined as the matrix whose rows are the columns of $\mathbf{A}$, i.e., $\mathbf{A}' = (a'(j, i))$, where $a'(j, i) = a(i, j)$, $1 \leqslant j \leqslant s$, $1 \leqslant i \leqslant r$. Clearly, $\mathbf{A}'$ is an $s \times r$ matrix. Notice that if $\mathbf{a}$ is an r-dimensional column vector, then $\mathbf{a}'$ is an r-dimensional row vector. To avoid any confusion and for typographical convenience we agree to consider any vector we deal with as a column vector. The above remark allowsus to write a row vector as the transpose of a column vector.

For example, if

$$\mathbf{A} = \begin{pmatrix} 2 & 1 \\ -1 & 0 \\ 3 & 2 \end{pmatrix}, \qquad \mathbf{a} = \begin{pmatrix} 1 \\ 0 \\ -2 \end{pmatrix},$$

then

$$\mathbf{A}' = \begin{pmatrix} 2 & -1 & 3 \\ 1 & 0 & 2 \end{pmatrix}, \qquad \mathbf{a}' = (1,\ 0,\ -2).$$

1.6.3. By *submatrix* of a matrix $\mathbf{A}$ we shall mean any matrix that either coincides with $\mathbf{A}$ or can be obtained from $\mathbf{A}$ by suppressing rows and/or columns.

For example, the matrix

$$\begin{pmatrix} 1 & 2 & 3 \\ 9 & 8 & 7 \end{pmatrix}$$

has the following submatrices

$$\begin{pmatrix} 1 & 2 & 3 \\ 9 & 8 & 7 \end{pmatrix}, \begin{pmatrix} 1 & 2 \\ 9 & 8 \end{pmatrix}, \begin{pmatrix} 1 & 3 \\ 9 & 7 \end{pmatrix}, \begin{pmatrix} 2 & 3 \\ 8 & 7 \end{pmatrix}, \begin{pmatrix} 1 \\ 9 \end{pmatrix}, \begin{pmatrix} 2 \\ 8 \end{pmatrix}, \begin{pmatrix} 3 \\ 7 \end{pmatrix},$$

$$(1,\ 2,\ 3),\ (9,\ 8,\ 7),\ (1,\ 2),\ (1,\ 3),\ (2,\ 3),\ (9,\ 8),\ (9,\ 7),\ (8,\ 7),$$
$$(1),\ (2),\ (3),\ (9),\ (8),\ (7).$$

1.6.4. Two $r \times s$ matrices $\mathbf{A} = (a(i, j))$ and $\mathbf{B} = (b(i, j))$ are said to be *equal* ($\mathbf{A} = \mathbf{B}$) if and only if $a(i, j) = b(i, j)$, $1 \leqslant i \leqslant r$, $1 \leqslant j \leqslant s$. More generally, for matrices with real entries, we write $\mathbf{A} > \mathbf{B}$, respectively $\mathbf{A} \geqslant \mathbf{B}$, if and only if $a(i, j) > b(i, j)$, respectively $a(i, j) \geqslant b(i, j)$, $1 \leqslant i \leqslant r$, $1 \leqslant j \leqslant s$.

A matrix $\mathbf{A}$ is called *symmetric* if and only if $\mathbf{A} = \mathbf{A}'$, i.e., if and only if it is identical to its transpose. Clearly, only square matrices can be symmetric. The simplest example of a symmetric matrix is a *diagonal* matrix for which all nonzero elements lie on its main diagonal (left upper corner—right lower corner).

For example, if

$$\mathbf{A}=\begin{pmatrix}1 & 3 & 1\\ 3 & 2 & 0\\ 1 & 0 & -2\end{pmatrix}, \qquad \mathbf{B}=\begin{pmatrix}-1 & 0 & 0\\ 0 & 2 & 0\\ 0 & 0 & -3\end{pmatrix},$$

then $\mathbf{A}$ is symmetric, $\mathbf{B}$ is diagonal, and $\mathbf{A}\geqslant\mathbf{B}$.

If $\mathbf{A}$ is a square matrix we will denote by $\mathbf{A}_{dg}$ the matrix resulting from setting all the entries off the main diagonal equal to 0.

For the previous example

$$\mathbf{A}_{dg}=\begin{pmatrix}1 & 0 & 0\\ 0 & 2 & 0\\ 0 & 0 & -2\end{pmatrix}, \qquad \mathbf{B}_{dg}=\begin{pmatrix}-1 & 0 & 0\\ 0 & 2 & 0\\ 0 & 0 & -3\end{pmatrix}.$$

Clearly, a matrix $\mathbf{A}$ is diagonal if and only if $\mathbf{A}=\mathbf{A}_{dg}$ (as is the case for the matrix $\mathbf{B}$ above).

1.6.5. The diagonal matrix of order r for which all main diagonal entries are equal to 1 is called the *unit* (or *identity*) matrix of order r and is denoted by $\mathbf{I}_r$ (the subscript will be omitted whenever the order is clear from the context).

For example, we have

$$\mathbf{I}_2=\begin{pmatrix}1 & 0\\ 0 & 1\end{pmatrix}, \qquad \mathbf{I}_3=\begin{pmatrix}1 & 0 & 0\\ 0 & 1 & 0\\ 0 & 0 & 1\end{pmatrix}.$$

1.6.6. The $r\times s$ matrix with all entries equal to zero will be denoted by $\mathbf{0}_{r\times s}$ and called the $r\times s$ *zero* matrix. The $r\times s$ matrix with all entries equal to 1 will be denoted by $\mathbf{E}_{r\times s}$.

In particular, for (column) vectors, corresponding to the case $s=1$, we shall write $\mathbf{0}_{r\times 1}=\mathbf{0}_r$, $\mathbf{E}_{r\times 1}=\mathbf{e}_r$. The subscripts will be omitted when the dimensions r and s are clear from the context.

For example, we have

$$\mathbf{0}_{3\times 2}=\begin{pmatrix}0 & 0\\ 0 & 0\\ 0 & 0\end{pmatrix}, \qquad \mathbf{E}_{3\times 2}=\begin{pmatrix}1 & 1\\ 1 & 1\\ 1 & 1\end{pmatrix},$$

$$\mathbf{0}_3=\begin{pmatrix}0\\ 0\\ 0\end{pmatrix}, \qquad \mathbf{e}_3=\begin{pmatrix}1\\ 1\\ 1\end{pmatrix}.$$

A matrix $\mathbf{A}$ with real entries is said to be *positive* or *nonnegative* according as $\mathbf{A}>\mathbf{0}$ or $\mathbf{A}\geqslant\mathbf{0}$.

1.7 OPERATIONS WITH MATRICES

1.7.1. The product $b\mathbf{A}$ of the matrix $\mathbf{A} = (a(i, j))$ by the real or complex number b is defined as the matrix $b\mathbf{A} = (ba(i, j))$. In particular, the product $(-1)\mathbf{A}$ will be denoted by $-\mathbf{A}$ and, consequently, $-\mathbf{A} = (-a(i, j))$. Clearly, $1\mathbf{A} = \mathbf{A}$ and $0\mathbf{A} = \mathbf{0}$. It is immediate that $(bc)\,\mathbf{A} = b(c\mathbf{A})$.

The sum $\mathbf{A} + \mathbf{B}$ of two $r \times s$ matrices $\mathbf{A} = (a(i, j))$ and $\mathbf{B} = (b(i, j))$ is defined as the matrix $\mathbf{A} + \mathbf{B} = (a(i, j) + b(i, j))$. This definition is extended in an obvious manner to any finite number of $r \times s$ matrices. Clearly, $\mathbf{A} + \mathbf{0} = \mathbf{0} + \mathbf{A} = \mathbf{A}$. In particular, $\mathbf{A} - \mathbf{B} = \mathbf{A} + (-\mathbf{B}) = (a(i, j) - b(i, j))$, therefore $\mathbf{A} - \mathbf{A} = -\mathbf{A} + \mathbf{A} = \mathbf{0}$.

For example, if

$$\mathbf{A} = \begin{pmatrix} 1 & 3 & 2 \\ 3 & -2 & 0 \\ 1 & 0 & -2 \end{pmatrix}, \quad \mathbf{B} = \begin{pmatrix} -1 & 0 & 2 \\ 0 & 2 & 1 \\ 1 & 2 & 3 \end{pmatrix},$$

then

$$3\mathbf{A} = \begin{pmatrix} 3 & 9 & 6 \\ 9 & -6 & 0 \\ 3 & 0 & -6 \end{pmatrix}, \quad \mathbf{A} + \mathbf{B} = \begin{pmatrix} 0 & 3 & 4 \\ 3 & 0 & 1 \\ 2 & 2 & 1 \end{pmatrix}.$$

Matrix addition is an associative and commutative operation, i.e.

$$(\mathbf{A} + \mathbf{B}) + \mathbf{C} = \mathbf{A} + (\mathbf{B} + \mathbf{C}), \ \mathbf{A} + \mathbf{B} = \mathbf{B} + \mathbf{A}$$

(here $\mathbf{C}$ is also an $r \times s$ matrix). Associativity of matrix addition allows us to omit the parantheses and write $\mathbf{A} + \mathbf{B} + \mathbf{C}$ instead of $(\mathbf{A} + \mathbf{B}) + \mathbf{C}$ or $\mathbf{A} + (\mathbf{B} + \mathbf{C})$. Next, multiplication by a number is distributive with respect to addition of matrices and of numbers, i.e.,

$$a(\mathbf{A} + \mathbf{B}) = a\mathbf{A} + a\mathbf{B}, \ (a + b)\mathbf{A} = a\mathbf{A} + b\mathbf{A}.$$

The following properties are immediate:

$$(\mathbf{A} + \mathbf{B})' = \mathbf{A}' + \mathbf{B}', \ (a\mathbf{A})' = a\mathbf{A}'.$$

1.7.2. Multiplication of two matrices is defined when the number of columns of the first factor equals the number of rows of the second one, i.e., for an $r \times s$ matrix $\mathbf{A} = (a(i, j))$ and an $s \times t$ matrix $\mathbf{B} = (b(j, k))$ (therefore, in particular, for two square matrices of the same order). The product $\mathbf{AB}$ of the above matrices $\mathbf{A}$ and $\mathbf{B}$ is defined as the $r \times t$ matrix.

$$\mathbf{AB} = \left(\sum_{j=1}^{s} a(i, j)\, b(j, k) \right)_{1 \leqslant i \leqslant r,\ 1 \leqslant k \leqslant t.}$$

This definition is extended in an obvious manner to any finite number of matrices.

For matrices **A** and **B** with real elements the *ik*th entry of the product **AB** is called the *inner product**) of the *s*-dimensional vector whose components are the elements of the *i*th row of the matrix **A** and the *s*-dimensional vector whose components are the elements of the *k*th column of the matrix **B**.

For example, if

$$\mathbf{A} = \begin{pmatrix} 1 & 2 & 1 \\ 2 & 1 & 3 \end{pmatrix}, \qquad \mathbf{B} = \begin{pmatrix} 1 & 0 \\ 0 & 2 \\ 1 & 3 \end{pmatrix},$$

then

$$\mathbf{AB} = \begin{pmatrix} 1\cdot 1 + 2\cdot 0 + 1\cdot 1 & 1\cdot 0 + 2\cdot 2 + 1\cdot 3 \\ 2\cdot 1 + 1\cdot 0 + 3\cdot 1 & 2\cdot 0 + 1\cdot 2 + 3\cdot 3 \end{pmatrix} = \begin{pmatrix} 2 & 7 \\ 5 & 11 \end{pmatrix}.$$

For the product **BA** to be defined we should have $r=t$, in which case **BA** is an $s \times s$ matrix (i.e., a (square) matrix of order s). For the previous example this condition if fulfilled and we have

$$\mathbf{BA} = \begin{pmatrix} 1\cdot 1 + 0\cdot 2 & 1\cdot 2 + 0\cdot 1 & 1\cdot 1 + 0\cdot 3 \\ 0\cdot 1 + 2\cdot 2 & 0\cdot 2 + 2\cdot 1 & 0\cdot 1 + 2\cdot 3 \\ 1\cdot 1 + 3\cdot 2 & 1\cdot 2 + 3\cdot 1 & 1\cdot 1 + 3\cdot 3 \end{pmatrix} = \begin{pmatrix} 1 & 2 & 1 \\ 4 & 2 & 6 \\ 7 & 5 & 10 \end{pmatrix}.$$

Therefore, matrix multiplication is not a commutative operation. However it is an associative one, i.e., $(\mathbf{AB})\mathbf{C} = \mathbf{A}(\mathbf{BC})$. This allows us to omit the parantheses and write **ABC** instead of $(\mathbf{AB})\mathbf{C}$ or $\mathbf{A}(\mathbf{BC})$. It is also immediate that $\mathbf{A}(a\mathbf{B}) = (a\mathbf{A})\mathbf{B} = a(\mathbf{AB})(=a\mathbf{AB})$. Next, matrix multiplication is distributive with respect to matrix addition, i.e., $(\mathbf{A} + \mathbf{B})\mathbf{C} = \mathbf{AC} + \mathbf{BC}$, $\mathbf{A}(\mathbf{B} + \mathbf{C}) = \mathbf{AB} + \mathbf{AC}$. We note also the fact that the transpose of a product of matrices is equal to the product of the transposed matrices in reversed order, i.e., $(\mathbf{AB})' = \mathbf{B}'\mathbf{A}'$. Finally, if **A** is an $r \times s$ matrix, then $\mathbf{0}_{t\times r}\mathbf{A} = \mathbf{0}_{t\times s}$, $\mathbf{A}\,\mathbf{0}_{s\times t} = \mathbf{0}_{r\times t}$, $\mathbf{I}_r\mathbf{A} = \mathbf{AI}_s = \mathbf{A}$.

An important special case where matrix multiplication is commutative is that of powers of a square matrix **A**. Let $\mathbf{A}^n$ be the matrix resulting from multiplication of **A** by itself n times. It is easily seen that $\mathbf{A}^n\mathbf{A}^m = \mathbf{A}^m\mathbf{A}^n = \mathbf{A}^{m+n}$ whatever the nonnegative integers m and n. (We define $\mathbf{A}^\circ = \mathbf{I}$.)

1.7.3. Let det(**A**) stand for the *determinant* of the square matrix $\mathbf{A} = (a(i,j))$ of order r defined by the expression

$$\det(\mathbf{A}) = \Sigma \pm a(1,j_1)\, a(2,j_2) \ldots a(r,j_r),$$

where the summation extends over all permutations $(j_1 \ldots, j_r)$ of $(1, \ldots, r)$ and the plus sign is chosen if the permutation is even and the minus

*) If $\mathbf{a} = (a(i))$ and $\mathbf{b} = (b(i))$ are s-dimensional vectors their inner product $(\mathbf{a}, \mathbf{b})$ is the number $\sum_{i=1}^{s} a(i)\, \overline{b(i)} (= \mathbf{a}'\overline{\mathbf{b}})$, where $\overline{b(i)}$ is the complex conjugate of $b(i)$.

sign if it is odd. A square matrix $\mathbf{A}$ is said to be *singular* or *nonsingular* according as $\det(\mathbf{A}) \neq 0$ or $\det(\mathbf{A}) = 0$. It can be shown that if $\mathbf{A}$ and $\mathbf{B}$ are square matrices of the same order then $\det(\mathbf{AB}) = \det(\mathbf{A}) \det(\mathbf{B})$.

For a square matrix $\mathbf{A} = (a(i, j))$ of order r let $\mathbf{A}_{ij}$ be the submatrix of $\mathbf{A}$ formed by deleting the ith row and the jth column of $\mathbf{A}$. Define the *cofactor* $A(i, j)$ of $a(i, j)$ as the number $A(i, j) = (-1)^{i+j} \det(\mathbf{A}_{ij})$. It can be shown that $\det(\mathbf{A}) = \sum_{k=1}^{r} a(i, k)A(i, k) = \sum_{l=1}^{r} a(l, j)\ A(l, j)$ whatever $1 \leqslant i, j \leqslant r$.

The square matrix adj$\mathbf{A}$ defined by

$$\operatorname{adj}\mathbf{A} = (A(i, j))'$$

is called the *adjoint**) of $\mathbf{A}$.

For example if

$$\mathbf{A} = \begin{pmatrix} 1 & 2 & 3 \\ 0 & 1 & 2 \\ 2 & 1 & 3 \end{pmatrix},$$

we have

$$A(1,1) = (-1)^{1+1} \begin{vmatrix} 1 & 2 \\ 1 & 3 \end{vmatrix} = 1, \quad A(1,2) = (-1)^{1+2} \begin{vmatrix} 0 & 2 \\ 2 & 3 \end{vmatrix} = 4,$$

$$A(1,3) = (-1)^{1+3} \begin{vmatrix} 0 & 1 \\ 2 & 1 \end{vmatrix} = -2, \quad A(2,1) = (-1)^{2+1} \begin{vmatrix} 2 & 3 \\ 1 & 3 \end{vmatrix} = -3,$$

$$A(2,2) = (-1)^{2+2} \begin{vmatrix} 1 & 3 \\ 2 & 3 \end{vmatrix} = -3, \quad A(2,3) = (-1)^{3+2} \begin{vmatrix} 1 & 2 \\ 2 & 1 \end{vmatrix} = 3,$$

$$A(3,1) = (-1)^{3+1} \begin{vmatrix} 2 & 3 \\ 1 & 2 \end{vmatrix} = 1, \quad A(3,2) = (-1)^{3+2} \begin{vmatrix} 1 & 3 \\ 0 & 2 \end{vmatrix} = -2,$$

$$A(3,3) = (-1)^{3+3} \begin{vmatrix} 1 & 2 \\ 0 & 1 \end{vmatrix} = 1.$$

Hence

$$\operatorname{adj}\mathbf{A} = \begin{pmatrix} 1 & -3 & 1 \\ 4 & -3 & -2 \\ -2 & 3 & 1 \end{pmatrix}.$$

It can be shown that for a square matrix $\mathbf{A}$ the existence of a matrix $\mathbf{A}^{-1}$ called the *inverse* of $\mathbf{A}$ such that $\mathbf{AA}^{-1} = \mathbf{A}^{-1}\mathbf{A} = \mathbf{I}$ is equivalent to nonsingularity of $\mathbf{A}$, i.e., $\det(\mathbf{A}) \neq 0$ and

$$\mathbf{A}^{-1} = \frac{1}{\det(\mathbf{A})} \operatorname{adj}\mathbf{A}.$$

*) It is convenient to define the adjoint of any 1×1 matrix to be the 1×1 matrix (1).

For the example considered

$$\det(\mathbf{A}) = \begin{vmatrix} 1 & 2 & 3 \\ 0 & 1 & 2 \\ 2 & 1 & 3 \end{vmatrix} = 1 \cdot A(1,1) + 0 \cdot A(2,1) + 2 \cdot A(3,1) = 3,$$

therefore

$$\mathbf{A}^{-1} = \frac{1}{3}\begin{pmatrix} 1 & -3 & 1 \\ 4 & -3 & -2 \\ -2 & 3 & 1 \end{pmatrix} = \begin{pmatrix} 1/3 & -1 & 1/3 \\ 4/3 & -1 & -2/3 \\ -2/3 & 1 & 1/3 \end{pmatrix}.$$

Clearly, $\mathbf{I}_r^{-1} = \mathbf{I}_r$ whatever the positive integer r.

It is easily seen that for nonsingular matrices $\mathbf{A}$ and $\mathbf{B}$ of the same order we have

$$(\mathbf{AB})^{-1} = \mathbf{B}^{-1}\mathbf{A}^{-1},$$

$$(\mathbf{A}')^{-1} = (\mathbf{A}^{-1})'.$$

1.7 4. The operations previously introduced allow us to define, by analogy with functions of a real or complex variable, certain functions whose arguments are matrices. The simplest case is that of a polynomial in a matrix. If

$$f(x) = a_0x^n + a_1x^{n-1} + \dots + a_{n-1}x + a_n$$

is a polynomial with real or complex coefficients and $\mathbf{A}$ a matrix of order r, then we define $f(\mathbf{A})$ as the matrix

$$f(\mathbf{A}) = a_0\mathbf{A}^n + a_1\mathbf{A}^{n-1} + \dots + a_{n-1}\mathbf{A} + a_n\mathbf{I}_r.$$

The same idea may be used to define functions of a matrix that in the case of a real or complex variable are defined in terms of series. Let us first make precise the concept of convergence of a sequence or a series of matrices. A sequence of $r \times s$ matrices $\mathbf{A}_n = (a_n(i, j))$, $n \geqslant 1$, is said to be convergent to the matrix $\mathbf{A} = (a(i, j))$ if and only if $a_n(i, j)$ converges to $a(i, j)$ as $n \to \infty$, $1 \leqslant i \leqslant r$, $1 \leqslant j \leqslant s$. Next, the matrix series $\sum_{n\geqslant 1} \mathbf{A}_n$ converges to the matrix $\mathbf{A}$ if and only if the sequence of partial sums $\mathbf{A}_1 + \dots + \mathbf{A}_n$, $n \geqslant 1$, converges to $\mathbf{A}$.

By analogy with the series expansion

$$e^x = \exp(x) = 1 + \sum_{n\geqslant 1} \frac{x^n}{n!}$$

we define $e^{\mathbf{A}} = \exp(\mathbf{A})$ for a square matrix $\mathbf{A} = (a(i, j))$ of order r as the sum of the series

$$\mathbf{I}_r + \sum_{n\geqslant 1} \frac{\mathbf{A}^n}{n!}.$$

It remains to be proved that it is convergent. To proceed, let us remark that setting $\delta = \max_{1\leqslant i,j\leqslant r} |a(i, j)|$, and $\mathbf{A}^n = (a^{(n)}(i, j))$, then $|a^{(2)}(i, j)| \leqslant$

$\leqslant r\,\delta^2$, and, by induction, $|a^{(n)}(i, j)| \leqslant r^{n-1}\delta^n$, $1 \leqslant i, j, \leqslant r$. The convergence of the series now follows, since

$$\sum_{n\geqslant 1} \frac{|a^{(n)}(i, j)|}{n!} \leqslant \sum_{n\geqslant 1} \frac{r^{n-1}\delta^n}{n!} = \frac{1}{r}\,(e^{r\delta} - 1).$$

It is easily verified that if **A** and **B** are square matrices of order r such that $\mathbf{AB} = \mathbf{BA}$, then

$$\exp(\mathbf{A} + \mathbf{B}) = \exp(\mathbf{A}) \exp(\mathbf{B}).$$

To conclude we consider the well known series expansion

$$\frac{1}{1 - x} = 1 + x + x^2 + \ldots$$

valid for $|x| < 1$. It suggests the matrix equation

$$(\mathbf{I} - \mathbf{A})^{-1} = \mathbf{I} + \mathbf{A} + \mathbf{A}^2 + \ldots$$

We shall prove

Theorem 1.7. *If* **A** *is a square matrix such that* $\mathbf{A}^n \to \mathbf{0}$ *(the null matrix) as* $n \to \infty$, *then the inverse* $(\mathbf{I} - \mathbf{A})^{-1}$ *exists and is given by*

$$(\mathbf{I} - \mathbf{A})^{-1} = \sum_{n\geqslant 0} \mathbf{A}^n.$$

Proof. Notice that

$$(\mathbf{I} - \mathbf{A})(\mathbf{I} + \mathbf{A} + \ldots + \mathbf{A}^{n-1}) = \mathbf{I} - \mathbf{A}^n.$$

By hypothesis the right side converges to the unit matrix **I** that has determinant 1. Therefore, for n large enough, $\det(\mathbf{I} - \mathbf{A}^n) \neq 0$. As the determinant of the product of two matrices is the product of their determinants we deduce that $\det(\mathbf{I} - \mathbf{A}) \neq 0$, hence $(\mathbf{I} - \mathbf{A})^{-1}$ does exist. Multiplying both sides of the above identity by $(\mathbf{I} - \mathbf{A})^{-1}$ yields

$$\mathbf{I} + \mathbf{A} + \ldots + \mathbf{A}^{n-1} = (\mathbf{I} - \mathbf{A})^{-1}(\mathbf{I} - \mathbf{A}^n).$$

Hence, letting $n \to \infty$, the proof is complete.

1.7.5. A sequence of $r \times s$ matrices is a special case of a matrix function, i.e., a matrix whose entries are functions of one or more variables. If the variables are real or complex then it makes sense to speak of continuity, absolute continuity, differentiability or integrability of a matrix function. These are defined component by component. For instance, if $\mathbf{A}(t) = (a(t, i, j))$ is a matrix function of a real variable then $\mathbf{A}(\cdot)$ is continuous at $t = t_0$ if and only if the $a(\cdot, i, j)$ are all continuous at $t = t_0$. Next, $\mathbf{A}(\cdot)$ is differentiable at $t = t_0$ if and only if the $a(\cdot, i, j)$ are all differentiable at $t = t_0$, and we write *) $\mathbf{A}'(t_0) = (a'(t_0, i, j))$.

*) The derivative $\mathbf{A}'(t_0)$ should not be confused with the transpose $(\mathbf{A}(t_0))'$.

Finally, $\mathbf{A}(\cdot)$ is integrable over an interval $[u, v]$ if and only if the $a(\cdot,i,j)$ are all integrable over $[u, v]$, and we write

$$\int_u^v \mathbf{A}(t)\,dt = \left(\int_u^v a(t, i, j)\,dt\right).$$

1.8 r-DIMENSIONAL SPACE

1.8.1. The set of all r-dimensional (column) vectors with complex components endowed with the operations of addition and multiplication by complex numbers and with the inner product is called the *r-dimensional (complex) space C^r*. The elements of C^r will be also called (r-dimensional) points.

Let us remember that the inner product of two vectors $\mathbf{a} = (a(i))$, $\mathbf{b} = (b(i)) \in C^r$ is defined as the complex number

$$(\mathbf{a}, \mathbf{b}) = \sum_{i=1}^{r} a(i)\,\overline{b(i)}.$$

By means of the operations with vectors and the inner product all the concepts of Euclidean r-dimensional geometry can be developed.

1.8.2. The vectors $\mathbf{a}_1, \ldots, \mathbf{a}_l \in C^r$ are said to be *linearly independent* if and only if the equation

$$c_1\mathbf{a}_1 + \ldots + c_l\mathbf{a}_l = \mathbf{0},$$

where $c_1, \ldots, c_l$ are complex numbers, is true only for $c_1 = \ldots = c_l = 0$. For examples, the vectors

$$\xi_1 = \begin{pmatrix} 1 \\ 0 \\ \vdots \\ 0 \end{pmatrix}, \quad \xi_2 = \begin{pmatrix} 0 \\ 1 \\ \vdots \\ 0 \end{pmatrix}, \quad \ldots, \quad \xi_r = \begin{pmatrix} 0 \\ 0 \\ \vdots \\ 1 \end{pmatrix}$$

are, clearly, linearly independent. If $\mathbf{a} = (a(i)) \in C^r$ then we have

$$\mathbf{a} = a(1)\,\xi_1 + \ldots + a(r)\,\xi_r.$$

Hence, in C^r there can exist at most r linearly independent vectors. Any system of r linearly independent vectors is called a *basis* of C^r. Any vector $\mathbf{a} \in C^r$ has a unique expression as a linear combination of the vectors $\mathbf{b}_1, \ldots, \mathbf{b}_r$ forming a basis of C^r, i.e., there exists a unique system of complex numbers $c_1, \ldots, c_r$ such that

$$\mathbf{a} = c_1\mathbf{b}_1 + \ldots + c_r\mathbf{b}_r.$$

The numbers $c_1, \ldots, c_r$ are called the *coordinates* of the vector $\mathbf{a}$ in the basis $\mathbf{b}_1, \ldots, \mathbf{b}_r$.

In the basis $\xi_1, \ldots, \xi_r$ the coordinates of **a** coincide with its components. Notice that

$$(\xi_i, \xi_j) = \delta(i, j) = \begin{cases} 1, & \text{if } i = j, \\ 0, & \text{if } i \neq j. \end{cases}$$

Two vectors $\mathbf{a}, \mathbf{b} \in C^r$ are said to be *orthogonal* if and only if $(\mathbf{a}, \mathbf{b}) = 0$. The quantity $(\mathbf{a}, \mathbf{a})^{1/2}$ is called the *length* of the vector **a**. In particular, all the vectors $\xi_1, \ldots, \xi_r$ have length 1. On account of the above properties the basis $\xi_1, \ldots, \xi_r$ is said to be *orthonormal.*

1.8.3. Analogous results hold for any *subspace* L of C^r, i.e., a subset L of C^r such that if $\mathbf{a}, \mathbf{b} \in L$ then $c_1\mathbf{a} + c_2\mathbf{b} \in L$ whatever the complex numbers c_1 and c_2. A subspace L is characterized by an integer $0 \leqslant q \leqslant r$ called the *dimension* of L which equals the largest number of linearly independent vectors in L. Any system of q linearly independent vectors in L is called a *basis* of L. Any vector $\mathbf{a} \in L$ has a unique expression as a linear combination of the vectors forming a basis of L. The extreme value $q = 0$ corresponds to the case where L consists of just the null vector $\mathbf{0}_r$. The other extreme value $q = r$ corresponds to the case where L coincides with C^r.

1.8.4. To conclude, we shall note that the set of all r-dimensional vectors with real components endowed with the operations of addition and multiplication by a real number and the inner product is called the *r-dimensional Euclidean space* R^r. The concepts defined for the space C^r are easily transferred to the space R^r by considering only real numbers instead of complex ones.

1.9 EIGENVALUES AND EIGENVECTORS

1.9.1. Let $\mathbf{A} = (a(i, j))$ be a square matrix of order r. A complex number λ is said to be an *eigenvalue* of the matrix **A** if and only if there exists a vector $\mathbf{u} \neq \mathbf{0}_r$ such that $\mathbf{u}'\mathbf{A} = \lambda\mathbf{u}'$. If λ is an eigenvalue of the matrix **A**, then the set $\mathfrak{U}_\lambda$ of all the vectors **u** satisfying the equation $\mathbf{u}'\mathbf{A} = \lambda\mathbf{u}'$ (the null vector $\mathbf{0}_r$ included) is called the *left eigenspace* corresponding to λ. It is immediate that $\mathfrak{U}_\lambda$ is a subspace of C^r. The dimension of $\mathfrak{U}_\lambda$ is called the *geometric multiplicity* of the eigenvalue λ. The eigenvalues of the matrix **A** are the roots of the rth degree algebraic equation (called the *characteristic equation*)

$$\det(\lambda\mathbf{I}_r - \mathbf{A}) = 0.$$

Indeed, if $\mathbf{u} = (u(i))$, then the equation $\mathbf{u}'\mathbf{A} = \lambda\mathbf{u}'$ is equivalent to the homogeneous system of linear equations

$$\sum_{i=1}^{r} u(i)\, a(i, j) = \lambda u(j), \qquad 1 \leqslant j \leqslant r,$$

which can be also written as

$$\sum_{i=1}^{r} u(i)\,(\lambda\delta(i,j) - a(i,j)) = 0, \qquad 1 \leqslant j \leqslant r. \qquad (1.10)$$

But a necessary and sufficient condition for a homogeneous system of linear equations to possess a nontrivial solution (i.e., such that at least one of the unknowns be nonzero) is that its determinant be zero. We are therefore led to the characteristic equation.

1.9.2. The characteristic equation has r complex roots, distinct or not. Therefore the matrix $\mathbf{A}$ has at most r eigenvalues and at most r left eigenspaces. It can be shown that the geometric multiplicity of an eigenvalue cannot exceed its algebraic multiplicity (as a root of the characteristic equation). Clearly, in the case where the algebraic multiplicity is 1, the geometric multiplicity is also 1. The converse is not true as we shall see from an example later on.

1.9.3. The values λ such that $\mathbf{Av} = \lambda\mathbf{v}$ for vectors $\mathbf{v} \neq \mathbf{0}_r$ are identical with the eigenvalues of the matrix $\mathbf{A}$ as defined above. Indeed, if $\mathbf{v} = (v(i))$, then the equation $\mathbf{Av} = \lambda\mathbf{v}$ is equivalent to the homogeneous system of linear equations

$$\sum_{j=1}^{r} (a(i,j) - \lambda\delta(i,j))\, v(j) = 0, \qquad 1 \leqslant i \leqslant r, \qquad (1.11)$$

whose determinant $\det(\mathbf{A} - \lambda\mathbf{I}_r)$ equals $(-1)^r \det(\lambda\mathbf{I}_r - \mathbf{A})$. It can be shown that if we denote by $\mathfrak{B}_\lambda$ the set of all the vectors satisfying the equation $\mathbf{Av} = \lambda\mathbf{v}$ (the *right eigenspace* corresponding to λ) then the dimension of $\mathfrak{B}_\lambda$ is precisely the geometric multiplicity of λ.

1.9.4. The nonnull vectors belonging to $\mathfrak{U}_\lambda$ are called *left eigenvectors* corresponding to λ. Similarly, the nonnull vectors belonging to $\mathfrak{B}_\lambda$ are called *right eigenvectors* corresponding to λ.

It should be stressed that there are no general methods for effectively determining the eigenvectors and the eigenvalues of a given matrix. Instead, we have at our disposal some general theoretical results. For example, it is known that the eigenvalues of a symmetric matrix with real entries are real and that the algebraic and geometric multiplicites of any eigenvalue are equal.

1.9.5. It can be shown that if $\lambda_1, \ldots, \lambda_l$ are pairwise distinct eigenvalues then, whatever the eigenvectors $\mathbf{u}_1 \in \mathfrak{U}_{\lambda_1} (\mathbf{v}_1 \in \mathfrak{B}_{\lambda_1}), \ldots, \mathbf{u}_l \in \mathfrak{U}_{\lambda_l} (\mathbf{v}_l \in \mathfrak{B}_{\lambda_l})$, they are linearly independent. Therefore, the largest number of linearly independent left (right) eingenvectors cannot exceed the sum of the geometric multiplicities of the distinct eigenvalues. This sum can be less than r, since it may happen that the geometric multiplicity of an eigenvalue is less than its algebraic multiplicity. As a simple example justifying this assertion, let us consider the matrix

$$\mathbf{A} = \begin{pmatrix} 1 & 1 \\ 0 & 1 \end{pmatrix}$$

whose characteristic equation is

$$\begin{vmatrix} \lambda - 1 & -1 \\ 0 & \lambda - 1 \end{vmatrix} = (\lambda - 1)^2 = 0,$$

which has the double root $\lambda = 1$. The left eigenvectors are found from the equation

$$u(1) + u(2) = u(1)$$
$$u(2) = u(2),$$

whence $u(2) = 0$, and $u(1)$ is arbitrary. Then the eigenspace $\mathfrak{U}_1$ consists of the vectors $\mathbf{u} = a(1, 0)'$, where a is an arbitrary complex number, and is therefore one-dimensional. Thus the geometric multiplicity of the eigenvalue $\lambda = 1$ is 1, although its algebraic multiplicity is 2.

1.9.6. A square matrix $\mathbf{A}$ is said to be *diagonalizable* if and only if there exists a nonsingular square matrix $\mathbf{U}$ of the same order such that $\mathbf{UAU}^{-1}$ is a diagonal matrix. It is immediate that the main diagonal elements of the latter matrix are eigenvalues of $\mathbf{A}$ and that the rows of $\mathbf{U}$ (the columns of $\mathbf{U}^{-1}$) are left (right) eigenvectors corresponding to these eigenvalues. Hence a necessary and sufficient condition for a square matrix of order r to be diagonalizable is that the sum of the geometric multiplicities of its distinct eigenvalues equals r (equivalently: the algebraic and geometric multiplicities are equal for any eigenvalue).

Let us consider in more detail the special case where the eigenvalues $\lambda_1, \ldots, \lambda_r$ are distinct. In this case for any $1 \leqslant i \leqslant r$ there exists a left (right) eigenvector $\mathbf{u}_i(\mathbf{v}_i)$ corresponding to the eigenvalue λ_i hence $\mathbf{u}_i'\mathbf{A} = \lambda_i\mathbf{u}_i'$, $\mathbf{A}\mathbf{v}_i = \lambda_i\mathbf{v}_i$. We have $\mathbf{u}_i'\mathbf{v}_j = 0$, $i \neq j$. Indeed, the above equations imply that $\lambda_i\mathbf{u}_i'\mathbf{v}_j = \mathbf{u}_i'\mathbf{A}\mathbf{v}_j = \lambda_j\mathbf{u}_i'\mathbf{v}_j$, whence $(\lambda_i - \lambda_j)\,\mathbf{u}_i'\mathbf{v}_j = 0$. As by hypothesis $\lambda_i \neq \lambda_j$, $i \neq j$, we get $\mathbf{u}_i'\mathbf{v}_j = 0$, $i \neq j$. Furthermore, assume that the vectors $\mathbf{u}_i$, $\mathbf{v}_i$ are chosen so that $\mathbf{u}_i'\mathbf{v}_i = 1$, $1 \leqslant i \leqslant r$, and denote by $\mathbf{U}(\mathbf{V})$ the matrix whose rows (columns) are the vectors $\mathbf{u}_i(\mathbf{v}_i)$, $1 \leqslant i \leqslant r$. Then we have $\mathbf{UV} = \mathbf{I}_r$. Hence the matrices $\mathbf{U}$ and $\mathbf{V}$ are nonsingular and $\mathbf{U} = \mathbf{V}^{-1}$, $\mathbf{V} = \mathbf{U}^{-1}$. This shows that the vectors $\mathbf{u}_1, \ldots, \mathbf{u}_r$ as well as the vectors $\mathbf{v}_1, \ldots, \mathbf{v}_r$ are linearly independent *). Therefore we have $\mathbf{UA} = (\lambda_i\delta(i, j))\mathbf{U}$, $\mathbf{AV} = \mathbf{V}(\lambda_i\delta(i, j))$, whence

$$\mathbf{UAU}^{-1} = \mathbf{V}^{-1}\mathbf{AV} = (\lambda_i\delta(i, j)),$$

$$\mathbf{A} = \mathbf{U}^{-1}(\lambda_i\delta(i, j))\,\mathbf{U} = \mathbf{V}(\lambda_i\delta(i, j))\mathbf{U},$$

---

*) This property follows also from 1.9.5.

that is

$$\mathbf{A} = \sum_{i=1}^{r} \lambda_i \mathbf{v}_i \mathbf{u}_i'. \tag{1.12}$$

The importance of this representation (called the *spectral representation*) mainly lies in the fact it implies that

$$\mathbf{A}^n = \sum_{i=1}^{r} \lambda_i^n \mathbf{v}_i \mathbf{u}_i'. \tag{1.12'}$$

Therefore once the spectral representation of a matrix is known its powers are easy to compute.

Remark. There exists a classical formula allowing one to compute the powers of a square matrix **A** of order r in terms of its distinct eigenvalues $\lambda_1, \ldots, \lambda_q$, $q \leqslant r$, of (algebraic) multiplicities $m_1, \ldots, m_q$, respectively, $m_1 + \ldots + m_q = r$. This is the Perron formula

$$\mathbf{A}^n = \sum_{h=1}^{q} \frac{1}{(m_h - 1)!} \left[\frac{d^{m_h - 1}}{d\lambda^{m_h - 1}} \left\{ \frac{\lambda^n \operatorname{adj}(\lambda \mathbf{I}_r - \mathbf{A})}{\prod_{i \neq h} (\lambda - \lambda_i)^{m_i}} \right\} \right]_{\lambda = \lambda_h}.$$

It can be shown that in the case where $q = r$, $m_1 = \ldots = m_r = 1$, the Perron formula reduces to equation (1.12'). A very simple proof of the Perron formula has been given in [64].

1.9.7. There exists a representation called the *Jordan canonical form* for an arbitrary square matrix. Let us denote by $\mathbf{L}_k(\lambda)$ the square matrix of order k given by

$$\mathbf{L}_k(\lambda) = \begin{pmatrix} \lambda & 1 & 0 \ldots 0 & 0 \\ 0 & \lambda & 1 \ldots 0 & 0 \\ \cdot & \cdot & \cdot \ldots \cdot & \cdot \\ 0 & 0 & 0 \ldots 1 & 0 \\ 0 & 0 & 0 \ldots \lambda & 1 \\ 0 & 0 & 0 \ldots 0 & \lambda \end{pmatrix}$$

with the convention $\mathbf{L}_1(\lambda) = (\lambda)$. It can be proved that for a given square matrix of order r there exists a nonsingular matrix **U** of the same order such that

$$\mathbf{A} = \mathbf{U} \begin{pmatrix} \mathbf{L}_{k_1}(\lambda_1) & & & \mathbf{0} \\ & \cdot & & \\ & & \cdot & \\ \mathbf{0} & & \cdot & \mathbf{L}_{k_s}(\lambda_s) \end{pmatrix} \mathbf{U}^{-1},$$

where $\lambda_1, \ldots, \lambda_s$ are the eigenvalues of $\mathbf{A}$ (not necessarily distinct) and $k_1 + \ldots + k_s = r$. The number of the $\mathbf{L}_k$ matrices in which the same eigenvalue λ occurs is equal to the geometric multiplicity of λ. On the other hand, the sum of the orders of the $\mathbf{L}_k$ matrices in which the same eigenvalue λ occurs is equal to the algebraic multiplicity of λ.

The spectral representation (1.12) is obviously a special case of the Jordan canonical form in which all the $\mathbf{L}_k$ matrices are of order 1.

Notice that the matrix occurring in the example considered in 1.9.5 is already represented in Jordan canonical form (with $\mathbf{U} = \mathbf{I}_2$).

1.10 NONNEGATIVE MATRICES. THE PERRON-FROBENIUS THEOREMS

1.10.1. For square matrices with nonnegative entries precise results concerning their eigenvalues and eigenvectors are known. They were obtained by O. Perron and G. Frobenius in the period 1907—1912.

Let $\mathbf{A} = (a(i, j))$ be a nonnegative square matrix. Then $\mathbf{A}$ is said to be *regular* if and only if there exists a natural number k such that $\mathbf{A}^k$ is a positive matrix.

Theorem 1.8. *Let* $\mathbf{A}$ *be a regular matrix. Then there exists a real eigenvalue* $\lambda_1 > 0$ *which is simple (i.e., of algebraic multiplicity* 1) *and which exceeds the absolute values of all other eigenvalues of* $\mathbf{A}$. *With* λ_1 *can be associated positive left and right eigenvectors* $\mathbf{u}_1$ *and* $\mathbf{v}_1$ *which are unique up to constant multiples. If* $\mathbf{u}_1$ *and* $\mathbf{v}_1$ *are chosen so that* $\mathbf{u}_1'\mathbf{v}_1 = 1$*), *then*

$$\mathbf{v}_1\mathbf{u}_1' = \frac{\operatorname{adj}(\lambda_1\mathbf{I} - \mathbf{A})}{f'(\lambda_1)}, \tag{1.13}$$

where $f(\lambda) = \det(\lambda\mathbf{I} - \mathbf{A})$. *The eigenvalue* λ_1 *is the only eigenvalue of* $\mathbf{A}$ *with which positive eigenvectors can be associated.*

Theorem 1.9. *Assume that the distinct eigenvalues of a regular matrix* $\mathbf{A}$ *of order* r *are* $\lambda_1, \lambda_2, \ldots, \lambda_q$, $q \leqslant r$, *where* $\lambda_1 > |\lambda_2| \geqslant |\lambda_3| \geqslant \ldots \geqslant |\lambda_q|$. *If* $|\lambda_2| = |\lambda_3|$ *let us stipulate that the algebraic multiplicity* m_2 *of* λ_2 *is at least as great as that of* λ_3 *and of any other eigenvalue having the same modulus as* λ_2. *If* $\mathbf{u}_1$ *and* $\mathbf{v}_1$ *have the same meanings as in the previous theorem (with* $\mathbf{u}_1'\mathbf{v}_1 = 1$), *then* **)

$$\mathbf{A}^n = \lambda_1^n\mathbf{v}_1\mathbf{u}_1' + O(n^{m_2-1}|\lambda_2|^n).$$

---

*) This is always possible because $\mathbf{u}_1$ and $\mathbf{v}_1$ are unique only up to constant multiples.

**) For a sequence $(a_n)_{n\geqslant 1}$ of real numbers, the Landau symbol $O(a_n)$ denotes a quantity which in absolute value does not exceed $c\,|a_n|$ for some $c > 0$ and for all sufficiently large n.

For example, if

$$\mathbf{A} = \begin{pmatrix} 0 & 6 & 6 \\ 4 & 3 & 5 \\ 8 & 3 & 1 \end{pmatrix},$$

then $\mathbf{A}$ is regular since $\mathbf{A}^2$, as is easily verified, is positive. We have

$$\det(\lambda \mathbf{I}_3 - \mathbf{A}) = \lambda^3 - 4\lambda^2 - 84\lambda - 144 = (\lambda - 12)(\lambda + 2)(\lambda + 6).$$

Therefore $\lambda_1 = 12$, $\lambda_2 = -6$, $\lambda_3 = -2$. The eigenvectors are found from systems (1.10) and (1.11) and we can take

$$\mathbf{u}_1 = \frac{1}{3}\begin{pmatrix} 1 \\ 1 \\ 1 \end{pmatrix}, \quad \mathbf{v}_1 = \begin{pmatrix} 1 \\ 1 \\ 1 \end{pmatrix}, \quad \mathbf{u}_2 = \frac{1}{12}\begin{pmatrix} 2 \\ -1 \\ -1 \end{pmatrix}, \quad \mathbf{v}_2 = \begin{pmatrix} 4 \\ 1 \\ -5 \end{pmatrix},$$

$$\mathbf{u}_3 = \frac{1}{4}\begin{pmatrix} -2 \\ 3 \\ -1 \end{pmatrix}, \quad \mathbf{v}_3 = \begin{pmatrix} 0 \\ 1 \\ -1 \end{pmatrix}.$$

Thus Theorem 1.8 is verified (the verification of equation (1.13) is left to the reader). At the same time, notice that this example also illustrates the spectral representation (1.12) and, implicitly, Theorem 1.9.

1.10.2. Theorem 1.8 can be generalized to a larger class of nonnegative matrices. A square matrix $\mathbf{A}$ (with real or complex entries) is said to be *reducible* if and only if, by applying the same permutation *) to its rows and columns it can be brought into the form

$$\mathbf{A} = \left(\begin{array}{c|c} \mathbf{A}_{11} & 0 \\ \hline \mathbf{A}_{21} & \mathbf{A}_{22} \end{array}\right),$$

where $\mathbf{A}_{11}$ and $\mathbf{A}_{22}$ are square matrices. A square matrix $\mathbf{A}$ is said to be *irreducible* if and only if it is not reducible. It can be shown that a necessary and sufficient condition for a square matrix $\mathbf{A} = (a(i, j))$ of order r to be irreducible is as follows: whatever the indices i, j, $1 \leqslant i, j \leqslant r$, either $a(i, j) \neq 0$ or there exists indices $i_0 = i, j_0, j_1, \ldots, j_n = j$, $n \geqslant 1$, such that $a(i_0, j_0)\, a(j_0, j_1) \ldots a(j_{n-1}, j_n) \neq 0$. Clearly, a regular matrix is irreducible. An irreducible matrix $\mathbf{A}$ is said to be *periodic*

*) It is immediate that such an operation does not alter the eigenvalues of the matrix as the characteristic equation remains invariant.

of period $d > 1$ if and only if by applying the same permutation to its rows and columns it can be brought into the form

$$\mathbf{A} = \begin{pmatrix} \mathbf{0} & \mathbf{A}_0 & \mathbf{0}\ldots & \mathbf{0} \\ \mathbf{0} & \mathbf{0} & \mathbf{A}_1\ldots & \mathbf{0} \\ \mathbf{0} & \mathbf{0} & \mathbf{0}\ldots & \mathbf{0} \\ . & . & \ldots\ldots & \\ \mathbf{0} & \mathbf{0} & \mathbf{0}\ldots & \mathbf{A}_{d-2} \\ \mathbf{A}_{d-1} & \mathbf{0} & \mathbf{0}\ldots & \mathbf{0} \end{pmatrix},$$

where the blocks along the diagonal are square. An irreducible matrix is said to be *aperiodic* (or of period $d = 1$) if and only if it is not periodic. It can be shown that a nonnegative matrix $\mathbf{A}$ is aperiodic if and only if there exist a natural number k such that all the entries of $\mathbf{A}^k$ are positive. Therefore, for nonnegative matrices, "aperiodic" is equivalent to "regular".

For example the matrix

$$\mathbf{A} = \begin{array}{c|cccccc} & 1 & 2 & 3 & 4 & 5 & 6 \\ \hline 1 & 1 & 1 & 1 & 2 & 2 & 0 \\ 2 & 0 & 2 & 0 & 3 & 0 & 0 \\ 3 & 6 & 0 & 0 & 1 & 1 & 4 \\ 4 & 0 & 0 & 0 & 1 & 0 & 0 \\ 5 & 2 & 1 & 7 & 1 & 5 & 1 \\ 6 & 2 & 2 & 0 & 0 & 4 & 3 \end{array}$$

is reducible since it can be written as

$$\mathbf{A} = \begin{array}{c|cc:cccc} & 2 & 4 & 3 & 5 & 6 & 1 \\ \hline 2 & 2 & 3 & 0 & 0 & 0 & 0 \\ 4 & 0 & 1 & 0 & 0 & 0 & 0 \\ \hdashline 3 & 2 & 1 & 0 & 1 & 4 & 6 \\ 5 & 1 & 1 & 7 & 5 & 1 & 2 \\ 6 & 2 & 0 & 0 & 4 & 3 & 2 \\ 1 & 1 & 2 & 1 & 2 & 0 & 1 \end{array}.$$

On the other hand, the matrix

$$\mathbf{A} = \begin{array}{c|cccccc} & 1 & 2 & 3 & 4 & 5 & 6 \\ \hline 1 & 0 & 1 & 0 & 0 & 0 & 0 \\ 2 & 0 & 0 & 2 & 3 & 0 & 1 \\ 3 & 1 & 0 & 0 & 0 & 4 & 0 \\ 4 & 1 & 0 & 0 & 0 & 0 & 0 \\ 5 & 0 & 2 & 0 & 0 & 0 & 0 \\ 6 & 0 & 0 & 0 & 0 & 3 & 0 \end{array}$$

is irreducible of period $d = 3$, since it can be written as

$$\mathbf{A} = \begin{array}{c} \\ 1 \\ 5 \\ 2 \\ 3 \\ 4 \\ 6 \end{array} \begin{array}{c} \begin{array}{cccccc} 1 & 5 & 2 & 3 & 4 & 6 \end{array} \\ \begin{pmatrix} 0 & 0 & 1 & 0 & 0 & 0 \\ 0 & 0 & 2 & 0 & 0 & 0 \\ 0 & 0 & 0 & 2 & 3 & 1 \\ 1 & 4 & 0 & 0 & 0 & 0 \\ 1 & 0 & 0 & 0 & 0 & 0 \\ 0 & 3 & 0 & 0 & 0 & 0 \end{pmatrix} \end{array}.$$

Theorem 1.10. *Let* $\mathbf{A}$ *be a nonnegative irreducible matrix. Then all of the assertions of Theorem 1.8 hold save that for the eigenvalue* λ_1 *one can only assert that it is greater than or equal to the absolute values of the remaining eigenvalues of* $\mathbf{A}$. *More precisely, if* $\mathbf{A}$ *has period* $d \geqslant 1$, *then there exist exactly* d *distinct eigenvalues of modulus* λ_1, *namely* $\lambda_1 \exp(2\pi i k/d)$, $0 \leqslant k \leqslant d-1$, *i.e., the* d *roots of the equation* $\lambda^d - \lambda_1^d = 0$. *Conversely, if a nonnegative irreducible matrix* $\mathbf{A}$ *has* $d \geqslant 1$ *eigenvalues of equal moduli exceeding the absolute values of the remaining eigenvalues then* $\mathbf{A}$ *is periodic of period* d.

In the case of the irreducible matrix in the previous example, the characteristic equation is $\lambda^6 - 27\lambda^3 = 0$ and has roots $\lambda_1 = 3$, $\lambda_2 = 3(1 + i\sqrt{3})/2$, $\lambda_3 = 3(1 - i\sqrt{3})/2$, $\lambda_4 = 0$ (of algebraic multiplicity 3; what is the geometric multiplicity of this eigenvalue ?). It is easily found that we can take

$$\mathbf{u}_1 = \frac{1}{27}\begin{pmatrix} 5 \\ 11 \\ 9 \\ 6 \\ 9 \\ 3 \end{pmatrix}, \qquad \mathbf{v}_1 = \frac{1}{9}\begin{pmatrix} 3 \\ 6 \\ 9 \\ 9 \\ 1 \\ 6 \end{pmatrix}.$$

1.11 STOCHASTIC MATRICES. ERGODICITY COEFFICIENTS

1.11.1. A very important class of nonnegative matrices is represented by the so-called stochastic matrices which are of basic importance in the theory of Markov chains and processes.

A nonnegative matrix $\mathbf{A}$ is said to be *stochastic* if and only if the sum of the entries in any row of $\mathbf{A}$ is 1. A stochastic matrix is said to be *doubly stochastic* if and only if the sum of the entries in any column of $\mathbf{A}$ is 1 also. Therefore, any row of a stochastic matrix is a so-called *probability vector* (a synonym for discrete probability distribution — see 1.4.2). In a doubly stochastic matrix both the rows and the columns are probability vectors.

For example, if

$$\mathbf{A} = \begin{pmatrix} 1/4 & 3/4 \\ 1/2 & 1/2 \\ 1/3 & 2/3 \end{pmatrix}, \qquad \mathbf{B} = \begin{pmatrix} 1/2 & 1/2 & 0 \\ 1/4 & 1/2 & 1/4 \\ 1/4 & 0 & 3/4 \end{pmatrix},$$

then **A** is stochastic and **B** doubly stochastic.

The definition and the example considered show that the concept of a stochastic matrix is not restricted to square matrices. However, unless stated otherwise, in what follows the adjective "stochastic" will be attached to square matrices. From now on to denote stochastic matrices we shall use the letter **P**(the reason is quite obvious).

1.11.2. Let us show that the eigenvalues of a stochastic matrix are in absolute value at most equal to 1. Indeet, if λ is an eigenvalue of the stochastic matrix $\mathbf{P} = (p(i,j))$ of order r and $\mathbf{u} = (u(i))$ a left eigenvector corresponding to λ, we have

$$\lambda u(j) = \sum_{i=1}^{r} u(i)p(i,j),\ 1 \leqslant j \leqslant r,$$

whence

$$\sum_{j=1}^{r} |\lambda\, u(j)| = \sum_{j=1}^{r} \Big|\sum_{i=1}^{r} u(i)\, p(i,j)\Big| \leqslant \sum_{i=1}^{r} |u(i)| \sum_{j=1}^{r} p(i,j) = \sum_{i=1}^{r} |u(i)|,$$

that is

$$|\lambda| \sum_{j=1}^{r} |u(j)| \leqslant \sum_{i=1}^{r} |u(i)|,$$

which implies $|\lambda| \leqslant 1$.

Let us notice that since $\mathbf{P}\mathbf{e}_r = \mathbf{e}_r$ (we remind the reader that $\mathbf{e}_r$ is the r-dimensional (column) vector with all its components equal to 1), $\lambda = 1$ is an eigenvalue of the stochastic matrix **P** of order r and $\mathbf{e}_r$ is a right eigenvector corresponding to it.

A. N. Kolmogorov raised in 1938 the problem of determining the set Λ_r of all the points of the unit disk $|\lambda| \leqslant 1$ which are the eigenvalues of some stochastic matrix of order r. This problem was completely solved in 1951 by F.I. Karpelevič. The boundary of the set Λ_r is constituted from arcs successively joining the points $\exp(2\pi i a/b)$, $0 \leqslant a \leqslant b \leqslant r$, of the unit circle. For a precise formulation the reader should consult [193].

In the case $r = 2$, when the matrix **P** can be written as

$$\mathbf{P} = \begin{pmatrix} 1 - p(1,2) & p(1,2) \\ p(2,1) & 1 - p(2,1) \end{pmatrix},$$

we easily find that $\lambda_1 = 1$, $\lambda_2 = 1 - p(1,2) - p(2,1)$. If $p(1,2), p(2,1) \neq 0$ (equivalently, $\mathbf{P} \neq \mathbf{I}_2$), then $\lambda_1 \neq \lambda_2$ and the left and right eigenvectors corresponding to λ_1 and λ_2 are

$$\mathbf{u}_1 = \begin{pmatrix} \pi(1) \\ \pi(2) \end{pmatrix}, \quad \mathbf{v}_1 = \begin{pmatrix} 1 \\ 1 \end{pmatrix}, \quad \mathbf{u}_2 = \begin{pmatrix} 1 \\ -1 \end{pmatrix}, \quad \mathbf{v}_2 = \begin{pmatrix} \pi(2) \\ -\pi(1) \end{pmatrix},$$

where $\pi(1) = p(2,1)/(1-\lambda_2)$, $\pi(2) = p(1,2)/(1-\lambda_2)$. Therefore, the spectral representation of $\mathbf{P}$ is given by

$$\mathbf{P} = \begin{pmatrix} \pi(1) & \pi(2) \\ \pi(1) & \pi(2) \end{pmatrix} + \lambda_2 \begin{pmatrix} \pi(2) & -\pi(2) \\ -\pi(1) & \pi(1) \end{pmatrix},$$

whence

$$\mathbf{P}^n = \begin{pmatrix} \pi(1) & \pi(2) \\ \pi(1) & \pi(2) \end{pmatrix} + \lambda_n^2 \begin{pmatrix} \pi(2) & -\pi(2) \\ -\pi(1) & \pi(1) \end{pmatrix}.$$

Finally, the set Λ_2 is the segment $[-1, +1]$.

1.11.3. In Markov chain theory it is especially important to study the conditions under which the nth power of a stochastic matrix $\mathbf{P}$ of order r converges as $n \to \infty$ to a *stable* stochastic matrix (i.e., a stochastic matrix whose rows are identical). This property, known as *ergodicity*, can be studied by means of the so called ergodicity coefficients.

By *ergodicity coefficient* we understand any $[0,1]$-valued function $\varepsilon(\cdot)$ defined and continuous on the set of stochastic matrices of order r (regarded as points in r^2-dimensional Euclidean space). An ergodicity coefficient is said to be *proper* if and only if the equation $\varepsilon(\mathbf{P}) = 1$ holds only for stable matrices $\mathbf{P}$. If $\mathbf{P} = (p(i, j))_{1 \leqslant i,j \leqslant r}$ we define

$$\mu(\mathbf{P}) = \max_{1 \leqslant j \leqslant r} (\min_{1 \leqslant i \leqslant r} p(i, j)), \quad \delta(\mathbf{P}) = \sum_{j=1}^{r} (\min_{1 \leqslant i \leqslant r} p(i, j)),$$

$$\alpha(\mathbf{P}) = \min_{1 \leqslant i,j \leqslant r} \sum_{k=1}^{r} \min(p(i, k), \; p(j, k)).$$

(The notation min (a, b) designates the smallest of the numbers a and b.) Clearly, μ, δ and α are ergodicity coefficients, the last two being proper, and

$$\mu(\mathbf{P}) \leqslant \delta(\mathbf{P}) \leqslant \alpha(\mathbf{P}). \tag{1.14}$$

In the case $r = 2$ (see 1.11.2) it is easily seen that

$$\mu(\mathbf{P}) = \frac{1}{2}(1 - |\lambda_2| + |p(1,2) - p(2,1)|), \quad \delta(\mathbf{P}) = \alpha(\mathbf{P}) = 1 - |\lambda_2|.$$

The coefficient μ was in fact used by A A. Markov himself, the coefficient δ by W. Doeblin [81] (cf. [358]), and the coefficient α by

R.L. Dobrušin [79]. Several other authors independently rediscovered and used these coefficients.

1.11.4. A stochastic matrix $\mathbf{P}$ is said to be a *Markov matrix* if and only if $\mu(\mathbf{P}) > 0$. This inequality is equivalent to the fact that the matrix $\mathbf{P}$ has a column whose entries are positive. Indeed, if

$$\mu(\mathbf{P}) = \max_{1 \leqslant j \leqslant r} (\min_{1 \leqslant i \leqslant r} p(i,j)) > 0,$$

then there exists j_0 such that $\min_{1 \leqslant i \leqslant h} p(i, j_0) > 0$, that is $p(i, j_0) > 0$ whatever $1 \leqslant i \leqslant r$. The converse is obvious. Notice that the inequality $\mu(\mathbf{P}) > 0$ is also equivalent to the inequality $\delta(\mathbf{P}) > 0$. Indeed, if $\mu(\mathbf{P}) > 0$ then $\delta(\mathbf{P}) > 0$ by virtue of inequalities (1.14). Conversely, if

$$\delta(\mathbf{P}) = \sum_{j=1}^{r} (\min_{1 \leqslant i \leqslant r} (p(i,j)) > 0,$$

then there exists j_0, such that $\min_{1 \leqslant i \leqslant r} p(i, j_0) > 0$ (since a positive sum of nonnegative numbers should contain a positive term), hence $\mu(\mathbf{P}) > 0$.

Therefore the Markov matrices — stochastic matrices possessing at least a column whose entries are positive — are characterized by the fact that both ergodicity coefficients δ and μ are positive. The main difference between them lies in the fact that the first one is proper while the second one is not.

1.11.5. A stochastic matrix $\mathbf{P}$ is said to be *scrambling* if and only if $\alpha(\mathbf{P}) > 0$. It follows from the very definition of the ergodicity coefficient α that $\alpha(\mathbf{P}) > 0$ if and only if any submatrix of $\mathbf{P}$ consisting of two rows of it possesses a column whose entries are positive. Clearly, a Markov matrix is scrambling, too. The converse is not true, as is shown by the matrix

$$\mathbf{P} = \begin{pmatrix} 0 & 1/2 & 1/2 \\ 1/3 & 2/3 & 0 \\ 1/4 & 0 & 3/4 \end{pmatrix}.$$

Although it is scrambling (we have $\alpha(\mathbf{P}) = 1/4$), it is not a Markov matrix.

Notice that as $\min(a, b) = (a + b - |a-b|)/2$, we have

$$\alpha(\mathbf{P}) = 1 - \frac{1}{2} \max_{1 \leqslant i,j \leqslant r} \sum_{k=1}^{r} |p(i, k) - p(j, k)|. \tag{1.15}$$

Setting

$$a^+ = \begin{Bmatrix} a, \text{ if } a \geqslant 0 \\ 0, \text{ if } a < 0 \end{Bmatrix}, \quad a^- = \begin{Bmatrix} 0, \text{ if } a \geqslant 0 \\ a, \text{ if } a < 0 \end{Bmatrix}$$

for any real number a, and taking into account that

$$\sum_{k=1}^{r} (p(i, k) - p(j, k)) = 0,$$

we get

$$\sum_{k=1}^{r} (p(i, k) - p(j, k))^+ = -\sum_{k=1}^{r} (p(i, k) - p(j, k))^- \qquad (1.16)$$

$$= \frac{1}{2} \sum_{k=1}^{r} |p(i, k) - p(j, k)|.$$

Hence

$$\max_{I} \sum_{k \in I} (p(i, k) - p(j, k)) = \frac{1}{2} \sum_{k=1}^{r} |p(i, k) - p(j, k)|$$

[the maximum is taken over all the subsets I of the set $(1, 2, \ldots, r)$], which allows us to write that

$$\alpha(\mathbf{P}) = 1 - \max_{1 \leqslant i, j \leqslant r} \max_{I} \sum_{k \in I} (p(i, k) - p(j, k)). \qquad (1.17)$$

We shall now prove the following fundamental property.

Theorem 1.11. *Whatever the stochastic matrices* $\mathbf{P}_1 = (p_1(i, j))$ *and* $\mathbf{P}_2 = (p_2(i, j))$ *of the same order* r *we have*

$$1 - \alpha(\mathbf{P}_1\mathbf{P}_2) \leqslant (1 - \alpha(\mathbf{P}_1))\,(1 - \alpha(\mathbf{P}_2)).$$

In the case $r = 2$ *the sign* ' $\leqslant$ ' *is to be replaced by the sign* ' $=$ '

Proof. From (1.17) we have

$$1 - \alpha(\mathbf{P}_1\mathbf{P}_2) = \max_{1 \leqslant i, j \leqslant r} \max_{I} \sum_{k \in I} \sum_{l=1}^{r} (p_1(i, l) - p_1(j, l))\, p_2(l, k).$$

Now, let us remark that whatever $1 \leqslant i,\ j \leqslant r$ and I

$$\sum_{k \in I} \sum_{l=1}^{r} (p_1(i, l) - p_1(j, l))\, p_2(l, k)$$

$$= \left(\sum_{l=1}^{r} (p_1(i, l) - p_1(j, l))^+ \sum_{k \in I} p_2(l, k) \right.$$

$$\left. + \sum_{l=1}^{r} (p_1(i, l) - p_1(j, l))^- \sum_{k \in I} p_2(l, k) \right)$$

$$\leqslant \frac{1}{2} \sum_{k=1}^{r} |p_1(i, k) - p_1(j, k)| \left(\max_{1 \leqslant l \leqslant r} \sum_{k \in I} p_2(l, k) - \min_{1 \leqslant l \leqslant r} \sum_{k \in I} p_2(l, k) \right)$$

(by (1.16))

$$= \frac{1}{2} \sum_{k=1}^{r} |p_1(i, k) - p_1(j, k)| \max_{1 \leqslant l', l'' \leqslant r} \sum_{k \in I} (p_2(l', k) - p_2(l'', k))$$

$$\leqslant (1 - \alpha(\mathbf{P}_1))\,(1 - \alpha(\mathbf{P}_2))$$

(by (1.15) and (1.17)).

Hence, i, j, and I being arbitrary, we get the stated inequality. The result for the case $r = 2$ follows by direct computation (see 1.11.3).

Corollary. *Whatever the stochastic matrices* $\mathbf{P}_1, \ldots, \mathbf{P}_n$ *of the same order we have*

$$1 - \alpha(\mathbf{P}_1 \ldots \mathbf{P}_n) \leqslant (1 - \alpha(\mathbf{P}_1)) \ldots (1 - \alpha(\mathbf{P}_n)).$$

The *proof* is immediate by induction on n.

Remark. Clearly, Theorem 1.11 is still true for an $r \times s$ stochastic matrix $\mathbf{P}_1$ and an $s \times t$ stochastic matrix $\mathbf{P}_2$. In particular, if $\mathbf{p}_1 = (p_1(i))$ and $p_2 = (\mathbf{p}_2(i))$ are two r-dimensional probability vectors and $\mathbf{P} = (p(i, j))$ a stochastic (square) matrix of order r, then by applying Theorem 1.11 to the matrices

$$\begin{pmatrix} p_1(1) \ldots p_1(r) \\ p_2(1) \ldots p_2(r) \end{pmatrix}$$

and $\mathbf{P}$ we get the inequality

$$\sum_{k=1}^{r} \left| \sum_{i=1}^{r} (p_1(i) - p_2(i))\, p(i, k) \right| \leqslant \sum_{k=1}^{r} | p_1(k) - p_2(k) | \, (1 - \alpha(\mathbf{P})), \tag{1.18}$$

which we shall use later, in 7.4.3.

CHAPTER 2

Fundamental Concepts in Homogeneous Markov Chain Theory

This chapter is meant to lay the foundations of a thorough study of homogeneous Markov chains as mathematical models of those random processes evolving in time that remember only the most recent past and whose conditional distributions are time invariant. As we shall see, the Markov property and time homogeneity which formalize these attributes allow us to develop a rich system of concepts and results.

2.1 THE MARKOV PROPERTY

2.1.1. The Markov chains in this chapter are constructed starting from three basic quantities: a finite set S whose elements, called *states*, will be assumed to be numbered in a definite manner, a probability distribution $\mathbf{p} = (p(i))_{i\in S}$ over S, whose components are called *initial probabilities*, and a stochastic matrix $\mathbf{P} = (p(i, j))_{i,j\in S}$, whose entries are called *transition probabilities*. Therefore

$$p(i) \geqslant 0, \quad i\in S, \quad \sum_{i\in S} p(i) = 1,$$

$$p(i, j) \geqslant 0, \quad i, j\in S, \quad \sum_{j\in S} p(i, j) = 1, \quad i\in S.$$

A sequence of S-valued random variables $(X(n))_{n\geqslant 0}$ is said to be a (*homogeneous finite*) *Markov chain* with *state space* S, *initial distribution* $\mathbf{p}$ and *transition matrix* $\mathbf{P}$ if and only if $\mathsf{P}(X(0) = i) = p(i)$, $i\in S$, and

$$\underline{\mathsf{P}(X(n+1) = i_{n+1} \mid X(n) = i_n, \ldots, X(0) = i_0) = \mathsf{P}(X(n+1) = i_{n+1} \mid X(n) = i_n)}$$
$$= p(i_n, i_{n+1})$$

for all $n \geqslant 0$ and $i_0, \ldots, i_{n+1}\in S$, whenever the left hand side is defined.

The underlined equality is known as the *Markov property*. Homogeneity is introduced by the second equality that tells us the conditional probability of state i_{n+1} at time $n+1$ given state i_n at time n has a prescribed value independent of n. (The nonhomogeneous case will be studied in Chapter 7.)

2.1.2. We shall first prove the existence of such a mathematical object. To this end let us define the joint probability distribution of the random variables $X(0), \ldots, X(n)$, $n > 0$, by the equation *)

$$\begin{aligned} \mathsf{P}(X(0) = i_0, \ldots, X(n-1) = i_{n-1}, X(n) = i_n) \\ = p(i_0)\, p(i_0, i_1) \ldots p(i_{n-1}, i_n). \end{aligned} \tag{2.1}$$

For $n = 0$ we put $\mathsf{P}(X(0) = i) = p(i)$, $i \in S$. Since the matrix $\mathbf{P}$ is stochastic, summing over $i_n \in S$, the right hand side of (2.1) yields $p(i_0)\, p(i_0, i_1) \ldots p(i_{n-2}, i_{n-1})$. But this value coincides with that assigned to $\mathsf{P}(X(0) = i_0, \ldots, X(n-1) = i_{n-1})$. Consequently, the joint probability distributions considered satisfy the consistency conditions (1.8). Thus, the existence of a probability P and of an infinite sequence of random variables $X(0), X(1), \ldots$, satisfying (2.1) for all natural numbers n is ensured (see 1.5.1). Let us prove that this sequence has the Markov property. We have

$$\mathsf{P}(X(n+1) = i_{n+1} \mid X(n) = i_n, \ldots, X(0) = i_0)$$

$$= \frac{\mathsf{P}(X(0) = i_0, \ldots, X(n) = i_n, X(n+1) = i_{n+1})}{\mathsf{P}(X(0) = i_0, \ldots, X(n) = i_n)}$$

(assuming the denominator to be nonzero)

$$= \frac{p(i_0)\, p(i_0, i_1) \ldots p(i_{n-1}, i_n)\, p(i_n, i_{n+1})}{p(i_0)\, p(i_0, i_1) \ldots p(i_{n-1}, i_n)} = p(i_n, i_{n+1}).$$

On the other hand

$$\mathsf{P}(X(n+1) = i_{n+1} \mid X(n) = i_n) = \frac{\mathsf{P}(X(n) = i_n, X(n+1) = i_{n+1})}{\mathsf{P}(X(n) = i_n)}$$

$$= \frac{\sum\limits_{i_0, \ldots, i_{n-1}} \mathsf{P}(X(0) = i_0, \ldots, X(n-1) = i_{n-1}, X(n) = i_n, X(n+1) = i_{n+1})}{\sum\limits_{i_0, \ldots, i_{n-1}} \mathsf{P}(X(0) = i_0, \ldots, X(n) = i_n)}$$

$$= \frac{\sum\limits_{i_0, \ldots, i_{n-1}} p(i_0)\, p(i_0, i_1) \ldots p(i_{n-1}, i_n)\, p(i_n, i_{n+1})}{\sum\limits_{i_0, \ldots, i_{n-1}} p(i_0)\, p(i_0, i_1) \ldots p(i_{n-1}, i_n)} = p(i_n, i_{n+1}),$$

and hence the Markov property holds.

In what follows the transition matrix of a Markov chain will be kept fixed but the initial distribution will be allowed to vary. In some cases,

---

*) This is a consequence of the definition of a homogeneous Markov chain. As we shall see it implies in turn the Markov property and homogeneity.

to emphasize its dependence on $\mathbf{p}$, the probability P will be written as $\mathsf{P}_{\mathbf{p}}$. Accordingly, the mean value operator under $\mathsf{P}_{\mathbf{p}}$ will be denoted by $\mathsf{E}_{\mathbf{p}}$. In particular, for an initial distribution concentrated in state i, i.e., for the case where $p(j) = \delta(i, j)$*), $j \in S$, the corresponding $\mathsf{P}_{\mathbf{p}}$ will be denoted by P_i and the corresponding Markov chain will be said to *start* in state i (or to have i as *initial* state). Acordingly, the mean value operator under P_i will be denoted by E_i.

Now a few remarks on the language we shall use. To make it intuitive, we shall say that the Markov chain $(X(n))_{n\geqslant 0}$ is in state i at time n (or at nth step), meaning that the event $\{X(n) = i\}$ occurs. Expressions like leaving, staying, passing, reaching, entering, returning, avoiding, moving, etc. are to be interpreted in the same spirit. Next, if n denotes the present moment, then $X(n)$ is the *present* state, the section $(X(0),\ldots X(n))$ is the *past* while $(X(n), X(n+1), \ldots)$ is the *future* of the Markov chain. Finally, the expression "it is possible" will be used in connection with Markov chains to mean "with positive probability".

2.1.3. A Markov chain with r states admits a concrete representation by means of an urn scheme as follows. There are r urns marked $1, \ldots, r$ each containing r types of balls also marked $1, \ldots, r$. The composition of the urns remains fixed but varies from urn to urn: for urn i the probability of drawing a ball marked j equals $p(i, j)$. At time 0 an urn is chosen according to the probability distribution $(p(i))_{1\leqslant i\leqslant r}$. From that urn a ball is drawn at random and then replaced. If it is marked i then the next drawing is made from urn i, etc. It is clear the sequence of types of balls successively drawn is a Markov chain with state space $(1, \ldots, r)$, initial distribution $p(i))_{1\leqslant i\leqslant r}$ and transition matrix $(p(i,j))_{1\leqslant i,j\leqslant r}$.

The above representation was very popular in the early days of Markov chain theory.

Notice that the urn scheme just described does not constitute a *proof* of the existence of finite state Markov chains, as this amounts to the existence of a probability P satisfying (2.1) whatever the natural number n.

2.1.4. In what follows we discuss a few immediate consequences of the Markov property.

A random event A is said to be *prior to time* $n \geqslant 0$ for a Markov chain $(X(n))_{n\geqslant 0}$ if and only if it is determined by the random variables $X(0), \ldots, X(n)$, i.e., it can be expressed as $A = \{(X(0), \ldots, X(n)) \in A^{(n+1)}\}$, where $A^{(n+1)}$ is a set of $(n+1)$-tuples of states of the chain. Similarly, a random event B is said to be *posterior to time* $n \geqslant 0$ for a Markov chain $(X(n))_{n\geqslant 0}$ if and only if it is determined by the random variables $X(n), X(n+1), \ldots$, i.e., it can be expressed as $B = \{(X(n), X(n+1), \ldots) \in B^{(\infty)}\}$, where $B^{(\infty)}$ is a set of infinite sequences of states of the chain. In other words, we can tell by examination of the first $n+1$ variables

---

*) $\delta(i, j)$ is Kronecker's symbol defined on p. 47.

of the chain whether or not an event prior to time n has occurred; for telling whether or not an event posterior to time n has occurred it is necessary to examine the random variables of the chain beginning with that of index $n+1$. For example, a random event A prior to time n may be a completely specified past history of the form $\{X(0) = i_0, \ldots, X(n) = i_n\}$, or a more general random event of the form $\{X(0) \in S_0, \ldots, X(n) \in S_n\}$, where $S_0, \ldots, S_n$ are subsets of the state space S. Some of these subsets may coincide with S so that no restriction is placed on the corresponding random variables. Thus $\{X(0) \in S_0, X(1) \in S, X(2) \in S_2, X(3) \in S\}$ is nothing but $\{X(0) \in S_0, X(2) \in S_2\}$. Analogous considerations can be made concerning random events posterior to some given time. It is easily proved that the sets of all the random events prior (respectively posterior) to a given time for a Markov chain are σ-algebras.

Now we shall prove that whatever the random event A prior to time n for a Markov chain $(X(n))_{n\geqslant 0}$ we have

$$\mathsf{P}(X(n+1) = i_{n+1} \mid X(n) = i_n, A) = \mathsf{P}(X(n+1) = i_{n+1} \mid X(n) = i_n), \quad (2.2)$$

whenever the left side is defined. Indeed

$$\mathsf{P}(X(n+1) = i_{n+1} \mid X(n) = i_n, A) = \frac{\mathsf{P}(X(n+1) = i_{n+1}, X(n) = i_n, A)}{\mathsf{P}(X(n) = i_n, A)}$$

(assuming the denominator to be nonzero)

$$= \frac{\sum\limits_{(i_0, \ldots, i_{n-1}) \in A^{(n)}} \mathsf{P}(X(n+1)=i_{n+1}, X(n)=i_n, X(n-1)=i_{n-1}, \ldots, X(0)=i_0)}{\sum\limits_{(i_0, \ldots i_{n-1}) \in A^{(n)}} \mathsf{P}(X(n) = i_n, X(n-1) = i_{n-1}, \ldots, X(0) = i_0)}$$

$$= \frac{\mathsf{P}(X(n+1)=i_{n+1} \mid X(n)=i_n) \sum\limits_{(i_0, \ldots i_{n-1}) \in A^{(n)}} \mathsf{P}(X(n)=i_n, X(n-1)=i_{n-1}, \ldots, X(0)=i_0)}{\sum\limits_{(i_0, \ldots, i_{n-1}) \in A^{(n)}} \mathsf{P}(X(n) = i_n, X(n-1) = i_{n-1}, \ldots, X(0) = i_0)}$$

$$= \mathsf{P}(X(n+1) = i_{n+1} \mid X(n) = i_n)$$

(for $n \geqslant 1$ we put $\{X(n)=i_n, A\}=\{X(n)=i_n, (X(0), \ldots, X(n-1)) \in A^{(n)}\}$).

In the above we used the Markov property by replacing the probability $\mathsf{P}(X(n+1) = i_{n+1}, X(n) = i_n, X(n-1) = i_{n-1}, \ldots, X(0) = i_0)$ by $\mathsf{P}(X(n+1) = i_{n+1} \mid X(n) = i_n)\, \mathsf{P}(X(n) = i_n, \ldots, X(0) = i_0)$.

Furthermore, we can extend (2.2) by introducing random events posterior to time n. For instance, we have

$$\begin{aligned} &\mathsf{P}(X(n+3)=l,\ X(n+2)=k,\ X(n+1)=j \mid X(n)=i,\ A) \\ &= \mathsf{P}(X(n+3)=l \mid X(n+2)=k,\ X(n+1)=j,\ X(n)=i,\ A) \times \\ &\times \mathsf{P}(X(n+2)=k \mid X(n+1)=j, X(n)=i, A)\, \mathsf{P}(X(n+1)=j \mid X(n)=i,\ A) \\ &= p(i,\ j)\, p(j,\ k)\, p(k,\ l) \\ &= \mathsf{P}(X(n+3)=l,\ X(n+2)=k,\ X(n+1)=j \mid X(n)=i)^{*)} \end{aligned}$$

(the justification of the last equality is left to the reader).

In general, whatever the random events A and $B=\{(X(n), X(n+1), \ldots) \in B^{(\infty)}\}$ prior and posterior to time n, respectively, we have

$$\begin{aligned} \mathsf{P}(B \mid X(n)=i_n,\ A) &= \mathsf{P}(B \mid X(n)=i_n) \\ &= \mathsf{P}((X(0),\ X(1), \ldots) \in B^{(\infty)} \mid X(0)=i_n) \end{aligned}$$

whenever the left hand side is defined.

Let us remark that the last equality allows us to write

$$\begin{aligned} \mathsf{P}(B \cap A \mid X(n)=i_n) &= \mathsf{P}(B \mid X(n)=i_n,\ A)\, \mathsf{P}(A \mid X(n)=i_n) \\ &= \mathsf{P}(B \mid X(n)=i_n)\, \mathsf{P}(A \mid X(n)=i_n). \end{aligned}$$

Therefore *for a Markov chain the past* (A) *and the future* (B) *are conditionally independent given the present* ($\{X(n)=i\}$).

This symmetrical extension of the Markov property suggests that *the Markov property still holds in reversed time*, i.e., whatever the random event B posterior to time n we have

$$\mathsf{P}(X(n-1)=i_{n-1} \mid X(n)=i_n,\ B)=\mathsf{P}(X(n-1)=i_{n-1} \mid X(n)=i_n) \tag{2.2'}$$

whenever the left hand side is defined. Indeed

$$\begin{aligned} &\mathsf{P}(X(n-1)=i_{n-1} \mid X(n)=i_n,\ B) = \frac{\mathsf{P}(X(n-1)=i_{n-1}, X(n)=i_n,\ B)}{\mathsf{P}(X(n)=i_n,\ B)} \\ &= \frac{\mathsf{P}(B \mid X(n)=i_n,\ X(n-1)=i_{n-1})\, \mathsf{P}(X(n)=i_n,\ X(n-1)=i_{n-1})}{\mathsf{P}(B \mid X(n)=i_n)\, \mathsf{P}(X(n)=i_n)} \\ &= \mathsf{P}(X(n-1)=i_{n-1} \mid X(n)=i_n) \end{aligned}$$

[since $\mathsf{P}(B \mid X(n)=i_n,\ X(n-1)=i_{n-1}) = \mathsf{P}(B \mid X(n)=i_n)$].

---

*) In general, the conditional probability of the successive occurrence of states $i_1, \ldots, i_n$ given the immediately preceding state i_0 equals $p(i_0, i_1)\, p(i_1, i_2), \ldots, p(i_{n-1}, i_n)$ regardless of the time at which i_0 occurs:

$$\mathsf{P}(X(m+n)=i_n, \ldots, X(m+1)=i_1 \mid X(m)=i_0) = p(i_0, i_1)\, p(i_1, i_2) \cdots p(i_{n-1}, i_n)$$

whatever $m \geqslant 0$. Homogeneity plays an essential part here

The *time reversibility* (2.2′) of the Markov property was first explicitly stated by G. Mihoc [269]. *)

It should be noted that the consequences just discussed of the Markov property are in fact statements equivalent to it, any of them being suitable as a definition. Our choice is a matter of tradition and simplicity.

Later on a more sophisticated extension of the Markov property, the so-called strong Markov property, will be considered.

2.1.5. The transition probabilities $p(i, j)$, $i, j \in S$, of a Markov chain govern the changes of state from one step to the next. Changes of state over $n > 1$ steps are ruled by probabilities simply expressed in terms of the $p(i, j)$. Let us denote by $p(n, i, j)$ the probability of transition from state i to state j in exactly n steps, namely

$$p(n, i, j) = \mathsf{P}(X(m+n) = j \mid X(m) = i),$$

whatever the integer $m \geq 0$. The right side does not depend on m its value being (see footnote on p. 64).

$$\sum_{i,\ldots,i_{n-1}\in S} \mathsf{P}(X(m+n)=j, X(m+n-1)=i_{n-1},\ldots,X(m+1)=i_1 \mid X(m)=i)$$

$$= \sum_{i,\ldots,i_{n-1}\in S} p(i, i_1)\, p(i_1, i_2) \ldots p(i_{n-1}, j), \; n > 2.$$

Clearly, $p(1, i, j)$ is really just $p(i, j)$. It is convenient to define

$$p(0, i, j) = \delta(i, j), \; i, j \in S.$$

On account of the above, $p(2, i, j)$ equals $\sum_{k\in S} p(i, k)\, p(k, j)$, in other words the matrix $(p(2, i, j))_{i,j\in S}$ is the square of the transition matrix $\mathbf{P} = (p(i, j))_{i,j\in S}$. In general, it is easily recognized the probabilities $p(n, i, j)$ (which are called *n-step transition probabilities*) are the entries of the nth power of the matrix $\mathbf{P}$. This is the starting point of the use of matrix theory in Markov chain theory.

The matrix $\mathbf{P}^n$ is called the *n-step transition matrix*, $n \geq 1$. For each $n \geq 1$ it is a stochastic matrix, hence

$$p(n, i, j) \geq 0, \quad i, j \in S, \; \sum_{j\in S} p(n, i, j) = 1, \; i \in S.$$

The identity $\mathbf{P}^{m+n} = \mathbf{P}^m\mathbf{P}^n$, $m, n \geq 0$ ($\mathbf{P}^0$ is, by definition, the unit matrix) is just the system of equations

$$p(m+n, i, j) = \sum_{k\in S} p(m, i, k)\, p(n, k, j), \quad i, j \in S,$$

known as the *Chapman—Kolmogorov equations*.

*) This is in fact the doctoral dissertation defended by Mihoc on 28 April 1934 at the Faculty of Sciences of Bucharest University.

To conclude we note that joint probabilities when some of the intermediate states are not specified can be expressed in terms of n-step transition probabilities. For instance

$$\mathsf{P}(X(10) = l,\ X(8) = k,\ X(5) = j \mid X(1) = i) =$$
$$= p(4,\ i,\ j)\ p(3,\ j,\ k)\ p(2,\ k,\ l).$$

The justification is left to the reader.

2.2 EXAMPLES OF HOMOGENEOUS MARKOV CHAINS

In this paragraph we consider a few Markov chains occurring as the result of mathematical modelling in various fields.

Example 1 (*Gambler's ruin problem*). Consider two persons playing a certain game repeatedly and starting off with k and $l-k$ units of capital. Assume that one unit is bet each time, the loser paying it to the winner, that the winning probabilities for the two gamblers are p and $q=1-p$, and that the outcomes of the successive games are (stochastically) independent. The distribution of the combined capital l among the two gamblers will go on varying until one of them is ruined, i.e., loses his last unit of capital. We are interested in the ruin probabilites of the two gamblers and the (mean) duration of the contest.

The evolution of the capital of the first gambler can be described by a Markov chain with state space $(0, \ldots, l)$ and transition matrix

$$\begin{array}{c} \\ 0 \\ 1 \\ \vdots \\ l-1 \\ l \end{array} \begin{array}{c} \begin{matrix} 0 & 1 & 2 & \ldots & l-2 & l-1 & l \end{matrix} \\ \begin{pmatrix} 1 & 0 & 0 & \ldots & 0 & 0 & 0 \\ q & 0 & p & \ldots & 0 & 0 & 0 \\ \cdot & \cdot & \cdot & \cdots & \cdot & \cdot & \cdot \\ \cdot & \cdot & \cdot & \cdots & \cdot & \cdot & \cdot \\ 0 & 0 & 0 & \ldots & q & 0 & p \\ 0 & 0 & 0 & \ldots & 0 & 0 & 1 \end{pmatrix} \end{array}$$

which starts in state k. Indeed, if just after game n the capital of the first gambler equals i units, $i \neq 0, l$, then, independently of its previous evolution, just after game $n+1$ it equals $i+1$ units or $i-1$ units with probabilities p and q, respectively. If $i = 0$ (the first gambler is ruined) or $i = l$ (the second gambler is ruined), no subsequent transition (which is fictitious as far as our problem is concerned) can change the value of the capital, that is $p(0, 0) = p(l, l) = 1$. States 0 and l are called *absorbing* since once entered they cannot be left. The gambler's ruin problem amounts, therefore, to computing the probabilities of reaching the absorbing states of a Markov chain and the mean time until an absorbing state is reached.

Our problem and the corresponding Markov chain admit an interpretation suggested by physical applications. Let us consider an infinite sequence of independent Bernoulli trials $E_1, E_2, \ldots$ each having two possible outcomes α and $\bar{\alpha}$ with probabilities p and q, respectively. Imagine a particle which may occupy the points of the abscissa $0, 1, \ldots, l$ on a directed axis. At any time $n \geqslant 0$, if at 0 or l, the particle remains where it is; otherwise, it moves one unit step to the right or to the left depending on whether E_{n+1} resulted in α or $\bar{\alpha}$. We say that the particle performs a ***random walk with absorbing barriers at*** 0 ***and*** l. It is clear that the evolution of the capital of the first gambler in the gambler's ruin problem is described by the motion of the particle performing the above random walk.

Various modifications of the barriers at 0 and l may be considered. For instance, they may be made *reflecting* which amounts to assuming that once in 0 or l the particle is "reflected" into 1 or $l-1$, respectively. This leads to a Markov chain for which $p(0,1) = p(l, l-1) = 1$, that is with transition matrix *)

$$\begin{array}{c} 0 \\ 1 \\ \cdot \\ \cdot \\ \cdot \\ l-1 \\ l \end{array}\overset{\begin{array}{ccccccc} 0 & 1 & 2 & \ldots & l-2 & l-1 & l \end{array}}{\begin{pmatrix} 0 & 1 & 0 & \ldots & 0 & 0 & 0 \\ q & 0 & p & \ldots & 0 & 0 & 0 \\ \cdot & \cdot & \cdot & \ldots & \cdot & \cdot & \cdot \\ \cdot & \cdot & \cdot & \ldots & \cdot & \cdot & \cdot \\ \cdot & \cdot & \cdot & \ldots & \cdot & \cdot & \cdot \\ 0 & 0 & 0 & \ldots & q & 0 & p \\ 0 & 0 & 0 & \ldots & 0 & 1 & 0 \end{pmatrix}}$$

More generally, the barriers may be made *elastic* implying a transition matrix of the form

$$\begin{array}{c} 0 \\ 1 \\ \cdot \\ \cdot \\ l-1 \\ l \end{array}\overset{\begin{array}{ccccccc} 0 & 1 & 2 & \ldots & l-2 & l-1 & l \end{array}}{\begin{pmatrix} r_0 & p_0 & 0 & \ldots & 0 & 0 & 0 \\ q & 0 & p & \ldots & 0 & 0 & 0 \\ \cdot & \cdot & \cdot & \ldots & \cdot & \cdot & \cdot \\ \cdot & \cdot & \cdot & \ldots & \cdot & \cdot & \cdot \\ 0 & 0 & 0 & \ldots & q & 0 & p \\ 0 & 0 & 0 & \ldots & 0 & q_l & r_l \end{pmatrix}}$$

where $r_0 + p_0 = q_l + r_l = 1$, $r_0, p_0, q_l, r_l > 0$.

One can also consider combinations of types of barriers (for instance, an absorbing barrier and an elastic one, etc.).

The study of Example 1 will be continued in 3.1.2, 3.2.7 and 4.3.5.

Example 2 (*Physical models of random walk type*). A more general random walk than the one from the previous example obtains by assuming that whenever the particle is in a position $i \neq 0, l$ it moves a unit step to the right or to the left with probabilities p_i and q_i, respectively,

---

*) For the evolution of the capital of the first gambler to be described by the Markov chain with this transition matrix one should make the convention that whenever one of the gamblers loses his last unit of capital it is replaced by his adversary so that the game can continue. Of course, under this regulation, one can no longer speak of gambler's ruin.

or remains where it is with probability r_i, $p_i + q_i + r_i = 1$. From position 0 the particle moves to position 1 with probability p_0, remaining unmoved with probability r_0; analogously, from position l the particle moves to position $l-1$ with probability q_l, remaining unmoved with probability r_l. Clearly, $p_0 + r_0 = q_l + r_l = 1$. The Markov chain describing the motion of the particle will have the transition matrix

$$\begin{array}{c} \\ 0 \\ 1 \\ \cdot \\ \cdot \\ \cdot \\ l-1 \\ l \end{array} \begin{array}{c} \begin{array}{ccccccc} 0 & 1 & 2 & \cdots & l-2 & l-1 & l \end{array} \\ \left(\begin{array}{ccccccc} r_0 & p_0 & 0 & \cdots & 0 & 0 & 0 \\ q_1 & r_1 & p_1 & \cdots & 0 & 0 & 0 \\ \cdot & \cdot & \cdot & \cdots & \cdot & \cdot & \cdot \\ \cdot & \cdot & \cdot & \cdots & \cdot & \cdot & \cdot \\ \cdot & \cdot & \cdot & \cdots & \cdot & \cdot & \cdot \\ 0 & 0 & 0 & \cdots & q_{l-1} & r_{l-1} & p_{l-1} \\ 0 & 0 & 0 & \cdots & 0 & q_l & r_l \end{array} \right) \end{array}$$

There are two celebrated models of physical phenomena which are random walks of this type (and which were conceived without reference to Markov chains).

The first model, proposed by Daniel Bernoulli in 1769, is intended to describe the flow of two incompressible liquids between two containers. These are represented by two urns containing a total of $2l$ balls among which l are white and l black. The balls are supposed to represent molecules of the two liquids. As such, the incompressibility assumption implies the number of balls in each urn remains constant (thus equal to l). The motion of balls between the two urns is imagined as follows: at any time one ball is chosen at random from each urn and then these two balls are interchanged. We say the system is in state i, $0 \leqslant i \leqslant l$, if and only if the first urn contains i white balls. Therefore if the state is i then the first urn contains $l-i$ black balls and the second one $l-i$ white balls and i black balls. It is easily verified that

$$p(i, i-1) = q_i = \frac{i}{l} \cdot \frac{i}{l} = \left(\frac{i}{l}\right)^2, \qquad 1 \leqslant i \leqslant l,$$

$$p(i, i+1) = p_i = \frac{l-i}{l} \cdot \frac{l-i}{l} = \left(1 - \frac{i}{l}\right)^2, \qquad 0 \leqslant i \leqslant l-1,$$

$$p(i, i) = r_i = \frac{i}{l} \cdot \frac{l-i}{l} + \frac{l-i}{l} \cdot \frac{i}{l} = \frac{2i}{l}\left(1 - \frac{i}{l}\right), \qquad 0 \leqslant i \leqslant l.$$

For instance, $p(i, i-1)$ has been computed as follows: if the first urn contains i white balls, then to have $i-1$ white balls in it after the first drawings one must draw a white ball from the first urn (the probability of this random event is i/l) and a black ball from the second one (the probability of this random event is also i/l). These two random events being independent, the value of $p(i, i-1)$ is found by multiplying their probabilities.

The study of the Bernoulli model will be continued in 4.2.2 and 4.5.2.

The second model, proposed by Tatiana and Paul Ehrenfest in 1907, has two physical interpretations. It describes, on the one hand, the exchange of heat between two isolated bodies of unequal temperatures from the point of view of the kinetic theory of matter. The temperatures are represented by the numbers of balls in two urns containing a total of $2l$ balls numbered consecutively from 1 to $2l$. The exchange of heat is imagined as follows: at any time an integer between 1 and $2l$ is chosen at random (i.e., according to the uniform distribution over $1, \ldots, 2l$ — see 1.4.2) and the ball whose number has been drawn is moved from the urn containing it to the other one. We say that the system is in state i, $0 \leqslant i \leqslant 2l$, if and only if the first urn contains i balls. Therefore, if the state is i, the second urn contains $2l - i$ balls. It is easily seen that $p(i, i) = r_i = 0$, $0 \leqslant i \leqslant 2l$, and

$$p(i, i-1) = q_i = \frac{i}{2l}, \qquad 1 \leqslant i \leqslant 2l,$$

$$p(i, i+1) = p_i = 1 - \frac{i}{2l}, \quad 0 \leqslant i < 2l.$$

On the other hand, the Ehrenfest model may be used for explaining reversibility in statistical mechanics. All these aspects will be taken up again in 4.5.2, 4.6, and 5.4.5.

The Markov chain associated with the Ehrenfest model can be also obtained as follows. Consider an urn containing $2l$ balls of two colours, white and black. At any time 0, 1, 2, ... a ball is drawn at random and a ball of opposite colour is replaced in the urn. It is not difficult to see that the evolution of the number of white balls in the urn is described by a Markov chain whose transition probabilities are those associated with the Ehrenfest model. This urn scheme was devised by O. Onicescu and G. Mihoc [294] (without reference to the Ehrenfest model) to illustrate an exceptional feature of probability theory. It can easily be shown that the relative frequency of occurrence of white balls in n drawings (the number of occurrences of white balls divided by n) converges to 1/2 as $n \to \infty$, the convergence taking place for *all* the possible sequences of drawings without exception. (Try to prove it!)

Example 3 (*A model of certain sports*). Random walks with two absorbing states also occur when modelling games played between two players (or teams) $\mathcal{A}$ and $\mathcal{B}$ which terminate when a certain score is reached. Typical examples are tennis and volley ball. A match between $\mathcal{A}$ and $\mathcal{B}$ consists of a number of basic units which may aggregate into compound units and it is won by the first contestant winning a certain number of such basic or compound units. A basic unit consists of contesting a number of points and it is won by the first player (team) winning at least, say, r points, but by a margin of at least two. It will be assumed that the probabilities of winning any given point for $\mathcal{A}$ and $\mathcal{B}$ are equal to p and q, respectively, $p + q = 1$. A similar rule applies to compound units. For tennis the basic unit is called a "game", $r = 4$, and the compound unit is called a "set", the match being won by the first contestant winning three (or two) sets. For volley-ball the basic unit is called a "set", $r = 15$, there are no compound units, the match being won by the first contestant winning three sets. For a basic unit there are the following $r^2 + 1$ possible scores: u-v, $0 \leqslant u$, $v \leqslant r-1$, $(u, v) \neq (r-2, r-2)$, $(r-2, r-1)$, $(r-1, r-1)$, $(r-1, r-2)$, Advantage $\mathcal{A}$ ($\mathcal{A}$ leads with $(u + 1)$ - u, $u \geqslant r - 2$), Advantage $\mathcal{B}$ ($\mathcal{B}$ leads with $(u + 1)$ - u, $u \geqslant r-2$), Deuce (the score is u - u, $u \geqslant r-2$), Unit $\mathcal{A}$, Unit $\mathcal{B}$.*) These possible scores will be considered as the states of a Markov chain whose transition probabilities are indicated in Fig. 2.1 for the case $r = 4$ (cf. [195], p. 161).

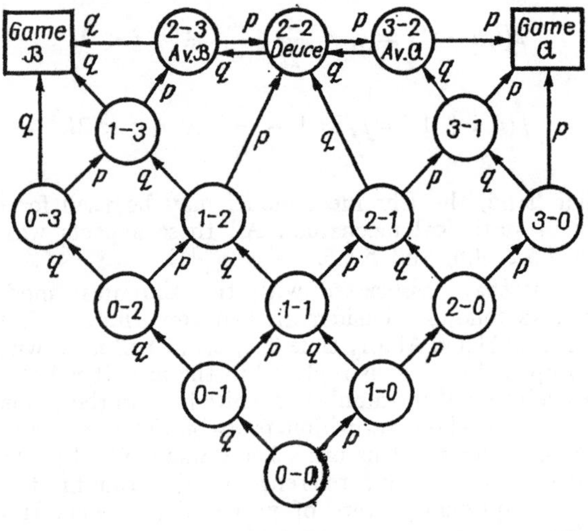

Fig. 2.1

A basic unit can be divided into two stages. At the beginning only the first r^2-4 states are involved (the preliminary stage of the basic unit), any transition from one state to another always taking place upwards (see Fig. 2.1 that represents the special case $r = 4$), and in r, $r + 1, \ldots,$ or $2r - 3$ steps one of the five states in the top row is reached. For convenience let us denote them by 0, 1, 2, 3, 4 from the left to the right. The only effect of the motion among the first r^2-4 states is to

*) For tennis the score is traditionally kept in the well known manner (with 15, 30, 40 corresponding to 1, 2, 3, respectively) which is, clearly, equivalent to the above one for $r = 4$.

assign well determined initial probabilities to states 0, 1, 2, 3, and 4, the evolution of the chain among these states (the final stage of the basic unit) being that of a random walk with absorbing barriers at 0 and 4. The initial probabilities of states 0, 1, 2, 3, 4 are easily obtained by simple computations. We have

$$p(0) = q^r\left(1 + \binom{r}{1}p + \binom{r+1}{2}p^2 + \dots + \binom{2r-4}{r-3}p^{r-3}\right),$$

$$p(1) = \binom{2r-4}{r-3}p^{r-2}q^{r-1}, \; p(2) = \binom{2r-4}{r-2}p^{r-2}q^{r-2},$$

$$p(3) = \binom{2r-4}{r-3}p^{r-1}q^{r-2},$$

$$p(4) = p^r\left(1 + \binom{r}{1}q + \binom{r+1}{2}q^2 + \dots + \binom{2r-4}{r-3}q^{r-3}\right).$$

The study of Example 3 will be continued in 3.2.11.

Example 4 (*A model for the spread of a contagious disease in a small community*). The setting in which the epidemic develops is a population whose size is assumed fixed and which comprises two categories of individuals: the infected individuals (infectives) and the uninfected ones (susceptibles). The progress of the epidemic is determined by the interactions between these two categories. In the Greenwood model (see, e.g., [118]) which we consider here, the evolution of the contagious disease is divided into a constant latent period (where the disease develops without clinical signs), which may be used as the time unit, and an infectious period contracted to a single point (when an infective becomes infectious to susceptibles). This means that, starting at time 0 with one or more infectives, the successive cases will occur in batches at time intervals equal to the latent period. Let us denote by r the initial number of uninfected susceptibles, by ξ_m, $m \geqslant 0$, the number of individuals getting infected at time m (equivalently: the number of infectious individuals at time $m+1$), and by $X(n)$, $n \geqslant 0$, the number of susceptibles still uninfected at time n. Clearly, $X(0) = r$, $X(n) = \xi_{n+1} + X(n+1)$ $n \geqslant 0$. Let us define the chance of contact $0 < p = 1-q < 1$ as the probability of contact at any time $n = 1, 2, \dots$ between any two individuals sufficient to produce a new infection if one of them is susceptible and the other is infectious *). Assuming the contacts to be (stochastically) independent, at any time $n = 1, 2, \dots$ the probability that among i susceptibles k of them ($k \leqslant i$) have contacts with infectious individuals equals $\binom{i}{k}p^k q^{i-k}$. (For, we are faced with i independent

*) It is implicitly assumed that the chance of contact does not depend on the number of infectious individuals present.

Bernoulli trials with outcomes α = contact takes place and $\bar{\alpha}$ = contact does not take place — see 1.3.2.) This allows us to write

$$\mathsf{P}(X(n+1) = j \mid X(n) = i) = \mathsf{P}(\xi_{n+1} = X(n) - j \mid X(n) = i)$$

$$= \mathsf{P}(\xi_{n+1} = i - j \mid X(n) = i) = \begin{cases} \binom{i}{j} p^{i-j} q^j, & \text{if } i \geqslant j, \\ 0, & \text{if } i < j, \end{cases}$$

and, obviously, the probability distribution of $X(n+1)$ is independent of the values assumed by the random variables $X(m)$, $m < n$. Therefore $(X(n))_{n \geqslant 0}$ is a Markov chain with states 0, 1, ..., r and transition matrix

$$\begin{array}{c} \\ 0 \\ 1 \\ 2 \\ \cdot \\ \cdot \\ \cdot \\ r \end{array} \begin{array}{c} \begin{array}{cccccc} 0 & 1 & 2 & & \dots & r \end{array} \\ \begin{pmatrix} 1 & 0 & 0 & \dots & 0 \\ p & q & 0 & \dots & 0 \\ p^2 & 2pq & q^2 & \dots & 0 \\ \cdot & \cdot & \cdot & \dots & \cdot \\ \cdot & \cdot & \cdot & \dots & \cdot \\ \cdot & \cdot & \cdot & \dots & \cdot \\ p^r & rp^{r-1}q & \binom{r}{2} p^{r-2}q^2 & \dots & q^r \end{pmatrix} \end{array}.$$

The duration of the epidemic is the smallest natural number τ such that $\xi_\tau = 0$, i.e., the moment τ when there are no further infectives. An equivalent definition obtains by using the equality $X(n) = \xi_{n+1} + X(n+1)$: the duration of the epidemic is the smallest natural number τ such that $X(\tau - 1) = X(\tau)$. Clearly, τ is a random variable. We can express in terms of τ the number of cases in the epidemic as $\xi_0 + X(0) - X(\tau) = \xi_0 + r - X(\tau)$.

Example 5 (*A model of schooling* [185]). Assume that a course of education consists of d stages each having a duration of one time unit (say, one year). At the end of each stage the promotion of a student to the next stage (or the completion of the course) is decided by an examination. It is also assumed that a student may leave the course, but that once he drops out he does not join the course again. Thus, at the end of a stage, a student's progress can be described by one of the following three possibilities: 1) he passes the examination and goes to the next stage; 2) he fails the examination and repeats the same stage; 3) he leaves the course altogether before the examination (if the student leaves the course immediately after the examination, he is considered to have dropped out during the next stage). Finally, assume that the probabilities of promotion to the next stage, repetition of the stage completed, and leaving the course do not depend on the previous performances of the student. Under these assumptions the progress of a student undergoing such a course of education until he either drops out or completes all the d stages can be described by a Markov chain with states 0, 1, ..., $d-1$, d, $d+1$. State 0 corresponds to a student entering the course in its first stage, state $d+1$ to a student leaving the course, and the intermediate states i, $1 \leqslant i \leqslant d$, to a student successfully completing the first i stages (i.e., in particular, state d corresponds to a student successfully completing the entire course). Clearly, states d and $d+1$ are absorbing. The transition matrix has the form

$$\begin{array}{c} \\ 0 \\ 1 \\ 2 \\ \cdot \\ \cdot \\ \cdot \\ d-1 \\ d \\ d+1 \end{array} \begin{array}{c} \begin{array}{cccccc} 0 & 1 & 2 \;\cdots\; d-1 & d & d+1 \end{array} \\ \left(\begin{array}{cccccc} r_1 & p_1 & 0 \;\cdots\; 0 & 0 & q_1 \\ 0 & r_2 & p_2 \cdots\; 0 & 0 & q_2 \\ 0 & 0 & r_3 \cdots\; 0 & 0 & q_3 \\ \cdot & \cdot & \cdot \;\cdots\; \cdot & \cdot & \cdot \\ \cdot & \cdot & \cdot \;\cdots\; \cdot & \cdot & \cdot \\ \cdot & \cdot & \cdot \;\cdots\; \cdot & \cdot & \cdot \\ 0 & 0 & 0 \;\cdots\; r_d & p_d & q_d \\ 0 & 0 & 0 \;\cdots\; 0 & 1 & 0 \\ 0 & 0 & 0 \;\cdots\; 0 & 0 & 1 \end{array} \right) \end{array}$$

the probabilities p_i, r_i, q_i, being associated with the possibilities 1) — 3) above at stage $i - 1$, $1 \leqslant i \leqslant d$.

The study of Example 5 will be continued in 3.1.3, 3.2.8, and 3.2.9.

Example 6 (*A simple waiting model*). A server (which may be an out-patient clinic, a ticket window, etc.) serves one customer (if any) at time instants 0, 1, 2, ... Assume that in time interval $(n, n+1)$ a number ξ_n of customers arrive, $n \geqslant 0$, where $(\xi_n)_{n\geqslant 0}$ is a sequence of independent identically distributed nonnegative integer valued random variables with $\mathsf{P}(\xi_0 = k) = p_k$, $k \geqslant 0$, $\Sigma_{k\geqslant 0} p_k = 1$, and that there is waiting room for at most m customers including the one being served. Customers who arrive to find the waiting room full leave without receiving service and do not return. Let $X(n)$ denote the number of customers present at time n, including the one being served. We shall prove that $(X(n))_{n\geqslant 0}$ is a Markov chain with states $0, 1, ..., m$.

Let us remark that the number of customers present at time $n+1$ equals the number of customers present at time n less the one served at time n (if any), plus the number of customers arriving in time interval $(n, n+1)$ if the result of this summation does not exceed m, and equals m otherwise. Therefore

$$X(n+1) = \min\,(X(n) + \delta((0, X(n)) - 1 + \xi_n, m)$$

$$= \begin{cases} X(n) - 1 + \xi_n, & \text{if } 1 \leqslant X(n) \leqslant m,\ 0 \leqslant \xi_n \leqslant m + 1 - X(n), \\ \xi_n, & \text{if } X(n) = 0,\ 0 \leqslant \xi_n \leqslant m - 1, \\ m, & \text{otherwise.} \end{cases}$$

Since $X(n+1)$ depends only on $X(n)$ and ξ_n and the random variables $X(n)$ and ξ_n are independent (why ?) the sequence $(X(n))_{n\geqslant 0}$ is a Markov chain. Let us find its transition matrix. Putting

$$P_l = p_l + p_{l+1} + ..., \quad 1 \leqslant l \leqslant m,$$

we have

$$p(0, j) = \mathsf{P}(X(n+1) = j \mid X(n) = 0)$$

$$= \begin{cases} \mathsf{P}(\xi_n = j \mid X(n) = 0) = p_j, \text{ if } 0 \leqslant j \leqslant m-1, \\ \mathsf{P}(\xi_n \geqslant m) = p_m + p_{m+1} + \ldots = P_m, \text{ if } j = m, \end{cases}$$

and for $1 \leqslant i \leqslant m$,

$$p(i, j) = \mathsf{P}(X(n+1) = j \mid X(n) = i)$$

$$= \begin{cases} \mathsf{P}(\xi_n = j+1-i) = p_{j-i+1}, \text{ if } i-1 \leqslant j \leqslant m-1, \\ \mathsf{P}(\xi_n \geqslant m+1-i) = P_{m+1-i}, \text{ if } j = m, \\ 0, \text{ otherwise.} \end{cases}$$

Therefore, the transition matrix is

$$\begin{array}{c} \\ 0 \\ 1 \\ 2 \\ \cdot \\ \cdot \\ \cdot \\ m-1 \\ m \end{array} \begin{array}{c} \begin{array}{cccccc} 0 & 1 & 2 & \ldots & m-1 & m \end{array} \\ \left(\begin{array}{cccccc} p_0 & p_1 & p_2 & \cdots & p_{m-1} & P_m \\ p_0 & p_1 & p_2 & \cdots & p_{m-1} & P_m \\ 0 & p_0 & p_1 & \cdots & p_{m-2} & P_{m-1} \\ \cdot & \cdot & \cdot & \cdots & \cdot & \cdot \\ \cdot & \cdot & \cdot & \cdots & \cdot & \cdot \\ \cdot & \cdot & \cdot & \cdots & \cdot & \cdot \\ 0 & 0 & 0 & \cdots & p_1 & P_2 \\ 0 & 0 & 0 & \cdots & p_0 & P_1 \end{array}\right) \end{array}$$

The study of Example 6 will be continued in 4.2.4.

Example 7 (*A simple storage model*). Consider a dam which can hold at most m units of water. Assume that during day n a quantity η_n of units of water flow into the dam, $n \geqslant 0$, where $(\eta_n)_{n \geqslant 0}$ is a sequence of independent identically distributed nonnegative integer valued random variables with $\mathsf{P}(\eta_0 = k) = p_k$, $k \geqslant 0$, $\sum_{k \geqslant 0} p_k = 1$. Any overflow is lost with no possibility of recovery. At the end of any day when the dam is not dry, one unit of water is released. Denoting by $X'(n)$ the content of the dam at the beginning of day n, it is easily seen that $(X'(n))_{n \geqslant 0}$ is a Markov chain identical to that arising in the preceding example. Let us show that, denoting by $X''(n)$ the content of the dam at the end of day n, the sequence $(X''(n))_{n \geqslant 0}$ is also a Markov chain (with states $0, 1, \ldots, m-1$). The reader will be able to verify without difficulty that

$$X''(n+1) = \min\left(X''(n) + \eta_{n+1} + \delta(0, X''(n) + \eta_{n+1}) - 1, m-1\right).$$

Hence, just as in the previous example, $(X''(n))_{n\geqslant 0}$ is a Markov chain. Its transition matrix is

$$\begin{array}{c} \\ 0 \\ 1 \\ 2 \\ \cdot \\ \cdot \\ \cdot \\ m-2 \\ m-1 \end{array}\begin{array}{c} \begin{array}{cccccc} 0 & 1 & 2 & & m-2 & m-1 \end{array} \\ \begin{pmatrix} p_0+p_1 & p_2 & p_3 & \cdots & p_{m-1} & P_m \\ p_0 & p_1 & p_2 & \cdots & p_{m-2} & P_{m-1} \\ 0 & p_0 & p_1 & \cdots & p_{m-3} & P_{m-2} \\ \cdot & \cdot & \cdot & \cdots & \cdot & \cdot \\ \cdot & \cdot & \cdot & \cdots & \cdot & \cdot \\ \cdot & \cdot & \cdot & \cdots & \cdot & \cdot \\ 0 & 0 & 0 & \cdots & p_1 & P_2 \\ 0 & 0 & 0 & \cdots & p_0 & P_1 \end{pmatrix} \end{array}$$

For instance

$$p(0, 0) = \mathsf{P}(X''(n+1) = 0 \mid X''(n) = 0)$$

$$= \mathsf{P}((\eta_{n+1} = 0) \cup (\eta_{n+1} = 1) \mid X''(n)) = p_0 + p_1,$$

(since the random variables $X''(n)$ and η_{n+1} are independent)

$$p(0, j) = \mathsf{P}(X''(n+1) = j \mid X''(n) = 0)$$

$$= \mathsf{P}(\eta_{n+1} = j+1 \mid X''(n) = 0) = p_{j+1}, \ 1 \leqslant j \leqslant m-2,$$

$$p(0, m-1) = \mathsf{P}(\eta_{n+1} \geqslant m \mid X''(n) = 0) = P_m,$$

$$p(1, 0) = \mathsf{P}(X''(n+1) = 0 \mid X''(n) = 1)$$

$$= \mathsf{P}(\eta_{n+1} = 0 \mid X''(n) = 1) = p_0, \ldots$$

For a systematic study of storage models the reader may consult the monograph [280] and the paper [115].

Example 8 (*A model of industrial accidents* [71]). Assume that in a certain industry a worker may meet with accidents of varying severity that need hospitalization for periods from 1 to $m-1$ time units (say, days) for which he remains immune to further accidents. Assume also that the probability that during any one day the worker does not meet with an accident is q, $0 < q < 1$, and the probability that during any one day the worker meets with an accident needing hospitalization for $m - r$ days is pp_{m-r}, $1 \leqslant r \leqslant m-1$, where $p = 1 - q$, $\sum_{r=1}^{m-1} p_{m-r} = 1$. We say that the worker is in state 0 at the end of a day if and only if he either did not meet with an accident during that day or he is at the end of a hospitalization period. We say that the worker is in state r at the end of a day if and only if he has to remain a further

$m-r$ days in hospital, $1 \leqslant r \leqslant m-1$. Clearly, the worker can be in state r at the end of a day either as a result of his meeting with an accident needing hospitalization for $m-r$ days or because he was in state $r-1$ at the end of the preceding day. It is plausible to assume that occurrences and non occurrences of accidents on different days are independent random events. Under these assumptions it is easily seen that the motion of the worker through states $0, \ldots, m-1$ is governed by a Markov chain with transition matrix

$$\begin{array}{c} 0 \\ 1 \\ \cdot \\ \cdot \\ \cdot \\ m-3 \\ m-2 \\ m-1 \end{array} \overset{\begin{array}{cccccc} 0 & 1 & 2 & \cdots & m-2 & m-1 \end{array}}{\begin{pmatrix} q & pp_{m-1} & pp_{m-2} & \cdots & pp_2 & pp_1 \\ 0 & 0 & 1 & \cdots & 0 & 0 \\ \cdot & \cdot & \cdot & \cdots & \cdot & \cdot \\ \cdot & \cdot & \cdot & \cdots & \cdot & \cdot \\ \cdot & \cdot & \cdot & \cdots & \cdot & \cdot \\ 0 & 0 & 0 & \cdots & 1 & 0 \\ 0 & 0 & 0 & \cdots & 0 & 1 \\ 1 & 0 & 0 & \cdots & 0 & 0 \end{pmatrix}}$$

The study of Example 8 will be continued in 4.2.3. and 4.5.2.

Example 9 (*An inventory model*). Assume that items of some merchandise are stocked in order to meet demand. The replenishment of stock is made at time instants 0, 1, 2, ... and the demand ξ_n during time interval $(n, n+1)$ is supposed to be a nonnegative integer valued random variable with probabillity distribution $\mathsf{P}(\xi_n = k) = p_k$, $k \geqslant 0$, $\sum_{k \geqslant 0} p_k = 1$, not depending on $n \geqslant 0$. Assume the random variables ξ_n, $n \geqslant 0$, are independent. The stocking policy is as follows. Let m and M be two given natural numbers, $m < M$. If at a time $n \geqslant 0$ the stock is below m items, then the replenishment of stock up to the level M is ordered instantaneously. No action is taken if the stock is at least m items.

Let us denote by $X(n)$ the number of items in stock just before the (possible) replenishment at time n. Assuming that any demand exceeding the stock available is met by a special instantaneous order, the possible values of $X(n)$ are $0, 1, \ldots, M$. Clearly, we have

$$X(n+1) = \begin{cases} \max\,(X(n) - \xi_n,\ 0), & \text{if } m < X(n) \leqslant M, \\ \max\,(M - \xi_n,\ 0), & \text{if } X(n) \leqslant m. \end{cases}$$

It follows as in Example 6 that $(X(n))_{n \geqslant 0}$ is a Markov chain. Finding out the transition probabilities is left as an exercise for the reader (one can start with the special case $M = m + 1$) who will get the transition matrix

$$\begin{array}{c} \\ 0 \\ 1 \\ \cdot \\ \cdot \\ \cdot \\ m \\ m+1 \\ m+2 \\ \cdot \\ \cdot \\ \cdot \\ M \end{array}
\begin{array}{c} \begin{array}{cccccccc} 0 & 1 & \cdots & m & m+1 & m+2 & \cdots & M \end{array} \\
\left(\begin{array}{cccccccc}
P_M & p_{M-1} & \cdots & p_{M-m} & p_{M-m-1} & p_{M-m-2} & \cdots & p_0 \\
P_M & p_{M-1} & \cdots & p_{M-m} & p_{M-m-1} & p_{M-m-2} & \cdots & p_0 \\
\cdot & \cdot & \cdots & \cdot & \cdot & \cdot & \cdots & \cdot \\
\cdot & \cdot & \cdots & \cdot & \cdot & \cdot & \cdots & \cdot \\
\cdot & \cdot & \cdots & \cdot & \cdot & \cdot & \cdots & \cdot \\
P_M & p_{M-1} & \cdots & p_{M-m} & p_{M-m-1} & p_{M-m-2} & \cdots & p_0 \\
P_{m+1} & p_m & \cdots & p_1 & p_0 & 0 & \cdots & 0 \\
P_{m+2} & p_{m+1} & \cdots & p_2 & p_1 & p_0 & \cdots & 0 \\
\cdot & \cdot & \cdots & \cdot & \cdot & \cdot & \cdots & \cdot \\
\cdot & \cdot & \cdots & \cdot & \cdot & \cdot & \cdots & \cdot \\
\cdot & \cdot & \cdots & \cdot & \cdot & \cdot & \cdots & \cdot \\
P_M & p_{M-1} & \cdots & p_{M-m} & p_{M-m-1} & p_{M-m-2} & \cdots & p_0
\end{array} \right) \end{array}$$

where $P_l = p_l + p_{l+1} + \ldots,\ m < l \leqslant M.$

2.3 STOPPING TIMES AND THE STRONG MARKOV PROPERTY

2.3.1. A random variable τ whose values are the nonnegative integers 0, 1, 2, ... and ∞ is said to be a *stopping time* (equivalently: *Markov time, optional, random time independent of the future*) for a Markov chain $(X(n))_{n \geqslant 0}$ if and only if whatever $k = 0, 1, 2, \ldots$ the random event $\{\tau = k\}$ is prior to time k for the chain considered, i.e., it is of the form $\{(X(0), \ldots, X(k)) \in A^{(k+1)}\}$, where $A^{(k+1)}$ is a set of $(k + 1)$ - tuples of states of the chain. In other words, we can tell by examination of the random variables $X(0), \ldots, X(k)$ whether or not τ assumes the value k. Clearly, a constant integer valued random variable $\tau = l =$ constant is a stopping time for any Markov chain. This is a trivial example of a stopping time (in fact not at all random). However, it should be noted that there are no concepts and results (including the Markov property itself) involving a non random time which cannot be generalized to a (random) stopping time.

A typical example of a stopping time is the *first passage time* to state i defined as $\tau_i = \min(n \geqslant 1 : X(n) = i)$. (Try to prove that τ_i is indeed a stopping time. Anyhow, the proof may be found on p. 86.) If we consider the sequence $X(\tau_i), X(\tau_i + 1), \ldots$ *) it seems reasonable to expect that it is a Markov chain with the same transition probabilities as the

*) The random variable $X(\tau_i)$ is defined as follows. If $\tau_i = k$ then $X(\tau_i) = X(k)$, $k = 1, 2, \ldots$; if $\tau_i = \infty$ (corresponding to the case where state i is never reached) then $X(\tau_i) = c$, where c is an arbitrary element not belonging to the state space S. We replace S by $S \cup (c)$ and define $p(c, c) = 1$, $p(c, i) = 0$, $i \in S$.

original chain $(X(n))_{n\geqslant 0}$. We might argue that if $\tau_i = k$ our sequence is $X(k)$, $X(k+1)$, ... which enjoys the stated property. However, τ_i being a random variable rather than a constant, a rigorous proof is needed. First, we introduce two new concepts that extend concepts already defined for a non random time (see 2.1.4).

2.3.2. Let τ be a stopping time for a Markov chain $(X(n))_{n\geqslant 0}$. A random event A is said to be *prior to* τ if and only if whatever $k = 0$, 1, 2, ..., the random event $A \cap \{\tau = k\}$ is prior to time k for the Markov chain considered. Equivalently, A is prior to τ if and only if it is determined by the random variables $X(0)$, ..., $X(\tau)$, i.e., it is of the form $A = \{X(0), \ldots, X(\tau)\} \in E$ where E is a subset of the set of all the finite sequences of states of the chain. Similarly, a random event B is *posterior to* τ if and only if it is determined by the random variables $X(\tau)$, $X(\tau+1)$, ..., i.e., it is of the form $B = \{(X(\tau), X(\tau+1), \ldots) \in B^{(\infty)}\}$ where $B^{(\infty)}$ is a set of infinite sequences of states of the chain. It can be shown that the sets of all the random events prior (respectively posterior) to a stopping time for a Markov chain are σ-algebras.

Typical examples of random events prior and posterior to a stopping time τ are $\{X(\min(\tau, m)) = i\}$ and $\{X(\tau+m) = i\}$, respectively, for any $m \geqslant 0$ and $i \in S$.

The concepts of a random event prior or posterior to a stopping time are not at all artificial. Everyday language contains phrases involving uncertain and therefore random dates. A few examples are "preoperative management", "post operative treatment", "the day before the earthquake". Also, when a roulette gambler decides to bet on black "after red has successively appeared four times", he is dealing with the random variable with subscript $\tau+1$ from a certain sequence of random variables, where the value of τ is a matter of chance. The last example illustrates convincingly the fact that the usefulness of the concept of a stopping time lies in that the determination of τ does not involve random events posterior to τ. Probability theory would be of no utility to a gambler able to foresee the future!

2.3.3. The theorem below shows that the Markov property holds for stopping times, too.

Theorem 2.1. *Let τ be a stopping time for a Markov chain $(X(n))_{n\geqslant 0}$. If A is a random event prior to τ then*

$$\mathsf{P}(X(\tau+1) = j \mid X(\tau) = i, A) = \mathsf{P}(X(\tau+1) = j \mid X(\tau) = i) = p(i, j)$$

whatever i, $j \in S$, whenever the left side is defined.

Proof. We can successively write

$$\mathsf{P}(X(\tau+1) = j, X(\tau) = i, A) = \sum_{n\geqslant 0} \mathsf{P}(X(\tau+1) = j, X(\tau) = i, \tau = n, A)$$

[since the random events $\{\tau = n\}$, $n \geq 0$, are disjoint]

$$= \sum_{n \geq 0} \mathsf{P}(X(n+1) = j,\ X(n) = i,\ \tau = n,\ A)$$

$$= \sum_{n \geq 0} \mathsf{P}(X(n) = i,\ \tau = n, A)\ \ \mathsf{P}(X(n+1) = j \mid X(n) = i,\ \tau = n, A)$$

[actually the sum is extended only over those n for which $\mathsf{P}(X(n) = i,$ $\tau = n,\ A) > 0$]

$$= \sum_{n \geq 0} \mathsf{P}(X(n) = i,\ \tau = n, A)\, p(i, j)$$

[by (2.2) taking into account that $A \cap \{\tau = n\}$ is a random event prior to time n]

$$= p(i, j) \sum_{n \geq 0} \mathsf{P}(X(\tau) = i,\ \tau = n,\ A) = p(i, j)\ \ \mathsf{P}(X(\tau) = i,\ A),$$

whence

$$\mathsf{P}(X(\tau + 1) = j \mid X(\tau) = i,\ A) = p(i,\ j).$$

This equality holds for any random event A prior to τ, in particular for $A =$ the sure event Ω. Therefore we have

$$\mathsf{P}(X(\tau + 1) = j \mid X(\tau) = i) = p(i,\ j)$$

and the proof is complete.

Corollary 1. *If A is a random event prior to τ, then*

$$\mathsf{P}(X(\tau + m) = i_m, \ldots,\ X(\tau + 1) = i_1 \mid X(\tau) = i_0,\ A)$$
$$= \mathsf{P}(X(\tau + m) = i_m, \ldots,\ X(\tau + 1) = i_1 \mid X(\tau) = i_0)$$
$$= p(i_0,\ i_1) \ldots p(i_{m-1},\ i_m)$$

for any $m > 1$ and $i_1, \ldots, i_m \in S$, whenever the left side is defined. In particular

$$\mathsf{P}(X(\tau + m) = i_m \mid X(\tau) = i_0,\ A) = \mathsf{P}(X(\tau + m) = i_m \mid X(\tau) = i_0)$$
$$= p(m,\ i_0,\ i_m).$$

Proof. We have

$$\mathsf{P}(X(\tau + m) = i_m, \ldots,\ X(\tau + 1) = i_1 \mid X(\tau) = i_0,\ A)$$
$$= \mathsf{P}(X(\tau + m) = i_m \mid X(\tau + m - 1) = i_{m-1}, \ldots,\ X(\tau + 1) = i_1, X(\tau) = i_0, A)$$
$$\ldots\ \mathsf{P}(X(\tau + 1) = i_1 \mid X(\tau) = i_0,\ A) = p(i_0,\ i_1) \ldots p(i_{m-1},\ i_m)$$

[on account of the fact that the random event $\{X(\tau + k) = i_k, \ldots,$ $X(\tau + 1) = i_1,\ X(\tau) = i_0,\ A\}$ is prior to the stopping time $\tau + k$,

$0 \leqslant k \leqslant m-1$, and using Theorem 2.1; here there are two assertions left for proof to the reader].

The second equality comes from taking $A = \Omega$. The special case follows summing over $i_1, \ldots, i_{m-1} \in S$.

Corollary 2. *If A and $B = \{(X(\tau), X(\tau+1), \ldots) \in B^{(\infty)}\}$ are random events prior and posterior to τ, respectively, then*

$$\mathsf{P}(B \mid X(\tau) = i, A) = \mathsf{P}(B \mid X(\tau) = i)$$

$$= \mathsf{P}(X(0), X(1), \ldots,) \in B^{(\infty)} \mid X(0) = i)$$

for any $i \in S$, whenever the left side is defined.

Proof. Immediate consequence of Corollary 1.

The first equality in the statement of Corollary 2 is known as the *strong Markov property*. It is, as we have just seen, implied by the Markov property and coincides with the latter in the case where τ is a constant. It follows that the two properties are equivalent. Until not long ago it was customary to use the strong Markov property after announcing the Markov property alone, with no mention as to the difference. Their equivalence ensured the correctness of the conclusions drawn in spite of some obscure points in the reasoning.

At the same time we should again emphasize the intuitively natural conceptual structure surrounding the strong Markov property, beginning with the concept of a stopping time itself. It is precisely the manner of defining the latter that ensures the validity of the strong Markov property. In this respect see Exercise 2.4.

2.4 CLASSES OF STATES

2.4.1. The states of a Markov chain may be grouped into classes according to whether or not other states of a class can be reached from any given state of the same class. We say "state i *leads to* state j"*), and write $i \to j$ if and only if there exists an $n \geqslant 1$ such that $p(n, i, j) > 0$. Next, we say "state i *communicates with* state j" and write $i \leftrightarrow j$ if and only if we have both $i \to j$ and $j \to i$, The relation "$\to$" is transitive, i.e., $i \to j$ and $j \to k$ imply $i \to k$. This is an immediate consequence of the inequality.

$$p(m+n, i, k) \geqslant p(m, i, j)\, p(n, j, k)$$

holding by the Chapman-Kolmogorov equations (see 2.1.5). The relation $\leftrightarrow$ is transitive, too, and is obviously symmetric, i.e., $i \leftrightarrow j$ implies $j \leftrightarrow i$.

---

*) Equivalent phrases are "state j is accessible from state i" and "state j is reachable from state i".

A *class of states* is a maximal *) subset of the state space such that any two states of it (distinct or not) communicate. Two not identical classes must be disjoint since the existence of a common element would imply all their states communicate (via that element), that is the two classes merge into one class.

The reader may be familiar with this kind of classification under the name of "equivalence classes" according to the relation "$\leftrightarrow$". It should however be noted that the relation $\leftrightarrow$ is not reflexive since there may be a state which does not communicate with itself and therefore with no other state. (Such a state is called *nonreturn.*) For instance, this is the case of state 1 in the Markov chain with the transition matrix

$$\begin{array}{c} \\ 1 \\ 2 \\ 3 \end{array} \begin{array}{c} \begin{array}{ccc} 1 & 2 & 3 \end{array} \\ \begin{pmatrix} 0 & 1/4 & 3/4 \\ 0 & 1/2 & 1/2 \\ 0 & 1/3 & 2/3 \end{pmatrix} \end{array}$$

Excepting the nonreturn states **) the relation $\leftrightarrow$ is an equivalence relation on the set of the remaining states and the equivalence classes according to it do coincide with the classes of states defined above.

2.4.2. A subset M of the state space is said to be *closed* if and only if $\sum_{j \in M} p(i, j) = 1$ for any $i \in M$, that is the submatrix $(p(i, j))_{i,j \in M}$ of the transition matrix $\mathbf{P}$ is stochastic [hence so is $(p(n, i, j))_{i,j \in M}$ for any $n \geqslant 1$]. In other words, M in closed if and only if it is not possible to leave M given the chain started in a state belonging to M. In particular, the concept of closedness applies to a class of states. A characterization of a closed class of states will be given later in Theorem 2.10. The simplest example is that of a closed class consisting of a single state i when $p(i, i) = 1$. Such a state is said to be *absorbing* (we have already encountered absorbing states in several examples from 2.2). The other extreme case is that where the whole state space S forms a (closed) class. In this case the state space (as well as the Markov chain) is said to be *irreducible.*

2.4.3. A property defined for all the states of a Markov chain is said to be a *class property* if and only if its possession for a state i implies its possession by all the states of the class containing i.

A first example of a class property occurs in connection with the concept of the period of a state. If i is a *return* state, i.e., $p(n, i, i) > 0$ for some $n \geqslant 1$, we define its *period* d_i as the greatest common divisor of all the natural numbers m such that $p(m, i, i) > 0$. A state i is said to be *periodic* or *aperiodic* according as $d_i > 1$ or $d_i = 1$.

---

*) The adjective "maximal" should be understood in the sense that no further state communicating with all the existing ones can be adjoined.

**) We may consider each nonreturn state as a class by itself.

Theorem 2.2. *Two distinct states belonging to the same class have the same period. In other words, the property of having period d is a class property.*

Proof. Let $i \neq j$ be two distinct states of period d_i and d_j, respectively. We should prove that $d_i = d_j$. Because $i \leftrightarrow j$ there exist natural numbers r and s such that $p(r, i, j) > 0$, $p(s, j, i) > 0$. If $p(m, i, i) > 0$ then $p(2m, i, i) \geqslant p(m, i, i)\, p(m, i, i) > 0$ and

$$p(m + r + s, j, j) \geqslant p(s, j, i)\, p(m, i, i)\, p(r, i, j) > 0,$$

$$p(2m + r + s, j, j) \geqslant p(s, j, i)\, p(2m, i, i)\, p(r, i, j) > 0$$

(the justification of the last two inequalities is left to the reader). Therefore d_j divides both $m + r + s$ and $2m + r + s$. Hence it also divides the difference $2m + r + s - (m+r+s) = m$. Thus d_j divides all the natural numbers m such that $p(m, i, i) > 0$. Since the greatest common divisor of these numbers is d_i we have $d_j \leqslant d_i$. By interchanging i and j we get $d_i \leqslant d_j$. Therefore $d_i = d_j$ and the proof is complete.

Theorem 2.2 allows us to speak of the period d of a class, and thus to qualify a class as periodic (cyclic) or aperiodic (acyclic) according as $d > 1$ or $d = 1$.

2.4.4. The motion of a Markov chain (i.e., the transitions from state to state) within a periodic closed class obeys a certain cyclic pattern we shall describe.

Let d be the period of the class $\mathcal{C}$ (not assumed for the moment to be closed). Consider two states $i, j \in \mathcal{C}$ and let r, s, t be natural numbers such that $p(r, i, j) > 0$, $p(s, i, j) > 0$, and $p(t, j, i) > 0$. Then $p(r+t, i, i) \geqslant p(r, i, j)\ p(t, j, i) > 0$ and $p(s + t, i, i) \geqslant p(s, i, j)\ p(t, j, i) > 0$. Hence d divides both $r + t$ and $s + t$, therefore their difference $s + t - (r + t) = s - r$. This means that if $r = ad + b$, with a and b nonnegative integers such that $0 \leqslant b \leqslant d - 1$, then $s = cd + b$ for some nonnegative integer c. Consequently, if i leads to j in n steps [that is $p(n, i, j) > 0$] then $n \equiv b \pmod d$, where $0 \leqslant b \leqslant d - 1$ depends on i and j but is independent of n. Now, for a fixed $i \in \mathcal{C}$, we may consider the subset $\mathcal{C}_b(i)$ of $\mathcal{C}$ that consists of those states corresponding to the same residue class b modulo d, namely

$$\mathcal{C}_b(i) = (j \in \mathcal{C} : p(n, i, j) > 0 \text{ implies } n \equiv b \pmod d),\ 0 \leqslant b \leqslant d - 1.$$

Clearly, the $\mathcal{C}_b(i)$ are pairwise disjoint and their union is $\mathcal{C}$. It is convenient to extend this notation to all the nonnegative integers u by setting $\mathcal{C}_u(i) = \mathcal{C}_v(i)$ if $u \equiv v \pmod d$. Let us remark that if $j \in \mathcal{C}_u(i)$ then $\mathcal{C}_v(j) = \mathcal{C}_{u+v}(i)$ for all the nonnegative integers u and v. Hence the sets $\mathcal{C}_b(i)$, $0 \leqslant b \leqslant d - 1$, do not depend on the choice of state i in $\mathcal{C}$ except for a cyclic permutation. Consequently, we may denote them simply by $\mathcal{C}_b$, $0 \leqslant b \leqslant d - 1$.

Theorem 2.3. *Let $\mathcal{C}$ be a closed class of period d. If $k \in \mathcal{C}_b$ and $p(k,j) > 0$, then $j \in \mathcal{C}_{b+1}$. Therefore, the Markov chain moves from $\mathcal{C}_0$ to $\mathcal{C}_1$, ..., from $\mathcal{C}_{d-2}$ to $\mathcal{C}_{d-1}$, from $\mathcal{C}_{d-1}$ back to $\mathcal{C}_0$, and so on.*

Proof. Let $n \geqslant 1$ be such that $p(n, i, k) > 0$. Then n is of the form $ad + b$. As $p(n+1, i, j) \geqslant p(n, i, k)\, p(k, j) > 0$, state i leads to state j. Hence $j \in \mathcal{C}$ since $\mathcal{C}$ is closed. But $n \equiv b \pmod{d}$, whence $n + 1 \equiv$ $\equiv b + 1 \pmod{d}$, so that $j \in \mathcal{C}_{b+1}$.

It follows from Theorem 2.3 that for a closed class $\mathcal{C}$ the subsets $\mathcal{C}_b$, $0 \leqslant b \leqslant d-1$, are not empty. They are called the (*cyclically moving*) *subclasses* of $\mathcal{C}$.

Clearly, the transition matrix of a Markov chain whose state space is a closed class of period $d > 1$ has the form

$$
\mathbf{P} = \begin{array}{c} \\ \mathcal{C}_0 \\ \mathcal{C}_1 \\ \cdot \\ \cdot \\ \cdot \\ \mathcal{C}_{d-1} \end{array}
\begin{array}{c} \begin{array}{ccccc} \mathcal{C}_0 & \mathcal{C}_1 & \mathcal{C}_2 & \cdots & \mathcal{C}_{d-1} \end{array} \\
\left(\begin{array}{ccccc}
0 & \times & 0 & \cdots & 0 \\
0 & 0 & \times & \cdots & 0 \\
 & & \cdot & & 0 \\
 & 0 & & \cdot & \\
 & & & & \cdot\times \\
\times & 0 & 0 & \cdots & 0
\end{array}\right) \end{array},
$$

where the sign $\times$ stands for stochastic matrices (in general rectangular), that is $\mathbf{P}$ is an irreducible matrix of period d. Hence, in particular, $\mathbf{P}^d$ has the diagonal form

$$
\mathbf{P}^d = \begin{array}{c} \\ \mathcal{C}_0 \\ \mathcal{C}_1 \\ \cdot \\ \cdot \\ \cdot \\ \mathcal{C}_{d-1} \end{array}
\begin{array}{c} \begin{array}{ccccc} \mathcal{C}_0 & \mathcal{C}_1 & & & \mathcal{C}_{d-1} \end{array} \\
\left(\begin{array}{ccccc}
\times & & & & \\
 & \times & & & 0 \\
 & & \cdot & & \\
 & & & \cdot & \\
 & 0 & & & \cdot \\
 & & & & \times
\end{array}\right) \end{array}.
$$

This means that

$$
\mathsf{P}(X(m+n) \in \mathcal{C}_{b+n} \mid X(m) = k) = 1,
$$

whatever the nonnegative integers m and n, $k \in \mathcal{C}_b$, $0 \leqslant b \leqslant d-1$. Hence given a closed class $\mathcal{C}$ of period $d > 1$ and an initial distribution $\mathbf{p} = (p(i))_{i \in S}$ on the state space S such that $p(i) = 0$ for $i \notin \mathcal{C}_b$, the sequence of random variables $(X(nd))_{n \geqslant 0}$ is a Markov chain with state space $\mathcal{C}_b$ and transition matrix $(p(d, i, j))_{i, j \in \mathcal{C}_b}$, $0 \leqslant b \leqslant d-1$. For this Markov chain the state space $\mathcal{C}_b$ is an aperiodic closed class.

2.4.5. An effective procedure for determining the period and the subclasses of a closed class is available (see [8], pp. 226—228). We shall describe it for the example of the Markov chain with transition matrix

$$\begin{array}{c} \\ 1 \\ 2 \\ 3 \\ 4 \\ 5 \\ 6 \\ 7 \end{array}\begin{array}{c} \begin{array}{ccccccc} 1 & 2 & 3 & 4 & 5 & 6 & 7 \end{array} \\ \left(\begin{array}{ccccccc} & & & \times & & & \\ & & & & & & \times \\ & \times & & & & & \\ & & \times & & \times & & \\ & \times & & & & & \\ \times & \times & & & & & \\ & & & & & \times & \end{array}\right) \end{array}$$

the only nonnegative entries of which are indicated by the sign $\times$. It is easily seen*) that all the states communicate, and therefore that the

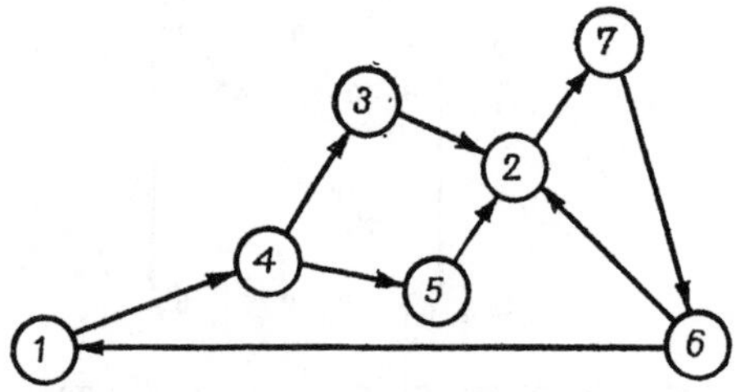

Fig. 2.2

state space forms a single closed class $\mathcal{C}$. Choose a state, say 1, and call $\mathcal{C}_0$ the subclass containing it. By Theorem 2.3 all the states which are accessible in one step from state 1 belong to $\mathcal{C}_1$; in our example $4 \in \mathcal{C}_1$. All the states which are accessible in one step from state 4 belong to $\mathcal{C}_2$; in our example $3{,}5 \in \mathcal{C}_2$. Going on in this way we obtain the following table

$\mathcal{C}_0$	$\mathcal{C}_1$	$\mathcal{C}_2$	$\mathcal{C}_3$	$\mathcal{C}_4$	$\mathcal{C}_5$	$\mathcal{C}_6$
1	4	3,5	2	7	6	1,2

whose construction is stopped when a state already encoutered reappears. It follows that $1 \in \mathcal{C}_0 \cap \mathcal{C}_6$, whence $\mathcal{C}_0 = \mathcal{C}_6$, and $2 \in \mathcal{C}_3 \cap \mathcal{C}_6$, whence $\mathcal{C}_3 = \mathcal{C}_6$, so that $\mathcal{C}_0 = \mathcal{C}_3 = \mathcal{C}_6$. Now, construct a new table starting with $\mathcal{C}_0$ = the subclass containing states 1 and 2. All the states which

*) A geometric representation (see Figure 2.2) that associates with the transition matrix a directed graph, whose vertices, corresponding to the states of the chain, are connected by oriented arcs ij if and only if $p(i, j) > 0$, offers a convenient intuitive image of the communication relation. For a systematic use of graph theory in the study of Markov chains the reader may consult the book [127]. See also the papers [20, 293, 348, 379, and 386].

are accessible in one step from either state 1 or state 2 belong to $\mathcal{C}_1$, etc. We get

$\mathcal{C}_0$	$\mathcal{C}_1$	$\mathcal{C}_2$	$\mathcal{C}_3$
1,2	4,7	3,5,6	1,2

Thus $\mathcal{C}_0 = \mathcal{C}_3$ (a fact already known) and since $\mathcal{C}_0 \cup \mathcal{C}_1 \cup \mathcal{C}_2 = \mathcal{C}$ we conclude that the period of $\mathcal{C}$ is 3 and the cyclically moving subclasses are $\mathcal{C}_0 = (1, 2)$, $\mathcal{C}_1 = (4,7)$, and $\mathcal{C}_2 = (3, 5, 6)$. By reordering the states in order to have the observed grouping into subclasses, the transition matrix can be written as

$$\mathbf{P} = \begin{array}{c} 1 \\ 2 \\ 4 \\ 7 \\ 3 \\ 5 \\ 6 \end{array} \overset{\begin{array}{ccccccc} 1 & 2 & 4 & 7 & 3 & 5 & 6 \end{array}}{\left(\begin{array}{cc|cc|ccc} & & \times & & & & \\ & & & \times & & & \\ \hline & & & & \times & \times & \\ & & & & & & \times \\ \hline & \times & & & & & \\ & \times & & & & & \\ \times & \times & & & & & \end{array} \right)}$$

or, schematically,

$$\mathbf{P} = \begin{array}{c} \mathcal{C}_0 \\ \mathcal{C}_1 \\ \mathcal{C}_2 \end{array} \overset{\begin{array}{ccc} \mathcal{C}_0 & \mathcal{C}_1 & \mathcal{C}_2 \end{array}}{\begin{pmatrix} & \times & \\ & & \times \\ \times & & \end{pmatrix}}.$$

Hence the schematic representation of $\mathbf{P}^2$ and $\mathbf{P}^3$ is as follows

$$\mathbf{P}^2 = \begin{array}{c} \mathcal{C}_0 \\ \mathcal{C}_1 \\ \mathcal{C}_2 \end{array} \overset{\begin{array}{ccc} \mathcal{C}_0 & \mathcal{C}_1 & \mathcal{C}_2 \end{array}}{\begin{pmatrix} & & \times \\ \times & & \\ & \times & \end{pmatrix}}, \quad \mathbf{P}^3 = \begin{array}{c} \mathcal{C}_0 \\ \mathcal{C}_1 \\ \mathcal{C}_2 \end{array} \overset{\begin{array}{ccc} \mathcal{C}_0 & \mathcal{C}_1 & \mathcal{C}_2 \end{array}}{\begin{pmatrix} \times & & \\ & \times & \\ & & \times \end{pmatrix}}.$$

The matrices $\mathbf{P}^{3n+1}$, $\mathbf{P}^{3n+2}$, $\mathbf{P}^{3n+3}$, $n = 1, 2, \ldots$, have the same form as $\mathbf{P}$, $\mathbf{P}^2$, $\mathbf{P}^3$, respectively (without coinciding numerically).

2.4.6. This subparagraph is devoted to reviewing the examples from 2.2 as regards the classes of states of the Markov chains, arising from them.

The Markov chain associated with the gambler's ruin problem has two absorbing states (0 and l) and a non closed class of period 2 consisting of states $1, \ldots, l-1$. In the variant where the barriers 0 and l are reflecting, the corresponding Markov chain is irreducible of period 2 and in the variant where the barriers 0 and l are elastic, the corresponding Markov chain is irreducible aperiodic.

The Markov chain associated with the Bernoulli model (Example 2) is irreducible aperiodic and that associated with the Ehrenfest model is irreducible of period 2.

In Example 3 the first $r^2 - 4$ states are nonreturn, and states 0, 1, 2, 3, 4 are, as noticed, the states of the Markov chain associated with the gambler's ruin problem for $l = 4$.

The Markov chain from Example 4 has an absorbing state (0) and r aperiodic non closed classes each consisting of a state.

The Markov chain from Example 5 has two absorbing states (d and $d+1$). Any state $0 \leqslant i \leqslant d-1$ for which $0 < r_i < 1$ forms an aperiodic non closed class. If $r_i = 0$ then state i is nonreturn.

If $0 < p_0 < 1$ and $P_m > 0$ then the Markov chain from Example 6 is irreducible aperiodic. Similarly, if $0 < p_0 + p_1 < 1$ and $P_m > 0$, the same is true of the Markov chain from Example 7. (These conditions are not necessary but just sufficient for irreducibility and aperiodicity.)

The Markov chain from Example 8 is irreducible aperiodic if and only if $p_{m-1} > 0$.

The discussion of Example 9 is left to the reader as an exercise.

2.5 RECURRENCE AND TRANSIENCE

2.5.1. According to whether or not a state occurs infinitely often two types of states will be distinguished. Let us first define the notation.

Let τ_j denote the first passage time to state j, defined as $\tau_j = \min\ (n \geqslant 1 : X(n) = j)$ (see 2.3.1). The probability distribution of τ_j is of paramount importance in what follows. Put*)

$$f(n, i, j) = \mathsf{P}_i(\tau_j = n), \quad n \geqslant 1,$$

and

$$f(i, j) = \sum_{n \geqslant 1} f(n, i, j) = \mathsf{P}_i(\tau_j < \infty),$$

whence

$$\mathsf{P}_i(\tau_j = \infty) = f(\infty, i, j) = 1 - f(i, j).$$

Therefore $(f(n, i, j),\ n = 1, 2, \ldots, \infty)$ is the probability distribution of τ_j under the probability P_i (i.e., given the Markov chain starts in state i).

The probabilities considered can be obviously expressed in a way that involves the Markov chain itself. We have

$$\{\tau_j = 1\} = \{X(1) = j\},$$

$$\{\tau_j = n\} = \{X(m) \neq j,\ 1 \leqslant m \leqslant n-1,\ X(n) = j\},\ n \geqslant 2,$$

$$\{\tau_j = \infty\} = \{X(m) \neq j,\ m \geqslant 1\},$$

$$\{\tau_j < \infty\} = \{X(m) = j \text{ for at least one value of } m \geqslant 1\} = \bigcup_{m \geqslant 1} \{X(m) = j\},$$

---

*) P_i has been defined on p. 62.

therefore

$$f(1, i, j) = \mathsf{P}_i(X(1) = j) = p(i, j),$$
$$f(n, i, j) = \mathsf{P}_i(X(m) \neq j,\ 1 \leqslant m \leqslant n-1,\ X(n) = j)$$
$$= \sum_{i_m \neq j (1 \leqslant m \leqslant n-1)} p(i, i_1)\, p(i_1, i_2) \ldots p(i_{n-1}, j),\ n \geqslant 2,$$
$$f(\infty, i, j) = \mathsf{P}_i(X(m) \neq j,\ m \geqslant 1),$$
$$f(i, j) = \mathsf{P}_i(X(m) = j \text{ for at least one value of } m \geqslant 1) = \mathsf{P}_i\Big(\bigcup_{m \geqslant 1} \{X(m) = j\}\Big).$$

The last equalities yield

$$\mathsf{P}_i(X(n) = j) \leqslant f(i, j) \leqslant \sum_{m \geqslant 1} \mathsf{P}_i(X(m) = j)$$

for all $n \geqslant 1$, whence

$$\sup_{n \geqslant 1} p(n, i, j) \leqslant f(i, j) \leqslant \sum_{m \geqslant 1} p(m, i, j)$$

for all states i and j.

It follows at once that $i \to j$ if and only if $f(i, j) > 0$ and that $i \leftrightarrow j$ if and only if $f(i, j)\, f(j, i) > 0$.

Notice that by homogeneity we also have

$$f(1, i, j) = \mathsf{P}(X(m+1) = j \mid X(m) = i),$$
$$f(n, i, j) = \mathsf{P}(X(m+l) \neq j,\ 1 \leqslant l \leqslant n-1,\ X(m+n) = j \mid X(m) = i),\ n \geqslant 2,$$

for any m for which the conditional probabilities are defined.

2.5.2. Now we are able to introduce the concepts of recurrence and transience.

Theorem 2.4. *If the Markov chain starts in state i then the probability of returning to i at least r times equals $[f(i, i)]^r$.*

Proof. We proceed by induction on r. For $r = 1$ the theorem is true by the very definition of $f(i, i)$. Assume it holds for $r = m - 1$. We have

$$\mathsf{P}_i(X(n) = i \text{ for at least } m \text{ values of } n \geqslant 1)$$
$$= \sum_{k \geqslant 1} \mathsf{P}_i(\tau_i = k,\ X(\tau_i + l) = i \text{ for at least } m-1 \text{ values of } l \geqslant 1)$$
$$= \sum_{k \geqslant 1} \mathsf{P}_i(\tau_i = k)\, \mathsf{P}_i(X(\tau_i + l) = i \text{ for at least } m-1 \text{ values of } l \geqslant 1 \mid \tau_i = k).$$

Next, taking into account that $\{X(\tau_i) = i\}$ is the sure event Ω, that τ_i is a stopping time for the Markov chain considered, and that the random event $\{\tau_i = k\}$ is prior to τ_i (why?), we have the following string of equalities:

$$\mathsf{P}_i(X(\tau_i + l) = i \text{ for at least } m - 1 \text{ values of } l \geqslant 1 \mid \tau_i = k)$$

$$= \mathsf{P}_i(X(\tau_i + l) = i \text{ for at least } m - 1 \text{ values of } l \geqslant 1 \mid X(\tau_i) = i, \tau_i = k)$$

$$= \mathsf{P}_i(X(\tau_i + l) = i \text{ for at least } m - 1 \text{ values of } l \geqslant 1 \mid X(\tau_i) = i)$$

(by Corollary 2 of Theorem 2.1)

$$= \mathsf{P}_i(X(l) = i \text{ for at least } m - 1 \text{ values of } l \geqslant 1)$$

(by the same corollary)

$$= [f(i, i)]^{m-1} \text{ (by the induction hypothesis).}$$

Therefore

$$\mathsf{P}_i(X(n) = i \text{ for at least } m \text{ values of } n \geqslant 1)$$

$$= \sum_{k \geqslant 1} f(k, i, i)\,[f(i, i)]^{m-1} = f(i, i)\,[f(i, i)]^{m-1},$$

which is the stated result.

Corollary. *Assume the Markov chain starts in state i. If $f(i, i) = 1$, then the probability of returning to i infinitely often is 1. If $f(i, i) < 1$, the probability of returning to i infinitely often is 0.*

Proof. The random event

$$A = \{\text{the Markov chain returns to } i \text{ infinitely often}\}$$

is the intersection of the decreasing sequence of random events

$$\{\text{the Markov chain returns to } i \text{ at least } r \text{ times}\},\ r \geqslant 1.$$

Therefore on account of (1.4) the probability of A is equal to

$$\lim_{m\to\infty} [f(i, i)]^m = \begin{cases} 1, & \text{if } f(i, i) = 1, \\ 0, & \text{if } f(i, i) < 1. \end{cases}$$

The above results justify the following

Definition. A state i is said to be *recurrent* or *transient* according as $f(i, i) = 1$ or $f(i, i) < 1$*).

2.5.3. Theorem 2.4 can be generalized as follows

Theorem 2.5. *If the Markov chain starts in state i then the probability that state j occurs at least r times equals $f(i, j)\,[f(j, j)]^{r-1}$.*

The *proof* is entirely similar to that of Theorem 2.4 using τ_j instead of τ_i.

---

*) For recurrent and transient one uses also the names *persistent* and *nonrecurrent* respectively.

Corollary. *Assume the Markov chain starts in state i. The probability that state j occurs infinitely often equals $f(i, j)$ or 0 according as j is recurrent or transient.*

Notice that in the case of a transient state j the above corollary holds whatever the initial distribution. Indeed

$$\mathsf{P}(\text{state } j \text{ occurs infinitely often})$$

$$= \sum_{i \in S} \mathsf{P}(X(0) = i)\, \mathsf{P}(\text{state } j \text{ occurs infinitely often} \mid X(0) = i)$$

$$= \sum_{i \in S} \mathsf{P}(X(0) = i) \cdot 0 = 0.$$

Moreover, the probability of remaining forever in the set T of transient states is 0 since

$$\mathsf{P}(X(n) \in T \text{ for all } n \geqslant 0) \leqslant \sum_{j \in T} \mathsf{P}(\text{ state } j \text{ occurs infinitely often}) = 0.$$

(The reader will be able to show that the random event $\{X(n) \in T$ for all $n \geqslant 0\}$ implies at least one of the random events {state j occurs infinitely often}, $j \in T$.)

2.5.4. Now we shall give the fundamental formula connecting the probabilities $f(n, i, j)$ and $p(n, i, j)$.

Theorem 2.6. (*First entrance theorem*). *Whatever the states i and j and the natural number n, we have*

$$p(n, i, j) = \sum_{m=1}^{n} f(m, i, j)\, p(n-m, j, j). \tag{2.3}$$

Proof. Intuitively we may argue as follows. To be in state j at the nth step, the Markov chain should reach that state for the first time at some time m, $1 \leqslant m \leqslant n$. After this happens, it should return to j in $n - m$ steps. The rigorous proof uses the strong Markov property. We have

$$p(n, i, j) = \mathsf{P}_i(X(n) = j) = \mathsf{P}_i(\tau_j \leqslant n, X(n) = j)$$

(since the random event $\{X(n) = j\}$ implies the random event $\{(\tau_j \leqslant n\})$

$$= \sum_{m=1}^{n} \mathsf{P}_i(\tau_j = m, X(n) = j)$$

(since $\{\tau_j \leqslant n\}$ is the union of the disjoint random events $\{\tau_j = m\}$, $1 \leqslant m \leqslant n$)

$$= \sum_{m=1}^{n} \mathsf{P}_i(\tau_j = m, X(\tau_j + n - m) = j)$$

(since $\{\tau_j = m\} = \{\tau_j + n - m = n\}$)

$$= \sum_{m=1}^{n} \mathsf{P}_i(\tau_j = m)\, \mathsf{P}_i(X(\tau_j + n - m) = j) \mid \tau_j = m)$$

$$= \sum_{m=1}^{n} \mathsf{P}_i(\tau_j = m)\, \mathsf{P}_i(X(\tau_j + n - m) = j \mid X(\tau_j) = j,\ \tau_j = m)$$

(since $\{X(\tau_j) = j\}$ is the sure event Ω)

$$= \sum_{m=1}^{n} \mathsf{P}_i(\tau_j = m)\, \mathsf{P}_i(X(\tau_j + n - m) = j \mid X(\tau_j) = j)$$

(by Corollary 1 of Theorem 2.1)

$$= \sum_{m=1}^{n} f(m, i, j)\, p(n - m, j, j).$$

2.5.5. The first entrance theorem allows us to express $f(i, j)$ in terms of the n step transition probabilities. We shall prove

Theorem 2.7. (*Doeblin's formula*). *Whatever the states i and j, we have*

$$f(i, j) = \lim_{s\to\infty} \frac{\sum_{n=1}^{s} p(n, i, j)}{1 + \sum_{n=1}^{s} p(n, j, j)}. \tag{2.4}$$

Proof. Equation (2.3) yields

$$\sum_{n=1}^{s} p(n, i, j) = \sum_{n=1}^{s} \sum_{m=1}^{n} f(m, i, j)\, p(n - m, j, j)$$

$$= \sum_{m=1}^{s} \left(f(m, i, j) \sum_{n=m}^{s} p(n - m, j, j) \right),$$

whence

$$\left(1 + \sum_{n=1}^{s} p(n, j, j)\right) \sum_{m=1}^{s} f(m, i, j) \geqslant \sum_{n=1}^{s} p(n, i, j)$$

$$\geqslant \left(1 + \sum_{n=1}^{s-s'} p(n, j, j)\right) \sum_{m=1}^{s'} f(m, i, j),$$

for all $s' < s$. For $1 + \sum_{n=1}^{s} p(n, j, j)$ dominates

$$\sum_{n=m}^{s} p(n - m, j, j) = 1 + \sum_{n=1}^{s-m} p(n, j, j)$$

and

$$\sum_{m=1}^{s}\left(f(m, i, j)\sum_{n=m}^{s} p(n-m, j, j)\right)$$

dominates

$$\sum_{m=1}^{s'}\left((f(m, i, j)\sum_{n=m}^{s-s'+m} p(n-m, j, j)\right)=\left(1+\sum_{n=1}^{s-s'} p(n, j, j)\right)\sum_{m=1}^{s'} f(m, i, j).$$

Dividing by $1+\sum_{n=1}^{s} p(n, j, j)$ and letting first $s \to \infty$ then $s' \to \infty$ yields equation (2.4).

In particular, for $j = i$ we have

$$f(i, i) = 1 - \lim_{s\to\infty}\frac{1}{1+\sum_{n=1}^{s} p(n, i, i)}.$$

Corollary 1. *State i is recurrent or transient according as the series $\sum_{n\geqslant 1} p(n, i, i)$ diverges or converges.*

Proof. By Doeblin's formula divergence of the series $\sum_{n\geqslant 1} p(n, i, i)$ implies $f(i, i) = 1$ and convergence of it implies $f(i, i) < 1$.

Corollary 2. *Let i be an arbitrary state and j a transient state. Then the series $\sum_{n\geqslant 1} p(n, i, j)$ converges, hence, in particular* $\lim_{n\to\infty} p(n, i, j) = 0$.

Proof. By Corollary 1 the series $\sum_{n\geqslant 1} p(n, j, j)$ converges. Then Doeblin's formula implies that the series $\sum_{n\geqslant 1} p(n, i, j)$ converges, too [and its sum is $f(i,j)\sum_{n\geqslant 0} p(n, j, j)$].

Remark. The fact that $\lim_{n\to\infty} p(n, i, j) = 0$ for a transient state j implies a geometric rate of convergence, i.e., there exist $a > 0$ and $0 < b < 1$ such that $p(n, i, j) \leqslant ab^n$. See Exercise 2.8.

Corollary 3. *The state space of a finite Markov chain contains at least one recurrent state.*

Proof. Assume for a contradiction that all the states are transient. Letting $n \to \infty$ in the equality $\sum_{j\in S} p(n, i, j) = 1$, $i \in S$, on account of Corollary 2 above we get

$$0 = \sum_{j\in S}\lim_{n\to\infty} p(n, i, j) = \lim_{n\to\infty}\sum_{j\in S} p(n, i, j) = 1,$$

i.e., the desired contradiction.

2.5.6. We are now able to study recurrence and transience within a class of states.

Theorem 2.8. *Recurrence and transience are class properties.*

Proof. Recurrence and transience being complementary properties we need only prove our assertion for the former. Let $\mathcal{C}$ be a class and $i \in \mathcal{C}$ a recurrent state. If $j \in \mathcal{C}$, $j \neq i$, then $i \leftrightarrow j$ and there exist natural numbers r and s such that $p(r, i, j) = a > 0$, $p(s, j, i) = b > 0$, whence

$$p(n + r + s, j, j) \geqslant p(s, j, i)\ p(n, i, i)\ p(r, i, j) = ab\, p(n, i, i),$$

what ever the nonnegative integer n. Since i is a recurrent state, by Corollary 1 of Theorem 2.7 the right hand side is the general term of a divergent series. Therefore, so is the left hand side and by the same corollary state j is recurrent, too.

Theorem 2.8 allows us to use the labels "recurrent" and "transient" for classes of states.

Theorem 2.9. *Let i be a recurrent state. If $i \rightarrow j$ then state j is recurrent, too, and $f(i, j) = f(j, i) = 1$. In other words, a transient state cannot be reached from a recurrent state.*

Proof. We shall prove that $j \rightarrow i$. It will then follow that i and j communicate, therefore that they are in the same class. An appeal to Theorem 2.8 will show state j is recurrent.

As we have previously noticed $j \rightarrow i$ is equivalent to $f(j, i) > 0$. We shall prove that we even have $f(j, i) = 1$.

Whatever the natural number n state i being recurrent we have

$$\begin{aligned}
1 &= \mathsf{P}_i \text{ (state } i \text{ occurs infinitely often)} \\
&\leqslant \mathsf{P}_i\ (X(l) = i \text{ for at least one value of } l > n) \\
&= \sum_{k \in S} \mathsf{P}_i(X(l) = i \text{ for at least one value of } l > n,\ X(n) = k) \\
&= \sum_{k \in S} \mathsf{P}_i(X(n) = k)\ \mathsf{P}_i(X(l) = i \text{ for at least one value of } l > n \mid X(n) = k) \\
&= \sum_{k \in S} p(n, i, k)\, f(k, i).
\end{aligned}$$

Since $i \rightarrow j$ there exists n such that $p(n, i, j) > 0$. So if $f(j, i) < 1$ we should have

$$1 \leqslant \sum_{k \in S} p(n, i, k)\, f(k, i) < \sum_{k \in S} p(n, i, k) = 1.$$

The contradiction reached shows we may only have $f(j, i) = 1$. By interchanging i and j we then get $f(i, j) = 1$.

Theorem 2.10. *A class is recurrent if and only if it is closed.*

Proof. Let $\mathcal{C}$ be a recurrent class. If $\mathcal{C}$ is not closed then there exists $i \in \mathcal{C}$ such that $\sum_{j \in \mathcal{C}} p(i, j) < 1$, and hence there exists $k \notin \mathcal{C}$ such that $p(i, k) > 0$, i.e., $i \rightarrow k$. State i being recurrent, by Theorem 2.9 we have $i \leftrightarrow k$, whence $k \in \mathcal{C}$. The contradiction reached shows that $\mathcal{C}$ is closed.

Next, let $\mathcal{C}$ be a closed class. By Corollary 3 of Theorem 2.7 [applied to the Markov chain with state space $\mathcal{C}$ and transition matrix $(p(i, j))_{i,j \in \mathcal{C}}$] there exists at least one recurrent state in $\mathcal{C}$. But recurrence is a class property, hence $\mathcal{C}$ is a recurrent class.

2.5.7. Consider a recurrent class $\mathcal{C}$ and let i be an arbitrary state of $\mathcal{C}$. Since i is recurrent it occurs infinitely often with probability 1. Let $\tau_i^{(1)} < \dots < \tau_i^{(l)} < \dots$ be the increasing sequence of all the values of $n \geqslant 1$ for which $X(n) = i$. Clearly, $\tau_i^{(l)}$, which is called the *moment of lth passage* to state i, is a random variable, $l = 1, 2, \dots$. The random variable $\tau_i^{(1)}$ is nothing but τ_i. Notice also that $X(\tau_i^{(l)}) = i$, $l = 1, 2, \dots$ by the very definition of the $\tau_i^{(l)}$. Finally, most importantly, $\tau_i^{(l)}$, $l = 1, 2, \dots$ are stopping times. The proof of this fact, which is immediate, is left to the reader.

Consider the random variables

$$\beta_i^{(l)} = (X(\tau_i^{(l)}),\ X(\tau_i^{(l)} + 1), \dots,\ X(\tau_i^{(l+1)} - 1)),\quad l = 1,\ 2, \dots$$

whose values are finite ordered sequences of states from $\mathcal{C}$, all beginning with i.

We shall prove

Theorem 2.11 (*W. Doeblin*). *The random variables* $\beta_i^{(l)}$, $l = 1, 2, \dots$ *are independent and identically distributed.*

Proof. The result is an immediate consequence of the strong Markov property. Indeed we have

$$\begin{aligned}
&\mathsf{P}(\beta_i^{(1)} = b_i^{(1)}, \dots, \beta_i^{(l-1)} = b_i^{(l-1)},\ \beta_i^{(l)} = b_i^{(l)}) \\
&= \mathsf{P}(\beta_i^{(1)} = b_i^{(1)}, \dots, \beta_i^{(l-1)} = b_i^{(l-1)}, X(\tau_i^{(l)}) = i,\ (X(\tau_i^{(l)} + 1), \dots, X(\tau_i^{(l+1)} - 1) = b'^{(l)}) \\
&= \mathsf{P}((X(\tau_i^{(l)} + 1), \dots, X(\tau_i^{(l+1)} - 1)) = b'^{(l)} \mid X(\tau_i^{(l)}) = i,\ \beta_i^{(l-1)} = b_i^{(l-1)}, \dots \\
&\quad \dots, \beta_i^{(1)} = b_i^{(1)}) \times \mathsf{P}(\beta_i^{(1)} = b_i^{(1)}, \dots, \beta_i^{(l-1)} = b_i^{(l-1)},\ X(\tau_i^{(l)}) = i),
\end{aligned}$$

where $b_i^{(1)}, \dots, b_i^{(l-1)}, b_i^{(l)}$ are finite ordered sequences of states from $\mathcal{C}$ beginning with i and $b_i^{(l)} = (i,\ b'^{(l)})$. By the strong Markov property the above conditional probability is equal to

$$\begin{aligned}
&\mathsf{P}((X(\tau_i^{(l)} + 1), \dots, X(\tau_i^{(l+1)} - 1)) = b'^{(l)} \mid X(\tau_i^{(l)}) = i) \\
&= \mathsf{P}((X(\tau_i^{(l)}),\ X(\tau_i^{(l)} + 1), \dots, X(\tau_i^{(l+1)} - 1)) = b_i^{(l)}) = \mathsf{P}(\beta_i^{(l)} = b_i^{(l)})
\end{aligned}$$

[on account of the fact that $\mathsf{P}(X(\tau_i^{(l)}) = i) = 1$]. Therefore

$$\mathsf{P}(\beta_i^{(1)} = b_i^{(1)}, \ldots, \beta_i^{(l-1)} = b_i^{(l-1)}, \beta_i^{(l)} = b_i^{(l)})$$
$$= \mathsf{P}(\beta_i^{(1)} = b_i^{(1)}, \ldots, \beta_i^{(l-1)} = b_i^{(l-1)})\, \mathsf{P}(\beta_i^{(l)} = b_i^{(l)}).$$

Hence

$$\mathsf{P}(\beta_i^{(1)} = b_i^{(1)}, \ldots, \beta_i^{(l-1)} = b_i^{(l-1)}, \beta_i^{(l)} = b_i^{(l)}) = \prod_{s=1}^{l} \mathsf{P}(\beta_i^{(s)} = b_i^{(s)}),$$

i.e., the random variables $\beta_i^{(l)}$, $l = 1, 2, \ldots$ are independent. Their common distribution has been already obtained. Again by the strong Markov property

$$\mathsf{P}(\beta_i^{(l)} = b_i^{(l)}) = \mathsf{P}((X(\tau_i^{(l)} + 1), \ldots, X(\tau_i^{(l+1)} - 1)) = b'^{(l)} \mid X(\tau_i^{(l)}) = i)$$
$$= \mathsf{P}_i((X(1), \ldots, X(\tau_i - 1)) = b'^{(l)}).$$

The proof is now complete.

R e m a r k. The random variable $\beta_i^{(0)} = (X(0), X(1), \ldots, X(\tau_i^{(1)} - 1))$ is also independent of the random variables $\beta_i^{(l)}$, $l \geqslant 1$, but its distribution is not identical, in general, to the common distribution of the latter. The distributions do coincide if $\mathsf{P} = \mathsf{P}_i$, i.e., if the Markov chain starts in i.

Corollary. *The random variables* $\tau_i^{(l+1)} - \tau_i^{(l)}$, $l = 1, 2, \ldots$ *are independent and identically distributed and*

$$\mathsf{P}(\tau_i^{(l+1)} - \tau_i^{(l)} = n) = f(n, i, i), \quad n = 1, 2, \ldots$$

Proof. We use the fact that functions of independent random variables are independent random variables. To obtain the corollary we take

$$F(\beta_i^{(l)}) = \text{number of components of } \beta_i^{(l)} = \tau_i^{(l+1)} - \tau_i^{(l)}.$$

Various other results can be derived from Theorem 2.11. One of them is the fact that the random variables $X(\tau_i^{(l)} + 1)$, $l \geqslant 1$, are independent.

2.6 CLASSIFICATION OF HOMOGENEOUS MARKOV CHAINS

2.6.1. The results obtained so far allow us to classify Markov chains according to the type of their states.

A finite Markov chain has at least one recurrent state. If there are no transient states, then renumbering the states in such a manner that those belonging to the same class have consecutive numbers, the transition matrix can be written as

$$\mathbf{P} = \begin{pmatrix} \mathbf{P}_1 & & & & \\ & \mathbf{P}_2 & & \mathbf{0} & \\ & & \cdot & & \\ & & & \cdot & \\ & \mathbf{0} & & \cdot & \\ & & & & \mathbf{P}_r \end{pmatrix}.$$

Here the stochastic matrices $\mathbf{P}_1, \ldots, \mathbf{P}_r$ are the stochastic matrices associated with the (recurrent) classes. If $r > 1$ no communication is possible between these classes. In fact we have r distinct Markov chains that may be studied separately.

A Markov chain without transient states will be called *recurrent*. A recurrent Markov chain whose states form only one class has been already called *irreducible* (see 2.4.2). An irreducible Markov chain will be called *cyclic* or *regular* according as its states are periodic or aperiodic. Irreducible Markov chains will be also called *ergodic*. The motivation of this nomenclature will appear from Chapter 4.

2.6.2. If the Markov chain also has transient states, then renumbering the states in such a manner that the transient states are placed after the recurrent ones and the latter are treated as in the preceding case, the transition matrix can be written as

$$\mathbf{P} = \begin{pmatrix} \mathbf{P}_1 & & & & \\ & \mathbf{P}_2 & & & \\ & & \cdot & \mathbf{0} & \\ & & & \cdot & \\ & \mathbf{0} & & \cdot & \\ & & & \mathbf{P}_r & \\ \mathbf{T}_1 & \mathbf{T}_2 & \cdots & \mathbf{T}_r & \mathbf{T} \end{pmatrix}.$$

where the matrices $\mathbf{T}_1, \ldots, \mathbf{T}_r, \mathbf{T}$ are associated with the transient states: the first r matrices comprise the transition probabilities from the transient states to the recurrent states and the last matrix the transition probabilities within the set of transient states. The powers of $\mathbf{P}$ are, clearly, of the same type. For instance

$$\mathbf{P}^2 = \begin{pmatrix} \mathbf{P}_1^2 & & & & \\ & \mathbf{P}_2^2 & & & \mathbf{0} \\ & & \cdot & & \\ & & & \cdot & \\ & \mathbf{0} & & \cdot & \\ & & & \mathbf{P}_r^2 & \\ \mathbf{T}_1\mathbf{P}_1 + \mathbf{T}\mathbf{T}_1 & \mathbf{T}_2\mathbf{P}_2 + \mathbf{T}\mathbf{T}_2 & \cdots & \mathbf{T}_r\mathbf{P}_r + \mathbf{T}\mathbf{T}_r & \mathbf{T}^2 \end{pmatrix}.$$

Hence for $i, j \in T$ the entries $p(m, i, j)$ of the matrix $\mathbf{P}^m$ coincide with the corresponding ones of the matrix $\mathbf{T}^m$. This is an immediate consequence of the fact that a transient state cannot be reached from a recurrent state. Notice that by Corollary 2 to Theorem 2.7 we have $\lim_{n\to\infty} \mathbf{T}^n = \mathbf{0}$.

It was proved (see 2.5.3) that if a Markov chain starts in a transient state then, with probability 1, it stays in the set T of transient states for just a finite number of steps after which it enters a recurrent class where it remains forever. On account of this property a Markov chain with transient states will be called *absorbing**). An absorbing Markov chain with only one aperiodic recurrent class will be called *indecomposable*. The motivation for this nomenclature will appear from the corollary to Theorem 5.7.

EXERCISES

2.1 Prove Kingman's inequality

$$p(m+n, i, i) \leqslant 1 + p(m, i, i)\, p(n, i, i) - \max(p(m, i, i), p(n, i, i)).$$

Deduce Davidson's inequality

$$1 - p(m+n, i, i) \geqslant p(m, i, i)(1 - p(n, i, i)).$$

2.2 Discuss similarly to the case of tennis (see Example 2) the case of volley-ball game $(r = 15)$.

2.3 Construct an absorbing Markov chain model for a chess match between two players $\mathcal{A}$ and $\mathcal{B}$ that ends when one player first wins r games. It is assumed that the probabilities are $p, d, q, p + d + q = 1$, and that any game results in a win for $\mathcal{A}$, a draw, and a win for $\mathcal{B}$, respectively.

2.4 Prove that $\tau = \tau_i - 1$ is not a Markov time by showing that Theorem 2.1 does not hold with τ so defined. Provide also a direct proof. (Hint: $X(\tau + 1) = X(\tau_i) = i$, hence $\{X(\tau + 1) = j\}$ is the impossible event if $j \neq i$.)

2.5 Prove that for a two state Markov chain the $f(n, i, j)$, $n \geqslant 1$, $i, j = 1, 2$, are given by

$$f(n, 1, 1) = \begin{cases} p(1, 1), \text{ if } n = 1, \\ p(1, 2)\, [p(2, 2)]^{n-2}\, p(2, 1), \text{ if } n \geqslant 2, \end{cases}$$

$$f(n, 1, 2) = [p(1, 1)]^{n-1} p(1, 2), \quad n \geqslant 1,$$

$$f(n, 2, 1) = [p(2, 2)]^{n-1} p(2, 1), \quad n \geqslant 1,$$

$$f(n, 2, 2) = \begin{cases} p(2, 2), \text{ if } n = 1, \\ p(2, 1)\, [p(1, 1)]^{n-2}\, p(1, 2), \text{ if } n \geqslant 2. \end{cases}$$

---

*) We draw to the reader's attention that it is usual to call absorbing a Markov chain all of whose recurrent states are absorbing. The nomenclature we propose is justified by the fact that absorption into the set of recurrent states is certain regardless of whether or not all these states are absorbing.

2.6 Define for $i \neq j$

$$g(1, i, j) = p(i, j),$$
$$g(n, i, j) = \mathsf{P}_i(X(m) \neq i,\ 1 \leqslant m \leqslant n-1,\ X(n) = j),\ n \geqslant 2,$$
$$g(i, j) = \sum_{n \geqslant 1} g(n, i, j).$$

Prove the *last exit formula*

$$p(n, i, j) = \sum_{m=0}^{n-1} p(m, i, i)\, g(n-m, i, j),\ i \neq j,\ n \geqslant 1.$$

Show that $i \to j$ if and only if $g(i, j) > 0$. Prove that $g(m, i, j) f(n, j, i) \leqslant f(m+n, i, i)$ and deduce that if $j \to i$ then $g(i, j) < \infty$.

2.7 Consider the generating functions

$$F_{ij}(z) = \sum_{n \geqslant 1} f(n, i, j)\, z^n,$$
$$G_{ij}(z) = \sum_{n \geqslant 1} g(n, i, j)\, z^n,$$
$$P_{ij}(z) = \sum_{n \geqslant 0} p(n, i, j)\, z^n,$$

where $|z| < 1$. Prove that

$$P_{ii}(z) = \frac{1}{1 - F_{ii}(z)},$$
$$P_{ij}(z) = F_{ij}(z)\, P_{jj}(z),$$
$$P_{ij}(z) = P_{ii}(z)\, G_{ij}(z),\ i \neq j.$$

(Hint: use the first entrance and the last exit formulas.)

2.8 Prove that if j is a transient state then $p(n, i, j)$ converges geometrically fast to 0 whatever the state i, that is there exist $a > 0$ and $0 < b < 1$ such that $p(n, i, j) \leqslant ab^n$. (Hint: by Corollary 2 to Theorem 2.7 there exist a natural number n_0 and a positive number $\eta < 1$ such that $\sum_{j \in T} p(n, i, j) < \eta$ for all $n \geqslant n_0$ and $i \in S$. Next, on account of Theorem 2.9

$$\sum_{j \in T} p((s+1)\, n_0, i, j) = \sum_{j, k \in T} p(sn_0, i, j)\, p(n_0, j, k) \leqslant \eta \sum_{j \in T} p(sn_0, i, j),$$

whence $\sum_{j \in T} p(sn_0, i, j) \leqslant \eta^s$. Finally, if $n = sn_0 + r$, $0 \leqslant r < n_0$, then

$$p(n, i, j) \leqslant \sum_{k \in T} p(sn_0, i, k)\, p(r, k, j) \leqslant \eta^s \leqslant \eta^{n/n_0 - 1} = ab^n,$$

where $a = \eta^{-1}$, $b = \eta^{1/n_0}$.)

2.9 *Regenerative phenomena.* Let $(\Omega, \mathcal{K}, \mathsf{P})$ be a probability space and $\mathcal{E} = (E_n)_{n\geqslant 0}$ a sequence of random events on it. $\mathcal{E}$ is said to be a *regenerative phenomenon**) (see [200 and 209]) if and only if $E_0 = \Omega$ and

$$\mathsf{P}\left(\bigcap_{r=1}^{s+1} E_{n_r}\right) = \mathsf{P}(E_{n_1}) \prod_{r=1}^{s} \mathsf{P}(E_{n_{r+1}-n_r})$$

whatever the natural number s and the increasing sequence $n_1 < n_2 < \ldots$ $\ldots < n_{s+1}$ of natural numbers.

Show that if $(X(n))_{n\geqslant 0}$ is a homogeneous finite Markov chain then $\mathcal{E} = (\{X(n) = i\})_{n\geqslant 0}$ is a regenerative phenomenon on $(\Omega, \mathcal{K}, \mathsf{P}_i)$.

2.10 *Continuation.* Put $p_n = \mathsf{P}(E_n)$, $n \geqslant 0$, $f_1 = \mathsf{P}(E_{n+1} \mid E_n)$, $f_r = \mathsf{P}(E_{n+r}\, E^c_{n+r-1} \ldots E^c_{n+1} \mid E_n)$, $r \geqslant 2$. Show that the f_m, $m \geqslant 1$, indeed do not depend on n. Prove that $p_0 = 1$, $0 \leqslant p_n \leqslant 1$, $n \geqslant 0$, $f_m \geqslant 0$, $m \geqslant 1$, $\sum_{m\geqslant 1} f_m \leqslant 1$, and

$$p_n = \sum_{m=1}^{n} f_m p_{n-m}, \quad n \geqslant 0.$$

Problem. Find the conditions the f_m should satisfy in order to represent $\mathcal{E}$ as the successive entrances of a homogeneous finite Markov chain into a state i such that $f(m, i, i) = f_m$, $m \geqslant 1$, $p(n, i, i) = p_n$, $n \geqslant 0$.

---

*) A *recurrent event* in Feller's terminology.

CHAPTER 3

Absorbing Markov Chains

The aim of this chapter is to study various aspects concerning the motion of an absorbing Markov chain until absorption (i.e., until reaching a recurrent state). We are interested in random variables such as the number of occurrences of a given transient state before absorption, the number of steps until absorption, the state in which absorption takes place. Clearly, when solving a problem concerning the motion of an absorbing Markov chain within the set of its transient states there is no loss of generality if all its recurrent states are assumed to be absorbing. Consequently, in such problems, the transition matrix of the Markov chain will be assumed to have the canonical form

$$\mathbf{P} = \begin{pmatrix} \mathbf{I} & \mathbf{0} \\ \mathbf{R} & \mathbf{T} \end{pmatrix}.$$

3.1 THE FUNDAMENTAL MATRIX

3.1.1. Let us denote by $\nu(j)$ the number of occurrences of a transient state j. Clearly, $\nu(j)$ is a random variable that by the corollary to Theorem 2.5 is finite with probability 1 (i.e., $\mathsf{P}(\nu(j) < \infty) = 1$). We can represent $\nu(j)$ as the sum of an infinite series by considering the random variables $u_m(j)$, $m \geqslant 0$, defined as

$$u_m(j) = \begin{cases} 1, & \text{if } X(m) = j, \\ 0, & \text{if } X(m) \neq j. \end{cases}$$

Clearly,

$$\nu(j) = \sum_{m \geqslant 0} u_m(j).$$

Notice that*) $\mathsf{E}_i(u_m(j)) = 1 \cdot \mathsf{P}_i(X(m) = j) + 0 \cdot \mathsf{P}_i(X(m) \neq j) = p(m, i, j)$. Therefore for all $i, j \in T$ we have

$$\mathsf{E}_i(\nu(j)) = \mathsf{E}_i\left(\sum_{m \geqslant 0} u_m(j)\right) = \sum_{m \geqslant 0} \mathsf{E}_i(u_m(j)) = \sum_{m \geqslant 0} p(m, i, j),$$

*) E_i denotes the expectation operator under the probability P_i (see 2.1.2).

which is finite by Corollary 2 to Theorem 2.7. (Here, though not difficult to prove it, we take for granted the second equality sign above.) In matrix notation

$$(\mathsf{E}_i(\nu(j)))_{i,j\in T} = \mathbf{I} + \mathbf{T} + \mathbf{T}^2 + \ldots = (\mathbf{I} - \mathbf{T})^{-1}$$

(see Theorem 1.7). The matrix $(\mathbf{I} - \mathbf{T})^{-1} = \mathbf{N}$ is called the *fundamental matrix* of the absorbing Markov chain we considered. In what follows we shall see this matrix is, indeed, of a paramount importance.

3.1.2. For Example 1 from 2.2 we have

$$\mathbf{T} = \begin{array}{c} \\ 1 \\ 2 \\ \cdot \\ \cdot \\ \cdot \\ l-2 \\ l-1 \end{array} \begin{array}{c} \begin{array}{cccccc} 1 & 2 & 3 & \cdots & l-2 & l-1 \end{array} \\ \begin{pmatrix} 0 & p & 0 & \cdots & 0 & 0 \\ q & 0 & p & \cdots & 0 & 0 \\ \cdot & \cdot & \cdot & \cdots & \cdot & \cdot \\ \cdot & \cdot & \cdot & \cdots & \cdot & \cdot \\ \cdot & \cdot & \cdot & \cdots & \cdot & \cdot \\ 0 & 0 & 0 & \cdots & 0 & p \\ 0 & 0 & 0 & \cdots & q & 0 \end{pmatrix} \end{array}$$

and it is easily verified by direct computation that the entries $n(i, j)$ of the fundamental matrix $\mathbf{N} = (\mathbf{I} - \mathbf{T})^{-1}$ are given by

$$n(i, j) = \frac{1}{(2p-1)(a^l - 1)} \cdot \begin{cases} (a^j - 1)(a^{l-i} - 1), & \text{if } j \leqslant i, \\ (a^i - 1)(a^{l-i} - a^{j-i}), & \text{if } j > i, \end{cases}$$

for $p \neq 1/2$, where $a = p/q$, and

$$n(i, j) = \frac{2}{l} \cdot \begin{cases} j(l-i), & \text{if } j \leqslant i, \\ i(l-j), & \text{if } j > i, \end{cases}$$

for $p = 1/2$. In particular, for $l = 4$ we have

$$\mathbf{N} = \frac{1}{p^2 + q^2} \begin{pmatrix} p + q^2 & p & p^2 \\ q & 1 & p \\ q^2 & q & q + p^2 \end{pmatrix}.$$

Let us also consider a variant of the classical gambler's ruin problem, namely the case of an elastic barrier at l with $q_l = q$, $r_l = p$ (0 remaining an absorbing barrier). In this case

$$\mathbf{T} = \begin{array}{c} \\ 1 \\ 2 \\ \cdot \\ \cdot \\ \cdot \\ l-1 \\ l \end{array} \begin{array}{c} \begin{array}{cccccc} 1 & 2 & 3 & \cdots & l-1 & l \end{array} \\ \begin{pmatrix} 0 & p & 0 & \cdots & 0 & 0 \\ q & 0 & p & \cdots & 0 & 0 \\ \cdot & \cdot & \cdot & \cdots & \cdot & \cdot \\ \cdot & \cdot & \cdot & \cdots & \cdot & \cdot \\ \cdot & \cdot & \cdot & \cdots & \cdot & \cdot \\ 0 & 0 & 0 & \cdots & 0 & p \\ 0 & 0 & 0 & \cdots & q & p \end{pmatrix} \end{array}$$

and the entries of the fundamental matrix are given by

$$n(i,j) = \frac{1}{2p-1} \cdot \begin{cases} a^j - 1, & \text{if } j \leqslant i, \\ a^j - a^{j-i}, & \text{if } j > i, \end{cases}$$

for $p \neq 1/2$ and

$$n(i,j) = \begin{cases} 2j, & \text{if } j \leqslant i, \\ 2i, & \text{if } j > i, \end{cases}$$

for $p = 1/2$, $1 \leqslant i, j \leqslant l$ *).

3.1.3. For Example 5 from 2.2 we have

$$\mathbf{T} = \begin{array}{c} \\ 0 \\ 1 \\ 2 \\ \cdot \\ \cdot \\ d-2 \\ d-1 \end{array} \begin{array}{c} \begin{array}{cccccc} 0 & 1 & 2 & \cdots & d-2 & d-1 \end{array} \\ \left(\begin{array}{cccccc} r_1 & p_1 & 0 & \cdots & 0 & 0 \\ 0 & r_2 & p_2 & \cdots & 0 & 0 \\ 0 & 0 & r_3 & \cdots & 0 & 0 \\ \cdot & \cdot & \cdot & \cdots & \cdot & \cdot \\ \cdot & \cdot & \cdot & \cdots & \cdot & \cdot \\ 0 & 0 & 0 & \cdots & r_{d-1} & p_{d-1} \\ 0 & 0 & 0 & \cdots & 0 & r_d \end{array} \right) \end{array}$$

and it is easily found that

$$n(i,i) = (1 - r_{i+1})^{-1}, \quad 0 \leqslant i \leqslant d-1,$$
$$n(i,j) = p_{i+1} \ldots p_j (1 - r_{i+1})^{-1} \ldots (1 - r_{j+1})^{-1}, \quad i < j,$$
$$n(i,j) = 0, \quad i > j.$$

3.2 APPLICATIONS OF THE FUNDAMENTAL MATRIX

3.2.1. To begin with, let us notice that the matrix $\mathbf{F} = (f(i,j))_{i,j \in T}$ can be expressed in terms of the fundamental matrix $\mathbf{N}$. By Doeblin's formula (2.4) we have

$$\mathbf{F} = (\mathbf{N} - \mathbf{I})\mathbf{N}_{dg}^{-1}.$$

Next, the distribution of $\nu(j)$ is given by

$$(\mathsf{P}_i(\nu(j) - \delta(i,j) = r))_{i,j \in T}$$
$$= \begin{cases} \mathbf{E} - \mathbf{F}, & \text{if } r = 0, \\ \mathbf{F}\mathbf{F}_{dg}^{r-1}(\mathbf{I} - \mathbf{F}_{dg}) = (\mathbf{N} - \mathbf{I})\,(\mathbf{N}_{dg}^{-1})^2\,(\mathbf{I} - \mathbf{N}_{dg}^{-1})^{r-1} & \text{if } r > 0. \end{cases}$$

Indeed, by Theorem 2.5, given the Markov chain starts in state i, the probability of exactly $r \geqslant 1$ occurrences of state j is

$$f(i,j)\,[f(j,j)]^{r-1} - f(i,j)\,[f(j,j)]^r = f(i,j)\,[f(j,j)]^{r-1}\,(1 - f(j,j)).$$

*) The reader should note that these values can be obtained from the entries of the fundamental matrix corresponding to the classical gambler's ruin problem by letting $l \to \infty$.

Obviously, the probability of no occurrence of state j is $1 - f(i, j)$.

Now we can obtain the generating function g_{ij} of the random variable $\nu(j) - \delta(i, j)$ under the probability P_i. By the very definition of the generating function (see 1.4.6)

$$\begin{aligned} g_{ij}(z) &= \mathsf{E}_i(z^{(\nu(j)-\delta(i,j))}) \\ &= 1 - f(i, j) + \sum_{r>0} f(i, j)\,[f(j, j)]^{r-1}(1 - f(j, j))\, z^r \\ &= 1 - f(i, j) + f(i, j)\,(1 - f(j, j)) \frac{z}{1 - f(j, j)\, z} \cdot \end{aligned}$$

(since $f(j, j) < 1$). In matrix notation

$$\begin{aligned} (g_{ij}(z))_{i, j\in T} &= \mathbf{E} - \mathbf{F} + z\,\mathbf{F}(\mathbf{I} - z\mathbf{F}_{\mathrm{dg}})^{-1}\,(\mathbf{I} - \mathbf{F}_{\mathrm{dg}}) \\ &= \mathbf{E} - \mathbf{F} + (\mathbf{N} - \mathbf{I})\,(\mathbf{N}_{\mathrm{dg}}^{-1})^2\,(z^{-1}\mathbf{I} - (\mathbf{I} - \mathbf{N}_{\mathrm{dg}}^{-1}))^{-1}. \end{aligned}$$

3.2.2. More generally, we can obtain the joint distribution of the random variables $\nu(j)$, $j \in T$, i.e.,

$$\mathsf{P}_i(\nu(j) = r_j, j \in T)$$

for all values $r_j \geqslant 0$, $j \in T$. To this end it is sufficient to remark that denoting by $\mathbf{D}(z)$ the diagonal matrix $(\delta(i, j)\; z_j)_{i,j\in T}, |z_j| \leqslant 1, j \in T$, then $\mathsf{P}_i(\nu(j) = r_j,\ j \in T)$ is nothing but the coefficient of $\prod_{j\in T} z_j^{r_j}$ in the product

$$(\delta(i, j)\;\; z_j)'_{j\in T}(\mathbf{TD}(z))^{\sum_{j\in T} r_j - 1}\;(\mathbf{I} - \mathbf{T})\mathbf{e}. \tag{3.1}$$

Indeed, $\mathsf{P}_i(\nu(j) = r_j,\ j \in T)$ is either the sum of the products of the form $p(i, i_1)\; p(i_1, i_2) \dots p(i_{n-1}, i_n)$ with $i_1, \dots, i_{n-1} \in T$, $i_n \in S - T$, where j occurs exactly r_j times among $i, i_1, \dots, i_{n-1}$ $\left(\text{if } n = \sum_{j\in T} r_j > 1\right)$, or the sum $\sum_{k\in S-T} p(i, k) \left(\text{if} \sum_{j\in T} r_j = 1\right)$. The sum of the products of the form $z_i p(i, i_1)\; z_{i_1} p(i_1, i_2) \dots z_{i_{n-1}} p(i_{n-1}, i_n)$ for any fixed $n > 1$ or the sum $\sum_{k\in S-T} z_i p(i, k)$ coincide with (3.1) according as $\sum_{j\in T} r_j = n$ or $\sum_{j\in T} r_j = 1$.

It follows from these considerations that

$$\begin{aligned} \left(\sum_{\substack{r_j\geqslant 0\\ j\in T}} \mathsf{P}_i(\nu(j) = r_j, j \in T) \prod_{j\in T} z_j^{r_j}\right)_{i\in T} &= \mathbf{D}(z) \sum_{n\geqslant 1} (\mathbf{TD}(z))^{n-1}\,(\mathbf{I} - \mathbf{T})\,\mathbf{e} \\ = \mathbf{D}(z)\,(\mathbf{I} - \mathbf{TD}(z))^{-1}\,(\mathbf{I} - \mathbf{T})\,\mathbf{e} &= (\mathbf{I} - \mathbf{D}(z)\mathbf{T})^{-1}\mathbf{D}(z)\,(\mathbf{I} - \mathbf{T})\,\mathbf{e}. \end{aligned} \tag{3.2}$$

The same trick leads to a yet more general result. Let us denote by $\nu(i, j)$ the number of values $m \geqslant 0$ for which $X(m) = i$ and $X(m + 1) = j$, $i \in T$, $j \in S$. Clearly, $\nu(i) = \sum_{j\in T} \nu(i, j)$, $i \in T$, and $\nu(k) =$

$= \sum_{i \in T} \nu(i, k) + \delta(k, X(0)),\ k \in T$. Putting $\mathbf{T}(z) = (p(i, k)\ z_{ik})_{i,\,k \in T}$, $\mathbf{R}(z) = (p(i, j)\ z_{ij})_{i \in T, j \in S-T}$, $|z_{ij}| \leqslant 1$, $i \in T$, $j \in S$, we gte

$$\left(\sum_{\substack{r_{ij} \geqslant 0 \\ i \in T, j \in S}} \mathsf{P}_k(\nu((i, j) = r_{ij},\ i \in T, j \in S) \prod_{i \in T, j \in S} z_{ij}^{r_{ij}} \right)_{k \in T}$$

$$= (\mathbf{I} - \mathbf{T}(z))^{-1}\, \mathbf{R}(z)\ \mathbf{e}. \tag{3.3}$$

It is easily verified that for $z_{ij} = z_i$, $i \in T$, $j \in S$, equation (3.3) reduces to equation (3.2).

3.2.3. The theorem below is a generalization of the result concerning the existence of the mean values $\mathsf{E}_i(\nu(j))$, $i, j \in T$ (see 3.1).

Theorem 3.1. *The moments of any order* $\mathsf{E}_i(\nu^s(j))$, $s = 1, 2, \ldots$ *of the random variable* $\nu(j)$ *are finite. The matrices* $\mathbf{N}_s = (\mathsf{E}_i(\nu^s(j)))_{i,j \in T}$ *are connected by the recursion* $\mathbf{N}_s = \mathbf{N}\left(\sum_{r=1}^{s-1} \binom{s}{r} (\mathbf{T}\mathbf{N}_{s-r})_{\mathrm{dg}} + \mathbf{I}\right)$, $s > 1$, *with* $\mathbf{N}_1 = \mathbf{N}$. *In particular*

$$\mathbf{N}_2 = \mathbf{N}(2(\mathbf{T}\mathbf{N})_{\mathrm{dg}} + \mathbf{I}) = \mathbf{N}(2\mathbf{N}_{\mathrm{dg}} - \mathbf{I}),$$

$$\mathbf{N}_3 = \mathbf{N}(3(\mathbf{T}\mathbf{N}_2)_{\mathrm{dg}} + 3(\mathbf{T}\mathbf{N})_{\mathrm{dg}} + \mathbf{I}) = \mathbf{N}(6\mathbf{N}_{\mathrm{dg}}^2 - 6\mathbf{N}_{\mathrm{dg}} + \mathbf{I}),$$

$$\mathbf{N}_4 = \mathbf{N}(4(\mathbf{T}\mathbf{N}_3)_{\mathrm{dg}} + 6(\mathbf{T}\mathbf{N}_2)_{\mathrm{dg}} + 4(\mathbf{T}\mathbf{N})_{\mathrm{dg}} + \mathbf{I})$$
$$= \mathbf{N}(24\mathbf{N}_{\mathrm{dg}}^3 - 36\mathbf{N}_{\mathrm{dg}}^2 + 14\mathbf{N}_{\mathrm{dg}} - \mathbf{I}).$$

Proof. The existence of the moments $\mathsf{E}_i(\nu^s(j))$ is ensured by the fact that the characteristic functions $\mathsf{E}_i(e^{it\nu(j)}) = g_{ij}(e^{it})\, e^{it\delta(i,j)}$ are infinitely differentiable (see 3.2.1 and 1.4.6). To derive the recursion connecting the matrices $\mathbf{N}_s$ we shall write

$$\mathsf{E}_i(\nu^s(j)) = \sum_{n \geqslant 0} n^s \mathsf{P}_i(\nu(j) = n)$$

$$= \delta(i, j)\, \mathsf{P}_i(\nu(j) = \delta(i, j)) + \sum_{n \geqslant \delta(i,j)+1} n^s \mathsf{P}_i(\nu(j) = n)$$

$$= \delta(i, j)\, \mathsf{P}_i(X(1) \notin T) + \sum_{n \geqslant \delta(i,j)+1} \sum_{k \in T} n^s \mathsf{P}_i(\nu(j) = n \mid X(1) = k)\, \mathsf{P}_i(X(1) = k)$$

$$= \delta(i,j) \sum_{l \notin T} p(i, l) + \sum_{k \in T} p(i, k) \sum_{n \geqslant 0} (n + \delta(i, j))^s \mathsf{P}_k(\nu(j) = n)$$

[since $\mathsf{P}_i(\nu(j) = n \mid X(1) = k) = \mathsf{P}_k(\nu(j) = n - \delta(i, j))$]

$$= \delta(i, j) \sum_{l \notin T} p(i, l) + \sum_{k \in T} p(i, k)\, \mathsf{E}_k((\nu(j) + \delta(i, j))^s)$$

$$= \delta(i, j) \sum_{l \notin T} p(i, l) + \sum_{k \in T} p(i, k) \sum_{r=0}^{s} \binom{s}{r} \mathsf{E}_k(\nu^{s-r}(j)\ \delta^r(i, j))$$

$$= \delta(i, j) + \sum_{k \in T} p(i, k)\, \mathsf{E}_k(\nu^s(j)) + \sum_{k \in T} p(i, k) \sum_{r=1}^{s-1} \binom{s}{r} \mathsf{E}_k(\nu^{s-r}(j))\ \delta(i, j).$$

In matrix notation

$$\mathbf{N}_s = \mathbf{T}\mathbf{N}_s + \sum_{r=1}^{s-1} \binom{s}{r} (\mathbf{T}\mathbf{N}_{s-r})_{\mathrm{dg}} + \mathbf{I},$$

$$(\mathbf{I} - \mathbf{T})\,\mathbf{N}_s = \sum_{r=1}^{s-1} \binom{s}{r} (\mathbf{T}\mathbf{N}_{s-r})_{\mathrm{dg}} + \mathbf{I},$$

whence our recursion follows at once. The special cases $s = 2, 3, 4$ follow without any difficulty. [It should be noted that $\mathbf{T}\mathbf{N} = \mathbf{T}(\mathbf{I} + \mathbf{T} + \mathbf{T}^2 + \ldots) = \mathbf{N} - \mathbf{I}$, whence $(\mathbf{T}\mathbf{N})_{\mathrm{dg}} = \mathbf{N}_{\mathrm{dg}} - \mathbf{I}$.]

3.2.4. Let ν denote the number of steps (including the starting position) in which the Markov chain is in a transient state. Clearly, $\nu = \sum_{j \in T} \nu(j)$. The random variable ν is called the *time to absorption*. The distribution of ν can be easily found as the random events $\{\nu > n\}$ and $\{X(n) \in T\}$ are equivalent. Indeed, absorption takes place at a time posterior to n if and only if at time n the Markov chain is still in the set T. On account of the fact that

$$\{\nu > n - 1\} = \{\nu > n\} \cup \{\nu = n\}, \quad n \geqslant 1,$$

we have

$$\mathsf{P}_i(\nu = n) = \mathsf{P}_i(\nu > n - 1) - \mathsf{P}_i(\nu > n)$$

$$= \mathsf{P}_i(X(n-1) \in T) - \mathsf{P}_i(X(n) \in T) = \sum_{j \in T} [p(n-1, i, j) - p(n, i, j)], \tag{3.4}$$

whence

$$(\mathsf{P}_i(\nu = n))_{i \in T} = (\mathbf{T}^{n-1} - \mathbf{T}^n)\,\mathbf{e} = \mathbf{T}^{n-1}(\mathbf{I} - \mathbf{T})\,\mathbf{e}.$$

Obviously, $\mathsf{P}_i(\nu = 0) = 0$, $i \in T$.

Equation (3.4) allows us to compute the mean vector $\mathbf{m} = (\mathsf{E}_i(\nu))_{i \in T}$. We have

$$\mathbf{m} = \left(\sum_{n \geqslant 0} n \mathsf{P}_i(\nu = n)\right)_{i \in T} = \sum_{n \geqslant 1} n(\mathbf{T}^{n-1} - \mathbf{T}^n)\,\mathbf{e} = \mathbf{N}\mathbf{e}.$$

The problem of determining the moments of ν is solved by

Theorem 3.2. *The moments of any order of the random variable ν are finite. The vectors $\mathbf{m}_s = (\mathsf{E}_i(\nu^s))_{i \in T}$ are connected by the recursion* $\mathbf{m}_s = \mathbf{N}\left(\mathbf{T}\sum_{r=1}^{s-1} \binom{s}{r} \mathbf{m}_{s-r} + \mathbf{e}\right)$, $s > 1$, *with* $\mathbf{m}_1 = \mathbf{m}$. *In particular*

$$\mathbf{m}_1 = \mathbf{N}\mathbf{e}, \quad \mathbf{m}_2 = (2\mathbf{N} - \mathbf{I})\,\mathbf{N}\mathbf{e}, \quad \mathbf{m}_3 = (6\mathbf{N}^2 - 6\mathbf{N} + \mathbf{I})\,\mathbf{N}\mathbf{e},$$

$$\mathbf{m}_4 = (24\mathbf{N}^3 - 36\mathbf{N}^2 + 14\mathbf{N} - \mathbf{I})\,\mathbf{N}\mathbf{e}.$$

Proof. The finiteness of the moments of any order of ν follows from Theorem 3.1 on account of the equation $\nu = \sum_{j \in T} \nu(j)$. A reasoning similar to that in the proof of Theorem 3.1 leads us to the equation

$$\mathsf{E}_i(\nu^s) = \sum_{l \notin T} p(i, l) + \sum_{k \in T} p(i, k)\, \mathsf{E}_k((\nu + 1)^s)$$

$$= \sum_{k \in T} p(i, k) \left(\mathsf{E}_k(\nu^s) + \sum_{r=1}^{s-1} \binom{s}{r} \mathsf{E}_k(\nu^{s-r}) \right) + 1.$$

In matrix notation

$$\mathbf{m}_s = \mathbf{T} \left(\mathbf{m}_s + \sum_{r=1}^{s-1} \binom{s}{r} \mathbf{m}_{s-r} \right) + \mathbf{e},$$

$$(\mathbf{I} - \mathbf{T})\, \mathbf{m}_s = \sum_{s=1}^{r-1} \binom{s}{r} \mathbf{m}_{s-r} + \mathbf{e},$$

whence our recursion follows at once. The vector $\mathbf{m}_1 = \mathbf{Ne}$ being known, the vectors $\mathbf{m}_2$, $\mathbf{m}_3$ and $\mathbf{m}_4$ can be derived without any difficulty.

3.2.5. Let ν' denote the number of distinct transient states that occur until absorption. Clearly, ν' is a random variable whose values are the natural numbers from 1 up to the order of the matrix $\mathbf{T}$ (the number of transient states).

The mean value of ν' can be obtained as follows. Let us define the random variables

$$u(j) = \begin{cases} 1, & \text{if } \tau_j < \infty, \\ 0, & \text{if } \tau_j = \infty, \end{cases} \quad j \in T,$$

where $\tau_j \geqslant 1$ is the moment of first passage to state j.

Whatever $i \in T$ we have

$$\nu' = \delta(X(0), i) + \sum_{j \neq i} u(j), \; i \in T,$$

and since

$$\mathsf{E}_i(u(j)) = \mathsf{P}_i(\tau_j < \infty) = f(i, j),$$

we get

$$\mathsf{E}_i(\nu') = 1 + \sum_{j \neq i} f(i, j).$$

Hence, in matrix notation,

$$(\mathsf{E}_i(\nu'))_{i \in T} = (\mathbf{F} + \mathbf{I} - \mathbf{F}_{\mathrm{dg}})\, \mathbf{e} = \mathbf{N}\mathbf{N}_{\mathrm{dg}}^{-1}\, \mathbf{e}.$$

3.2.6. Now we shall be concerned with the problem of computing the absorption probabilities.

Theorem 3.3. *Let $a(i, k)$ denote the probability that the Markov chain starting in transient state i ends up in absorbing state k. Then*

$$\mathbf{A} = (a(i, k))_{i \in T, k \in S-T} = \mathbf{NR}.$$

Proof. Notice that $a(i, k)$ is nothing but $f(i, k)$ (see 2.5.1) since transitions between absorbing states are impossible. As

$$f(i, k) = \mathsf{P}_i\Big(\bigcup_{n \geq 1} \{X(n) = k\}\Big)$$

and $\{X(n) = k\} \subset \{X(n+1) = k\}$, state k being absorbing, we have [see (1.4′)]

$$a(i, k) = \lim_{n \to \infty} \mathsf{P}_i(X(n) = k) = \lim_{n \to \infty} p(n, i, k). \tag{3.5}$$

Using the Chapman-Kolmogorov equation

$$p(n + 1, i, k) = \sum_{j \in S} p(i, j)\, p(n, j, k)$$

and taking into account the fact that only the probabilities $p(n, j, k)$, $j \in T$, can be different from 0 and that $p(n, k, k) = 1$, we get

$$p(n + 1, i, k) = p(i, k) + \sum_{j \in T} p(i, j)\, p(n, j, k).$$

On account of (3.5) letting $n \to \infty$ yields

$$a(i, k) = p(i, k) + \sum_{j \in S} p(i, j)\, a(j, k),$$

or, in matrix notation,

$$\mathbf{A} = \mathbf{R} + \mathbf{TA}, \quad (\mathbf{I} - \mathbf{T})\mathbf{A} = \mathbf{R},$$

whence the stated result.

Remark. On account of the fact that absorption represents the sure event (see 2.3.5) we should have $\sum_{j \in S-T} a(i, j) = 1$, $i \in T$, or, in matrix notation, $\mathbf{NRe} = \mathbf{e}$. The proof of this equation is left to the reader (see Exercise 3.4).

Corollary. *Consider the matrix*

$$\mathbf{A}^* = \begin{pmatrix} \mathbf{I} \\ \mathbf{A} \end{pmatrix}.$$

We have $\mathbf{PA}^* = \mathbf{A}^*$. *In other words, the columns of the matrix* $\mathbf{A}^*$ *are right eigenvectors of the transition matrix* $\mathbf{P}$ *corresponding to the eigenvalue* $\lambda = 1$.

Proof. We have

$$\mathbf{PA}^* = \begin{pmatrix} \mathbf{I} & \mathbf{0} \\ \mathbf{R} & \mathbf{T} \end{pmatrix} \begin{pmatrix} \mathbf{I} \\ \mathbf{A} \end{pmatrix} = \begin{pmatrix} \mathbf{I} \\ \mathbf{R} + \mathbf{TA} \end{pmatrix}.$$

But $\mathbf{R} + \mathbf{TA} = \mathbf{R} + \mathbf{TNR} = (\mathbf{I} + \mathbf{TN})\,\mathbf{R} = (\mathbf{I} + \mathbf{T} + \mathbf{T}^2 + \dots)\,\mathbf{R} = \mathbf{NR} = \mathbf{A}$, that is $\mathbf{PA}^* = \mathbf{A}^*$.

By the above corollary the geometric multiplicity of the eigenvalue $\lambda = 1$ equals its algebraic multiplicity (which is equal to the number r of the absorbing states). It is easily seen that the left eigenvectors corresponding to this eigenvalue are the unit vectors $(\delta(i, j))_{j \in S}$, $1 \leqslant i \leqslant r$. It follows that if the matrix $\mathbf{T}$ is diagonalizable then the matrix $\mathbf{P}$ itself is diagonalizable and then the spectral representation (1.12) shows that the convergence rate of $p(n, i, k)$, $i \in T$, $k \in S - T$, to $a(i, k)$ is of the order of a^n, where a is the maximum of the absolute values of the eigenvalues of $\mathbf{T}$.

Theorem 3.3 also allows us to solve the problem of computing the absorption probabilities in the case of a general absorbing chain for which the recurrent classes do not reduce to absorbing states. If $\mathcal{C}$ is a recurrent class then, clearly, the probability $a(i, \mathcal{C})$ that the Markov chain starting in transient state $i \in T$ reaches $\mathcal{C}$, equals the sum of absorbing probabilities $a(i, k)$ over all states $k \in \mathcal{C}$, that is

$$a(i, \mathcal{C}) = \sum_{k \in \mathcal{C}} a(i, k),$$

whence

$$(a(i, \mathcal{C}))_{i \in T} = \mathbf{NR}(\delta(k, \mathcal{C}))_{k \in S-T},$$

where

$$\delta(k, \mathcal{C}) = \begin{cases} 1, & \text{if } k \in \mathcal{C}, \\ 0, & \text{if } k \notin \mathcal{C}. \end{cases}$$

Next, if the recurrent classes of the Markov chain are $\mathcal{C}^1, \dots, \mathcal{C}^r$, then the vectors

$$\mathbf{v}_l = (a(j, \mathcal{C}^l))_{j \in S}, \quad 1 \leqslant l \leqslant r,$$

where $a(j, \mathcal{C}^l) = 1$ if $j \in \mathcal{C}^l$ and $a(j, \mathcal{C}^l) = 0$ if $j \in \mathcal{C}^{l'}$, $l \neq l'$, are a set of linearly independent right eigenvectors corresponding to the eigenvalue $\lambda = 1$. The problem of finding the left eigenvectors corresponding to this eigenvalue will be solved in 4.4.5.

3.2.7. Theorems 3.2 and 3.3 allow us to answer certain questions concerning the examples considered in 2.2.

For the gambler's ruin problem (Example 1) we have

$$\mathbf{R} = \begin{array}{c} \\ 1 \\ 2 \\ \cdot \\ \cdot \\ \cdot \\ l-2 \\ l-1 \end{array} \begin{array}{c} \begin{array}{cc} 0 & l \end{array} \\ \begin{pmatrix} q & 0 \\ 0 & 0 \\ \cdot & \cdot \\ \cdot & \cdot \\ \cdot & \cdot \\ 0 & 0 \\ 0 & p \end{pmatrix} \end{array}$$

and therefore

$$a(i, 0) = \sum_{j=1}^{l-1} n(i, j)\, p(j, 0) = \begin{cases} \dfrac{a^{l-i} - 1}{a^l - 1}, & \text{if } p \neq 1/2, \\ 1 - i/l, & \text{if } p = 1/2, \end{cases}$$

$$a(i, l) = 1 - a(i, 0), \quad 1 \leqslant i \leqslant l - 1.$$

These probabilities of absorption in 0, respectively l, are nothing but the ruin probabilities of the first, respectively of the second gambler.

The mean value $\mathsf{E}_i(\nu)$ represents the mean duration of the game. We have

$$\mathsf{E}_i(\nu) = \sum_{j=1}^{l-1} n(i, j) = \begin{cases} \dfrac{1}{2p - 1}\left(l\dfrac{a^l - a^{l-i}}{a^l - 1} - i\right), & \text{if } p \neq 1/2, \\ i(l - i), & \text{if } p = 1/2, \end{cases}$$

for $1 \leqslant i \leqslant l - 1$.

In the case of the variant of the gambler's ruin problem for which the barrier l is an elastic one (see 3.1.2) all the absorption probabilities $a(i, 0)$, $1 \leqslant i \leqslant l$, are equal to 1 since 0 is the only absorbing state and

$$\mathsf{E}_i(\nu) = \begin{cases} \dfrac{1}{2p - 1}\left(\dfrac{a^{l+1} - a^{l-i+1}}{a - 1} - i\right), & \text{if } p \neq 1/2, \\ (2l - i + 1)i, & \text{if } p = 1/2. \end{cases}$$

3.2.8. For Example 5 from 2.2 we have

$$\mathbf{R} = \begin{array}{c} \\ 0 \\ 1 \\ 2 \\ \cdot \\ \cdot \\ \cdot \\ d-2 \\ d-1 \end{array} \begin{array}{c} \begin{array}{cc} d & d+1 \end{array} \\ \begin{pmatrix} 0 & q_1 \\ 0 & q_2 \\ 0 & q_3 \\ \cdot & \cdot \\ \cdot & \cdot \\ \cdot & \cdot \\ 0 & q_{d-1} \\ p_d & q_d \end{pmatrix} \end{array}$$

and therefore

$$a(i, d) = \sum_{j=0}^{d-1} n(i, j)\, p(j, d) = p_{i+1} \dots p_d (1 - r_{i+1})^{-1} \dots (1 - r_d)^{-1},$$

$$a(i, d+1) = 1 - a(i, d), \quad 0 \leqslant i \leqslant d-1.$$

Notice that the absorption probabilities $a(0, d)$ and $a(0, d+1)$ represent the probabilities that a beginner successfully completes the entire course or leaves it, respectively.

Next,

$$\mathsf{E}_i(\nu) = \sum_{j=0}^{d-1} n(i, j) =$$

$$= \begin{cases} (1 - r_{i+1})^{-1} \left[1 + \sum_{j=i+1}^{d-1} p_{i+1} \dots p_j (1 - r_{i+2})^{-1} \dots (1 - r_{j+1})^{-1} \right], & \text{if } 0 \leqslant i < d-1, \\ (1 - r_d)^{-1}, \text{ if } i = d-1. \end{cases}$$

In particular $\mathsf{E}_0(\nu)$ represents the mean value of the duration of schooling for a beginner, ending by either successful completion or leaving of the course.

3.2.9. Consider the random event

$A = \{$absorption takes place in state $k\} = \{X(\nu) = k\}$,
where k is an absorbing state.

In some cases we may be interested in computing the conditional probabilities given A of various random events. For example we might ask ourselves which is the mean duration of schooling for a beginner conditional on the fact that he successfully completes the course.

We shall show that with respect to the conditional probability P_A the sequence $(X(n))_{n \geqslant 0}$ is still an absorbing Markov chain with state space $T \cup (k)$, the only recurrent (thus absorbing) state being k. Notice first that for the conditional probability

$$\mathsf{P}_A(X(n+1) = i_{n+1} \mid X(n) = i_n, \dots, X(0) = i_0) \tag{3.6}$$

to make sense it is necessary the sequence $i_0, \dots, i_n$ contains no recurrent state different from k, and if $i_m = k$ for some $m \leqslant n$ then $i_l = k$, $m \leqslant l \leqslant n$. This remark justifies the assertion concerning the state space. It remains to prove the Markov property. To this end fix $i_0, \dots, i_n \in T \cup (k)$ and put $B_0 = \{X(0) = i_0, \dots, X(n) = i_n\}$. It is easily seen that the conditional probability (3.6) equals

$$\frac{\mathsf{P}(A \cap \{X(n+1) = i_{n+1}\} \cap B_0)}{\mathsf{P}(A \cap B_0)}.$$

To begin with, assume that $i_0, \ldots, i_n \neq k$. Then for any random event $B_s = \{X(s) = i_s, \ldots, X(n) = i_n\}$, $0 \leqslant s \leqslant n$, we have (see 2.5.1)

$$\begin{aligned}\mathsf{P}(A \cap B_s) &= \mathsf{P}(\{X(\nu) = k\} \cap B_s)\\ &= \sum_{l \geqslant 1} \mathsf{P}(\{X(n+l) = k\} \cap \{X(u) \neq k,\ n < u < n+l\} \cap B_s)\\ &= \mathsf{P}(B_s) \sum_{l \geqslant 1} f(l, i_n, k) = \mathsf{P}(B_s)\, f(i_n, k).\end{aligned}$$

Similarly

$$\begin{aligned}&\mathsf{P}(A \cap \{X(n+1) = i_{n+1}\} \cap B_s)\\ &= \mathsf{P}(B_s)\, p(i_n, i_{n+1}) \Big(\delta(i_{n+1}, k) + \sum_{l \geqslant 1} (1 - \delta(i_{n+1}, k))\, f(l, i_{n+1}, k)\Big)\\ &= \mathsf{P}(B_s)\, p(i_n, i_{n+1})\, (\delta(i_{n+1}, k) + (1 - \delta(i_{n+1}, k))\, f(i_{n+1}, k)).\end{aligned}$$

Therefore the ratio $\mathsf{P}(A \cap \{X(n+1) = i_{n+1}\} \cap B_s)/\mathsf{P}(A \cap B_s)$ does not depend on s, $0 \leqslant s \leqslant n$, if $i_0, \ldots, i_n \neq k$. The same holds true if $i_n = k$ since then the random event $\{X(n) = i_n\}$ implies the random event A and the above ratio reduces to $\mathsf{P}(X(n+1) = i_{n+1} \mid X(n) = k)$ whatever s, $0 \leqslant s \leqslant n$. Consideration of the extreme cases $s = 0$ and $s = n$, that is the equation

$$\frac{\mathsf{P}(A \cap \{X(n+1) = i_{n+1}\} \cap B_0)}{\mathsf{P}(A \cap B_0)} = \frac{\mathsf{P}(A \cap \{X(n+1) = i_{n+1}\} \cap B_n)}{\mathsf{P}(A \cap B_n)},$$

concludes the proof of Markov property with respect to the conditional probability P_A.

The transition probabilities of the new Markov chain follow at once from the previous computations. We have for $i \neq k$

$$\begin{aligned}\mathsf{P}_A(X(n+1) = j \mid X(n) = i) &= \frac{p(i, j)\, (\delta(j, k) + (1 - \delta(j, k))\, f(j, k))}{f(i, k)}\\ &= \frac{p(i, j)\, a(j, k)}{a(i, k)}\end{aligned}$$

(see 3.2.6) with the convention $a(k, k) = 1$, and

$$\mathsf{P}_A(X(n+1) = k \mid X(n) = k) = 1.$$

Concerning the canonical form

$$\mathbf{P}_A = \begin{array}{c} \\ k \\ T \end{array}\!\!\begin{array}{c} \begin{array}{cc} k & T \end{array} \\ \begin{pmatrix} 1 & 0 \\ \mathbf{R}_A & \mathbf{T}_A \end{pmatrix} \end{array}$$

of the transition matrix, it is clear that $\mathbf{R}_A$ is the vector with entries $p(i, k)/a(i, k)$, $i \in T$, and $\mathbf{T}_A = \mathbf{D}_A^{-1}\mathbf{T}\mathbf{D}_A$, where $\mathbf{D}_A$ is a diagonal matrix with entries $a(i, k)$, $i \in T$, on the main diagonal. It follows then that $\mathbf{T}_A^n = \mathbf{D}_A^{-1}\mathbf{T}^n\mathbf{D}_A$, $n \geqslant 1$, $\mathbf{N}_A = \mathbf{I} + \mathbf{T}_A + \mathbf{T}_A^2 + \ldots = \mathbf{D}_A^{-1}\mathbf{N}\mathbf{D}_A$.

We shall apply the above results to Example 5 from 2.2 to compute the mean duration of schooling for a beginner conditional on the facts that he successfully completes the course (A) or leaves the course (A'). On account of 3.1.3 and 3.2.8 we have*)

$$\mathsf{E}(\nu|A) = \sum_{j=0}^{d-1} [a(0, d)]^{-1}\, n(0, j)\, a(j, d) = \sum_{i=1}^{d} (1 - r_i)^{-1},$$

$$\mathsf{E}(\nu|A') = \sum_{j=1}^{d-1} [a(0, d+1)]^{-1}\, n(0, j)\, a(j, d+1) = (1 - r_1)^{-1}$$

$$+ \sum_{j=1}^{d-1} \frac{p_1 \ldots p_j (1 - r_1)^{-1} \ldots (1 - r_{j+1})^{-1} [1 - p_{j+1} \ldots p_d (1 - r_{j+1})^{-1} \ldots (1 - r_d)^{-1}]}{1 - p_1 \ldots p_d (1 - r_1)^{-1} \ldots (1 - r_d)^{-1}}.$$

3.2.10. Until now we computed probabilities and mean values given the Markov chain started in a transient state $i \in T$. It is not difficult to obtain the corresponding results in the case of an arbitrary initial probability distribution $\mathbf{p} = (p(i))_{i \in S}$ over the whole state space S. To this end we notice that by (1.7) we have

$$\mathsf{P}_{\mathbf{p}}(A) = \sum_{i \in S} \mathsf{P}_{\mathbf{p}}(X(0) = i)\, \mathsf{P}_{\mathbf{p}}(A \mid X(0) = i) = \sum_{i \in S} p(i)\, \mathsf{P}_i(A),$$

whatever the random event A. In particular, if A is defined by the motion of the chain within set of its transient states, then $\mathsf{P}_i(A) = 0$ for any absorbing state i and we have

$$\mathsf{P}_{\mathbf{p}}(A) = \sum_{i \in T} p(i)\, \mathsf{P}_i(A).$$

These equations show that setting $\mathbf{p}_T = (p(i))_{i \in T}$ we have, for instance,

$$\mathsf{E}_{\mathbf{p}}(\nu(j))_{j \in T} = \mathbf{p}_T'\mathbf{N}, \quad \mathsf{E}_{\mathbf{p}}(\nu) = \mathbf{p}_T'\mathbf{N}\mathbf{e},$$

$$(\mathsf{P}_{\mathbf{p}} \text{ (absorption takes place in state } k))_{k \in S-T} = \mathbf{p}_T'\mathbf{N}\mathbf{R} + (p(k))_{k \in S-T}.$$

3.2.11. Let us apply the results we have obtained so far to Example 3 from 2.2. We shall consider only the case of tennis (the case of volleyball is left to the reader — see Exercise 3.7). As we have already seen the problem of winning a single game is in fact the gambler's ruin problem for the case $l = 4$ with given initial probabilities (corresponding to $r = 4$) as follows

$$p(0) = q^4(1 + 4p),\ p(1) = 4p^2q^3,\ p(2) = 6p^2q^2,$$
$$p(3) = 4p^3q^2,\ p(4) = p^4(1 + 4q).$$

*) The notation $\mathsf{E}(\nu|A)$ designates the mean value of ν under the probability P_A.

The fundamental matrix was explicitly computed in 3.1.2 so that we get immediately

$$(\mathsf{E}_i(\nu))_{1\leqslant i\leqslant 3} = \frac{1}{p^2+q^2}\begin{pmatrix} p^2+q^2+2p \\ 2 \\ p^2+q^2+2q \end{pmatrix},$$

$$(a(i,4))_{1\leqslant i\leqslant 3} = \frac{1}{p^2+q^2}\begin{pmatrix} p^3 \\ p^2 \\ p(p^2+q) \end{pmatrix}.$$

Therefore for the given initial probabilities we have

$$\mathsf{E}(\nu) = \frac{1}{p^2+q^2}\,[4p^2q^3(p^2+q^2+2p) + 12p^2q^2 + 4p^3q^2(p^2+q^2+2q)]$$

$$= \frac{8p^2q^2(2+pq)}{p^2+q^2},$$

$$\mathsf{P}(\text{Player } \mathcal{A} \text{ wins game}) = \mathsf{P}(\text{absorption takes place in state } 4)$$

$$= \frac{1}{p^2+q^2}\,[4p^5q^3 + 6p^4q^2 + 4p^4q^2(p^2+q)] + p^4(1+4q)$$

$$= \frac{p^4(1+2q)(1+4q^2)}{p^2+q^2}.$$

Notice that for $p = 1/2$ the probability P (Player $\mathcal{A}$ wins game) equals 1/2, a result that was expected by symmetry.

Notice that $\mathsf{E}(\nu)$ does not represent the mean length (expressed in points) of a game but the mean length of its final stage. To obtain the mean length of a game we should add to $\mathsf{E}(\nu)$ the mean length of its preliminary stage. In this stage either 4 or 5 points are disputed and the corresponding probabilities are $p^4 + q^4 + 6p^2q^2$ and $1 - (p^4 + q^4 + 6p^2q^2)$ so that its mean length equals

$$4(p^4+q^4+6p^2q^2) + 5(1-p^4-q^4-6p^2q^2)$$

$$= 5 - (p^4+q^4+6p^2q^2) = 4(1+pq-2p^2q^2).$$

It follows that the mean length of a game is

$$\frac{8p^2q^2(2+pq)}{p^2+q^2} + 4(1+pq-2p^2q^2) = \frac{4(6p^3q^3-pq+1)}{1-2pq}.$$

Thus, in the case where $p = 1/2$ (equally strong players) the mean length of a game is $27/4 = 6\,^3/_4$ points (the mean length of its final stage being $2\,^1/_4$ points). Notice that for two equally strong players the mean length of a game exceeds by almost 75% its minimum length (which is 4 points).

Now let us consider the problem of winning a set. This is in fact the same problem as above where p is to be replaced by the probability $p' = \mathsf{P}(\text{Player } \mathcal{A} \text{ wins game})$ and the initial probabilities (corresponding to $r = 6$) are as follows

$$p(0) = q'^6\left(1+\binom{6}{1}p'+\binom{7}{2}p'^2+\binom{8}{3}p'^3\right), \quad p(1) = \binom{8}{3}p'^4q'^5,$$

$$p(2) = \binom{8}{4}p'^4q'^4, \ p(3) = \binom{8}{3}p'^5q'^4,$$

$$p(4) = p'^6\left(1+\binom{6}{1}q'+\binom{7}{2}q'^2+\binom{8}{3}q'^3\right).$$

Then we have

$\mathsf{P}(\text{Player } \mathcal{A} \text{ wins set}) = \mathsf{P}(\text{absorption takes place in state } 4)$

$$= \frac{1}{p'^2+q'^2}[56p'^7q'^5+70p'^6q'^4+56p'^6q'^4(p'^2+q')]+p'^6(1+6q'+21q'^2+56q'^3)$$

$$= \frac{p'^6}{p'^2+q'^2}(1 + 4q' + 11\ q'^2 + 26\ q'^3 + 56\ q'^4 + 112\ q'^5),$$

where $q' = 1 - p'$.

The problem of winning the match is simpler. A match is won by the first player winning three sets. If we put $p'' = \mathsf{P}(\text{Player } \mathcal{A} \text{ wins set})$, then the probability p''' that Player $\mathcal{A}$ wins the match is

$$p''' = p''^3(1 + 3q'' + 6q''^2).$$

Here are a few illustrative figures:

Probability of winning point (p)	.5100	.5200	.5300	.5600	.5900
Probability of winning game (p')	.5250	.5499	.5746	.6468	.7145
Probability of winning set (p'')	.5734	.6443	.7103	.8652	.9502
Probability of winning match (p''')	.6357	.7559	.8503	.9802	.9989

Therefore the better player always has a good chance of winning the match. This "good chance" means "practically with certainty" in the case where the difference between the two player is substantial.

We should not believe that the model we studied is very realistic. It takes into account neither the various psychological elements that are very important in tennis nor the unavoidable variations of the fighting capacity of the players during the match. Nevertheless the model is to be considered as a starting point allowing further improvement.

3.3 EXTENSIONS AND COMPLEMENTS

3.3.1. The results obtained in the preceding paragraph can be applied to more general situations. A set of states $M \subset S$ is called *open* if and only if from every state of M a state of $S - M$ is accessible. It follows from the very definition that any subset of an open set of states is open, too. Clearly, the set of transient states T is open. We shall prove that $M \subset S$ is open if and only if no recurrent class is a subset of M. Indeed, if M contains a recurrent class $\mathcal{C}$, then from every state $i \in \mathcal{C}$ no state of $S - \mathcal{C} \supset S - M$ is accessible, that is M is not open. On the other hand, if M contains a recurrent state i without containing the whole class $\mathcal{C}$ of i, then from i all the states of $\mathcal{C} - M \subset S - M$ are accessible. Next, from any transient state of M (if any) a recurrent state (either of $S - M$ or of M) is accessible. Hence, if no recurrent class is a subset of M, then M is open.

The study of the motion of a Markov chain within an open set of states M reduces to the already investigated case $M = T$ on account of the following trick. Consider a new Markov chain with transition probabilities defined as

$$\tilde{p}(i, j) = \begin{cases} p(i, j), & \text{if } i \in M, \ j \in S, \\ \delta(i, j), & \text{if } i \in S - M, \ j \in S. \end{cases}$$

Therefore all the states of $S - M$ are made absorbing. For the new chain the set of transient states coincides with M and before leaving M the evolutions of the old and new chains are identical. Thus we can derive quantities such as the mean number of occurrences of a state of an open set of states before leaving that set, the mean number of steps in an open set of states, etc. For computing them we should use the fundamental matrix corresponding to the new chain, i.e. $(\mathbf{I} - \mathbf{M})^{-1}$, where $\mathbf{M} = (p(i, j))_{i, j \in M}$.

3.3.2. As a very simple application (which can also be dealt with in a direct manner) let us consider the case of a set M consisting of a single nonabsorbing state. Clearly, such a set is open. In this case the fundamental matrix $\mathbf{N}$ corresponding to the new chain has a unique entry equal to

$$1 + p(i, i) + p^2(i, i) + \dots = 1/(1 - p(i, i)).$$

If we denote by $\rho(i)$ the number of times the chain remains in state i once the state is entered (including the entering step), we then have (see 3.2.4)

$$\mathsf{P}_i(\rho(i) = r + 1) = (p(i, i))^r (1 - p(i, i)), \quad r \geqslant 0.$$

Consequently, under the probability P_i the random variable $\rho(i)$ has a geometric distribution with parameter $p(i, i)$ (see 1.4.2). Thus

$$\mathsf{E}_i(\rho(i)) = 1/(1 - p(i, i)),$$

$$\mathsf{E}_i(\rho^2(i)) = (1 + p(i, i))/(1 - p(i, i))^2.$$

Corresponding to the probability of absorption in $j \neq i$ for the new chain we get the value of the probability of reaching j after an unbroken sequence of occurrences of i (see 3.2.6):

$$\mathsf{P}_i(X(\rho(i)) = j) = \frac{p(i, j)}{1 - p(i, i)}. \tag{3.7}$$

3.3.3. Now we shall describe an application of equation (3.7). Denote by σ the number of changes of states in the trajectory of the Markov chain within the set T of its transient states. If, for instance, this trajectory is *iijijjii*, then the value of σ is 4. Defining a *run* of order $s \geqslant 1$ as any maximal segment of the trajectory consisting of the same state repeated s times, it is almost immediate that the total number of runs in the trajectory of the Markov chain in T equals $\sigma+1$ *). We want to compute the moments of the random variable σ. The problem will be reduced to one already solved. Define the random variables θ_0, θ_1, ... as follows

$$\theta_0 = 0,$$

$$\theta_1 = \begin{cases} 0, \text{ if } X(0) \text{ is an absorbing state,} \\ \text{the moment of the first change of state, if } X(0) \in T, \end{cases}$$

and, assuming $\theta_1, \ldots, \theta_n$ to be already defined,

$$\theta_{n+1} = \begin{cases} \theta_n, \text{ if } X(\theta_n) \text{ is an absorbing state,} \\ \text{the moment of the first change of state posterior} \\ \text{to } \theta_n, \text{ if } X(\theta_n) \in T. \end{cases}$$

Put $\widetilde{X}(n) = X(\theta_n)$, $n \geqslant 0$. Then $(\widetilde{X}(n))_{n\geqslant 0}$ is an absorbing Markov chain with the same state space as $(X(n))_{n\geqslant 0}$. The proof, which is based on the fact that $\theta_1, \theta_2, \ldots$ are stopping times for $(X(n))_{n\geqslant 0}$ and makes use of the strong Markov property, will be omitted. The transition probabilities of the associated Markov chain $(\widetilde{X}(n))_{n\geqslant 0}$ are as follows

$$\widetilde{p}(k, k) = \mathsf{P}_k(\widetilde{X}(1) = k) = 1, \quad k \in S - T,$$

$$\widetilde{p}(i, i) = 0, \; i \in T,$$

$$\widetilde{p}(i, j) = \mathsf{P}_i(\widetilde{X}(1) = j) = \mathsf{P}_i(X(\rho(i)) = j) = \frac{p(i, j)}{1 - p(i, i)}, \; i \in T, \; i \neq j,$$

[by equation (3.7)]. Therefore the transition matrix of the associated Markov chain is found from the transition matrix of the original Markov chain by setting equal to 0 the diagonal entries corresponding to the transient states and dividing each row by the sum of its elements.

---

*) For various results concerning runs the reader may consult [271, 320, 344, and 407].

Clearly, in the trajectory of $(\widetilde{X}(n))_{n\geqslant 0}$ within T there can occur no runs of order > 1, that is the associated Markov chain changes its state on every step. In other words, the random variable $\sigma+1$ for the original Markov chain coincides with the time to absorption ν for the associated Markov chain.

Notice that, once the mean value of σ is known, we can find the mean number of steps within T in which the Markov chain does not change its state. The latter is the difference between the mean time to absorption (for the original Markov chain) and the mean value of $\sigma+1$.

3.4 CONDITIONAL TRANSIENT BEHAVIOUR

3.4.1. The time to absorption ν may be sufficiently long for the probability distribution over the set T of transient states to approach a limiting probability distribution. Clarifying this statement, in what follows we shall prove that under suitable conditions the conditional probabilities

$$\mathsf{P}(X(n)=j\,|\,\nu>n),\qquad j\in T,$$

and

$$\mathsf{P}(X(m)=j\,|\,\nu>n),\qquad m<n,\ j\in T,$$

possess limits independent of the initial probability distribution over T. The study of this problem was initiated by M. S. Bartlett [15] and the basic results were obtained in [68] (see also [354]).

3.4.2. In what follows, in order to focus as much attention as possible on the conceptual rather than the technical aspects, we will assume the matrix $\mathbf{T}$ is irreducible*) aperiodic, thus regular (see 1.10.2).

Then it follows from the theory of nonnegative matrices (see 1.10) that the eigenvalues $\lambda_1, \lambda_2, \ldots, \lambda_t$ of $\mathbf{T}$ can be numbered in such a way that λ_1 is positive and simple (actually, as is easily seen using the argument from 1.11.2 we have $\lambda_1 < 1$) and $\lambda_1 > |\lambda_2| \geqslant |\lambda_3| \geqslant \ldots \geqslant |\lambda_t|$. If $|\lambda_2| = |\lambda_3|$ we stipulate that the (algebraic) multiplicity h of λ_2 is at least as great as the multiplicity of λ_3. Let $\mathbf{u} = (u(i))_{i\in T}$ and $\mathbf{v} = (v(i))_{i\in T}$ be the (positive) left and right eigenvectors corresponding to λ_1, chosen in such a way that

$$\sum_{i\in T} u(i) = 1,\quad \sum_{i\in T} u(i)\,v(i) = 1.$$

Then by Theorem 1.9 we have

$$\mathbf{T}^n = \lambda_1^n \mathbf{v}\mathbf{u}' + O(n^{h-1}\,|\lambda_2|^n). \tag{3.8}$$

* Some results in the case where $\mathbf{T}$ is reducible may be found in [256].

3.4.3. Returning to our problem let $\mathbf{p} = (p(i))_{i \in T}$ be an arbitrary probability distribution over T. Whatever state $j \in T$ we have

$$\mathsf{P}_{\mathbf{p}}(X(n)) = j \mid \nu > n) = \frac{\mathsf{P}_{\mathbf{p}}(X(n) = j)}{\mathsf{P}_{\mathbf{p}}(\nu > n)}$$

(since the random event $\{X(n) = j\}$ implies the random event $\{\nu > n\}$)

$$= \frac{\sum_{i \in T} p(i)\, p(n, i, j)}{\sum_{i \in T} p(i) \sum_{j \in T} p(n, i, j)}$$

[since $\mathsf{P}_i(\nu > n) = \mathsf{P}_i(X(n) \in T) = \sum_{j \in T} p(n, i, j)$ — see 3.2.4]

$$= \frac{\sum_{i \in T} p(i)\, \lambda_1^n v(i)\, u(j) + O(n^{h-1} \mid \lambda_2 \mid^n)}{\sum_{i \in T} p(i) \left(\sum_{j \in T} \lambda_1^n v(i)\, u(j) + O(n^{h-1} \mid \lambda_2 \mid^n) \right)}$$

[by (3.8)]

$$= u(j) + O(n^{h-1}[\mid \lambda_2 \mid / \lambda_1]^n).$$

Therefore

$$\lim_{n \to \infty} \mathsf{P}_{\mathbf{p}}(X(n) = j \mid \nu > n) = u(j), \; j \in T.$$

This can be expressed by saying that *whatever the initial distribution over T, for large n the probability distribution of $X(n)$ given that absorption has not yet taken place is approximately* $\mathbf{u}$.

Similarly, whatever $m < n$ and $j \in T$, we have

$$\mathsf{P}_{\mathbf{p}}(X(m) = j \mid \nu > n) = \frac{\mathsf{P}_{\mathbf{p}}(X(m) = j, \; \nu > n)}{\mathsf{P}_{\mathbf{p}}(\nu > n)}$$

$$= \frac{\mathsf{P}_{\mathbf{p}}(\nu > n \mid X(m) = j)\; \mathsf{P}_{\mathbf{p}}(X(m) = j)}{\mathsf{P}_{\mathbf{p}}(\nu > n)}$$

$$= \frac{\sum_{i \in T} p(i)\, p(m, \; i, \; j) \sum_{k \in T} p(n - m, j, k)}{\sum_{i \in T} p(i) \sum_{j \in T} p(n, i, j)}$$

$\Big[$since $\mathsf{P}_{\mathbf{p}}(\nu > n \mid X(m) = j) = \mathsf{P}_{\mathbf{p}}(X(n) \in T \mid X(m) = j) = \mathsf{P}_j(X(n - m) \in T) =$
$= \sum_{k \in T} p(n - m, j, k)\Big]$

$$= \frac{\sum_{i \in T} p(i)(\lambda_1^m v(i) u(j) + O(m^{h-1}|\lambda_2|^m)) \left(\sum_{k \in T} \lambda_1^{n-m} v(j) u(k) + O((n-m)^{h-1}|\lambda_2|^{n-m}) \right)}{\sum_{i \in T} p(i) \left(\sum_{j \in T} \lambda_1^n v(i)\, u(j) + O(n^{h-1} \mid \lambda_2 \mid^n) \right)}$$

$$= v(j)\, u(j) + O(m^{h-1}[\mid \lambda_2 \mid / \lambda_1]^m) + O((n - m)^{h-1}[\mid \lambda_2 \mid / \lambda_1]^{n-m}).$$

Therefore

$$\lim_{m\to\infty}\lim_{n\to\infty} \mathsf{P}_{\mathbf{p}}(X(m) = j \mid \{\nu > n\}) = v(j)\, u(j),\ j \in T.$$

This can be expressed by saying that *whatever the initial distribution over T, for large m the probability distribution of $X(m)$ given that absorption has not yet taken place and will not take place for a long time is approximately $(v(i)u(i))_{i\in T}$.*

It can also be shown that

$$\mathsf{E}_{\mathbf{p}}\left(\frac{\nu(j)}{\nu}\,\middle|\, \{\nu = n\}\right) = v(j)\, u(j) + O\left(\frac{1}{n}\right),$$

$$\mathsf{E}_{\mathbf{p}}\left(\frac{\sum_{m=0}^{n} u_m(j)}{n}\,\middle|\, \{\nu > n\}\right) = v(j)\, u(j) + O\left(\frac{1}{n}\right),\quad j \in T.$$

In other words, $v(j)u(j)$ *appears as both the expected proportion of time spent in state j before absorption given a long time to absorption and the expected relative frequency of state j in the first n steps given absorption takes place at a time posterior to n, for large n.*

3.4.4. Anticipating results from 4.2.1 the probability distribution $(v(i)u(i))_{i\in T}$ may be interpreted as the stationary distribution of a Markov chain with transition probabilities

$$\tilde{p}(i, j) = \frac{1}{\lambda_1}\,\frac{p(i, j)\, v(j)}{v(i)},\quad i, j \in T. \tag{3.9}$$

Under the assumed conditions this is a regular Markov chain. It is not difficult to prove (see Exercise 3.10) that

$$\lim_{m, n\to\infty} \mathsf{P}_{\mathbf{p}}(X(m + l) = i_l,\ 0 \leqslant l \leqslant s \mid X(m + n + s) \in T)$$

$$= v(i_0)\, u(i_0)\, \tilde{p}(i_0, i_1)\, \tilde{p}(i_1, i_2) \ldots \tilde{p}(i_{s-1}, i_s) \tag{3.10}$$

whatever $i_l \in T$, $0 \leqslant l \leqslant s$, and the initial distribution $\mathbf{p}$ over T. Therefore, given absorption has not taken place although a long time has elapsed, and that it will also not take place for a long time to come, the Markov chain behaves as a regular Markov chain with initial probabilities $v(i)\, u(i)$, $i \in T$, and transition probabilities $\tilde{p}(i, j)$, $i, j \in T$ (cf. [199]).

The result obtained justifies the nomenclature "quasi-stationary" for the probability distribution $(v(i)u(i))_{i\in T}$.

EXERCISES

3.1 Show [26] that

$$\frac{\mathsf{E}_k(\nu(i, j))}{\mathsf{E}_k(\nu(i))} = p(i, j), \quad i, k \in T, \; j \in S.$$

(Hint: use (3.3).)

3.2 Show [350] that whatever $i \in T$

$$\mathsf{E}_i(\nu(i, j)/\nu(i)) = \frac{p(i, j)}{1 - \sum_{k \in T} p(i, k)} \left(1 - c_i \sum_{k \in T} p(i, k)\right), \quad j \in T,$$

$$\mathsf{E}_i(\nu(i, j)/\nu(i)) = c_i p(i, j), \; j \in S - T,$$

where

$$c_i = \begin{cases} \dfrac{\ln n(i, i)}{n(i, i) - 1}, & \text{if } n(i, i) > 1, \\ 1, & \text{if } n(i, i) = 1. \end{cases}$$

Use the strong Markov property to deduce that whatever $k \neq i$

$$\mathsf{E}_k(\nu(i, j)/\nu(i)) = f(k, i) \; \mathsf{E}_i(\nu(i, j)/\nu(i)), \quad i, k \in T, \quad j \in S.$$

3.3 Prove [391] that whatever $i, j \in T$, $s \geqslant 1$,

$$\mathsf{E}_i((\nu^s(j)) = \begin{cases} \delta(i, j) + (1 - \delta(i, j)) f(i, j), & \text{if } f(j, j) = 0, \\ \dfrac{\delta(i, j) + (1 - \delta(i, j)) f(i, j)}{f(j, j)} \displaystyle\sum_{r=0}^{s} a_r^{(s)} n^r(j, j), & \text{if } f(j, j) \neq 0, \end{cases}$$

where $a_1^{(1)} = -1$, $a_0^{(s)} = 1$, $s \geqslant 1$, and

$$(a_l^{(s)})_{1 \leqslant l \leqslant s} = \mathbf{A}_s (a_l^{(s-1)})_{0 \leqslant l \leqslant s-1}, \quad s \geqslant 2,$$

with

$$\mathbf{A}_s = \begin{pmatrix} -1 & 2 & 0 & 0 & \cdots & 0 & 0 \\ 0 & -2 & 3 & 0 & \cdots & 0 & 0 \\ \cdot & \cdot & \cdot & \cdot & \cdots & \cdot & \cdot \\ \cdot & \cdot & \cdot & \cdot & \cdots & \cdot & \cdot \\ \cdot & \cdot & \cdot & \cdot & \cdots & \cdot & \cdot \\ 0 & 0 & 0 & 0 & \cdots & -s+1 & s \\ 0 & 0 & 0 & 0 & \cdots & 0 & -s \end{pmatrix}, \quad s \geqslant 2.$$

(H i n t: if $f(j, j) \neq 0$ then $\mathsf{E}_i(e^{it\nu(j)}) = g_{ij}(e^{it})e^{it\delta(i,j)} = c_{ij} + (\delta(i, j) + (1 - \delta(i, j))\, f(i, j))(1 - f(j, j))h_j(t) \,/\, f(j, j)$ with $h_j(t) = (1 - f(j, j)\, e^{it})^{-1}$. Use then the fact that $d^s h_j(t)/dt^s = (-\mathrm{i})^s \sum_{r=0}^{s} a_r^{(s)} h_j^s(t)$, and $d^{s+1}h_j(t)/dt^{s+1} = d(d^s h_j(t)/dt^s)/dt,\ s \geqslant 1$.)

3.4 Prove [391] that whatever $s \geqslant 1$

$$\mathbf{m}_s = \sum_{r=1}^{s} b_r^{(s)}\, \mathbf{N}^s \mathbf{e},$$

where $b_1^{(1)} = 1$ and

$$(b_l^{(s)})_{1 \leqslant l \leqslant s} = \mathbf{B}_s (b_l^{(s-1)})_{0 \leqslant l \leqslant s-1},\ s \geqslant 2,$$

with $b_0^{(s)} = 0,\ s \geqslant 1$, and

$$\mathbf{B}_2 = \begin{pmatrix} 0 & -1 \\ 0 & 2 \end{pmatrix},\ \mathbf{B}_s = \begin{pmatrix} 0 & -1 & 0 & 0 & \cdots & 0 & 0 \\ 0 & 2 & -2 & 0 & \cdots & 0 & 0 \\ \cdot & \cdot & \cdot & \cdot & \cdots & \cdot & \cdot \\ \cdot & \cdot & \cdot & \cdot & \cdots & \cdot & \cdot \\ \cdot & \cdot & \cdot & \cdot & \cdots & \cdot & \cdot \\ 0 & 0 & 0 & 0 & \cdots & s-1 & -(s-1) \\ 0 & 0 & 0 & 0 & \cdots & 0 & s \end{pmatrix},\ s \geqslant 3.$$

(H i n t: $(\mathsf{E}_i(e^{it\nu}))_{i \in T} = (e^{-it}\,\mathbf{I} - \mathbf{T})^{-1}\,\mathbf{N}^{-1}\mathbf{e}$ and

$$d^s(e^{-it}\,\mathbf{I} - \mathbf{T})^{-1}/dt^s = \mathrm{i}^s \sum_{r=1}^{s} b_r^{(s)} e^{-irt}(e^{-it}\,\mathbf{I} - \mathbf{T})^{-(r+1)}.)$$

3.5 Prove computationally that $\sum_{j \in S-T} a(i, j) = 1,\ i \in T$, or, in matrix notation, $\mathbf{Ae} = \mathbf{e}$. (H i n t: $\mathbf{Ae} = \mathbf{NRe}$, and $\mathbf{Re} = \mathbf{e} - \mathbf{Te}$.)

3.6 Prove that convergence in (3.5) is geometrically fast.

(H i n t: $p(m + n, i, k) = \sum_{j \in S} p(m, i, j)\ p(n, j, k)$

$$= p(m, i, k) + \sum_{j \in T} p(m, i, j)\ p(n, j, k),$$

whence $0 \leqslant a(i, k) - p(m, i, k) \leqslant \sum_{j \in T} p(m, i, j)$. Now use Exercise 2.8.)

3.7 Consider the formula

$$(\mathsf{P}_{\mathbf{p}}\ (\text{absorption takes place in state } k))_{k \in S-T} = \mathbf{p}'\,\mathbf{N}\,\mathbf{R}$$

where $\mathbf{p}$ is a probability distribution over T (see 3.2.10). Discuss the problem of finding $\mathbf{p}$ given the vector

$$(\mathsf{P}_{\mathbf{p}}\ (\text{absorption takes place in state } k))_{k \in S-T}.$$

(H i n t: [321].)

3.8 Discuss similarly to 3.2.11 the case of volley-ball game ($r = 15$ — see Exercise 2.2).

3.9 Show that under the assumptions of 3.4.2 the probability distribution of ν under P_u is geometric with parameter $1-\lambda_1$.

3.10 Show that under the assumptions of 3.4.2

$$\lim_{m\to\infty} \mathsf{P}_p(X(m) = j \mid \nu = m+s) = \frac{u(j)\,\mathsf{P}_j(\nu = s)}{\lambda_1^{s-1}(1-\lambda_1)}, \quad s > 1,$$

$$\lim_{s\to\infty} \mathsf{P}_p(X(m) = j \mid \nu = m+s) = \frac{\mathsf{P}_p(X(m) = j)\,v(j)}{\lambda_1^m \sum_{i\in T} p(i)\,v(i),}, \quad m \geqslant 0,$$

$$\lim_{m\to\infty}\lim_{s\to\infty} \mathsf{P}_p(X(m)=j \mid \nu = m+s)$$

$$=\lim_{s\to\infty}\lim_{m\to\infty} \mathsf{P}_p(X(m)=j \mid \nu = m+s) = v(j)\,u(j),\ j\in T.$$

(Hint:

$$\mathsf{P}_p(X(m)=j \mid \nu = m+s) = \frac{\sum_{i\in T} p(i)\,p(m,i,j) \sum_{k\in T}(p(s-1,j,k)-p(s,j,k))}{\sum_{i\in T} p(i) \sum_{j\in T}(p(m+s-1,i,j)-p(m+s,i,j))}.\Bigg)$$

3.11 Prove equation (3.10).

(Hint: the conditional probability there equals

$$\frac{\sum_{i\in T} p(i)\,p(m,i,i_0)\,p(i_0,i_1)\ldots p(i_{s-1},i_s) \sum_{k\in T} p(n,i_s,k)}{\sum_{i\in T} p(i) \sum_{j\in T} p(m+n+s,i,j)}.\Bigg)$$

CHAPTER 4

Ergodic Markov Chains

This chapter is devoted to the study of irreducible Markov chains. It will be seen that for such chains the n-step transition probabilities converge (either in the usual or in the Cesàro sense) as $n \to \infty$ to limits independent of the starting state. This property called "ergodicity", though not characteristic of irreducible Markov chains (see Theorem 4.3 and Exercise 5.1), has established the nomenclature "ergodic Markov chain" as a synonym for "irreducible Markov chain".

4.1 REGULAR MARKOV CHAINS

4.1.1. In 1.10.1 we introduced the concept of regularity of a square matrix with nonnegative entries. In particular, we may consider regular stochastic matrices: a stochastic matrix $\mathbf{P}$ is said to be *regular* if and only if there exists a natural number n_0 such that $\mathbf{P}^{n_0} > \mathbf{0}$. In what follows we first clarify the relationship between regular Markov chains and regular stochastic matrices.

Proposition 4.1. *A Markov chain is regular if and only if its transition matrix is regular.*

Proof. Clearly, the condition is sufficient: the inequality $\mathbf{P}^{n_0} > 0$ amounts to the fact that any state can be reached in n_0 steps regardless the starting state; hence there is only one class that cannot be periodic (see 2.4.4).

To prove the necessity of the condition, let us notice that the states being aperiodic, for all states i the greatest common divisor of the natural numbers n for which $p(n, i, i) > 0$ equals 1. At the same time, the set of these numbers is closed under addition since $p(n+n', i, i) \geqslant \geqslant p(n, i, i)\, p(n', i, i)$. A classical result in elementary number theory*) leads us to the conclusion that whatever the state i there exists a natural number n_i such that $p(n, i, i) > 0$ for $n \geqslant n_i$. On the other hand, since the states of the chain form a unique class, whatever the pair of states i and j there exists a natural number n_{ij} such that $p(n_{ij}, i, j) > 0$. Therefore,

---

* A set of natural integers that is closed under addition contains all but a finite number of multiples of its greatest common divisor.

on account of the fact that $p(n_{ij}+n_j, i, j) \geqslant p(n_{ij}, i, j)\, p(n_j, j, j) > 0$ we get $p(n_0, i, j) > 0$ whatever the states i and j, where $n_0 = \max\limits_{i,j} (n_{ij}+n_j)$.

4.1.2. Remember that the absolute values of the eigenvalues of a stochastic matrix are at most equal to 1 and that 1 is an eigenvalue (see 1.11.2). It follows by Theorem 1.8 that for a regular stochastic matrix $\lambda_1 = 1$ is that simple real eigenvalue which is strictly larger than the absolute values of all the other eigenvalues. Since as right eigenvector corresponding to $\lambda_1 = 1$ we can take the vector $\mathbf{v}_1$ whose entries are all equal to 1, we conclude that the left eigenvector $\mathbf{u}_1$ is a probability vector with nonnull entries ($\mathbf{u}_1$ and $\mathbf{v}_1$ were chosen such that $\mathbf{u}_1'\mathbf{v}_1 = 1$).

These considerations allow us to obtain a quick solution of the problem of the asymptotic behaviour of the powers of a regular stochastic matrix.

Theorem 4.2. *If* $\mathbf{P}$ *is a regular stochastic matrix then* $\mathbf{P}^n$ *converges as* $n \to \infty$ *to a positive stable stochastic matrix* $\mathbf{\Pi} = \mathbf{e}\boldsymbol{\pi}'$, *where* $\boldsymbol{\pi} = (\pi(i))_{i\in S}$ *is a probability vector with nonnull entries. If we agree to number the eigenvalues of* $\mathbf{P}$ *as in the statement of Theorem* 1.9, *then there exists a constant* $a > 0$ *such that*

$$|p(n, i, j) - \pi(j)| \leqslant a n^{m_2-1} |\lambda_2|^n,$$

whatever the states i *and* j *and the natural number* n.

Proof. Apply Theorem 1.9 to the regular matrix $\mathbf{P}$. Obviously, $\boldsymbol{\pi} = \mathbf{u}_1$, the left eigenvector corresponding to the eigenvalue $\lambda_1 = 1$.

Theorems 4.1 and 4.2 show that for a regular Markov chain, whatever the states i and j, the n step transition probability $p(n, i, j)$ converges as $n \to \infty$ to a limit $\pi(j)$ independent of i. For the effective computation of these values we can proceed as follows. Let D_i, $i \in S$, denote the cofactors of the main diagonal entries in the determinant of the matrix $\mathbf{I} - \mathbf{P}$. Equation (1.13) shows that the D_i are nonnull and have the sign of $f'(1)$. This is a positive quantity on account of the fact that 1 is in absolute value the largest root of the algebraic equation $f(\lambda)=0$. It follows that

$$\pi(i) = \frac{D_i}{\sum\limits_{i\in S} D_i}, \quad i \in S, \tag{4.1}$$

equations first obtained by Mihoc [269]*).

*) These equations have been subsequently rediscovered by many other authors. The most recent paper on this matter seems to be [104].

Equivalently, we can use the fact that, the eigenvalue $\lambda = 1$ being simple, $\boldsymbol{\pi}$ is the only probability vector for which $\boldsymbol{\pi}'\mathbf{P} = \boldsymbol{\pi}'$*). This equation leads to the system

$$\sum_{i \in S} \pi(i)\, p(i, j) = \pi(j), \quad j \in S.$$

Any equation of the above system can be obtained by summing the other ones. Thus the general solution of the system can be written as $\pi(j) = c_j \pi(i)$, $j \in S$, with $c_i = 1$, i being fixed. Since $\sum_{j \in S} \pi(j) = 1$ we shall have

$$\pi(j) = \frac{c_j}{\sum_{j \in S} c_j}, \quad j \in S. \tag{4.2}$$

Notice that the matrix $\mathbf{\Pi}$ satisfies the equations

$$\mathbf{\Pi P} = \mathbf{P\Pi} = \mathbf{\Pi} = \mathbf{\Pi}^2. \text{**)} \tag{4.3}$$

They come from writing $\mathbf{P}^{2n}\mathbf{P} = \mathbf{P}\mathbf{P}^{2n} = \mathbf{P}^{2n+1} = \mathbf{P}^n\mathbf{P}^{n+1}$ and letting then $n \to \infty$.

4.1.3. Theorem 4.2 can be proved directly by making use of the properties of the ergodicity coefficient α (see 1.11.5) without appealing to the theory of nonnegative matrices.

Note first that if $\mathbf{P}^{n_0} > \mathbf{0}$ then setting

$$\min_{i \in S} p(n_0, i, j) = \delta_j > 0, \quad j \in S,$$

we have

$$\begin{aligned}\sum_{k \in S} (p(n_0, i, k) - p(n_0, j, k))^+ &\leqslant \sum_{k \in S} (p(n_0, i, k) - \delta_k) \\ &= 1 - \sum_{k \in S} \delta_k < 1, \quad i, j \in S,\end{aligned}$$

whence

$$\alpha(\mathbf{P}^{n_0}) = 1 - \max_{i,j \in S} \sum_{k \in S} (p(n_0, i, k) - p(n_0, j, k))^+ > \sum_{k \in S} \delta_k > 0.$$

On account of Theorem 1.11 we have for $0 \leqslant m < n_0$, $s \geqslant 1$,

$$(1 - \alpha(\mathbf{P}^{sn_0 + m})) \leqslant (1 - \alpha(\mathbf{P}^{n_0}))^s\, (1 - \alpha(\mathbf{P}^m)).$$

---

*) Analogously, the right eigenvector corresponding to $\lambda_1 = 1$ being $\mathbf{v}_1 = \mathbf{e}$, the general solution of the equation $\mathbf{P}\boldsymbol{\rho} = \boldsymbol{\rho}$ is $\boldsymbol{\rho} = c\mathbf{e}$, with c an arbitrary constant. This remark will be used several times in what follows.

**) An explicit description of idempotent stochastic matrices can be found in [248].

Hence, setting

$$m_j(n) = \min_{i \in S} p(n, i, j), \quad M_j(n) = \max_{i \in S} p(n, i, j),$$

we get

$$M_j(sn_0 + m) - m_j(sn_0 + m) \leq (1 - \alpha(\mathbf{P}^{sn_0+m})) \leq (1 - \alpha(\mathbf{P}^{n_0}))^s \tag{4.4}$$

whatever $j \in S$.

If we remark that

$$m_j(n + 1) = \min_{i \in S} p(n + 1, i, j) = \min_{i \in S} \sum_{k \in S} p(i, k)\, p(n, k, j)$$

$$\geq \min_{i \in S} \sum_{k \in S} p(i, k)\, m_j(n) = m_j(n),$$

$$M_j(n + 1) = \max_{i \in S} p(n + 1, i, j) = \max_{i \in S} \sum_{k \in S} p(i, k)\, p(n, k, j)$$

$$\leq \max_{i \in S} \sum_{k \in S} p(i, k)\, M_j(n) = M_j(n),$$

so that

$$m_j(1) \leq m_j(2) \leq \ldots \leq M_j(2) \leq M_j(1),$$

it follows that $m_j(n)$ and $M_j(n)$ have the same limit as $n \to \infty$. Let us denote

$$\lim_{n\to\infty} m_j(n) = \lim_{n\to\infty} M_j(n) = \pi(j), \qquad j \in S.$$

Since

$$m_j(n) \leq p(n, i, j) \leq M_j(n), \tag{4.5}$$

it follows that

$$\lim_{n\to\infty} p(n, i, j) = \pi(j), \qquad i, j \in S.$$

As $m_j(n_0) = \delta_j > 0$, we shall have $\pi(j) > 0$, $j \in S$. Finally, it follows from (4.4) and (4.5) that

$$|p(n, i, j) - \pi(j)| \leq (1 - \alpha(\mathbf{P}^{n_0}))^{n/n_0 - 1},$$

whatever the states i and j and the natural number n.

Notice that in the above proof of Theorem 4.2 we only made use of the fact that $\alpha(\mathbf{P}^{n_0}) > 0$. It is clear that the transition matrix of a Markov chain having a periodic recurrent class or several recurrent classes cannot verify this inequality. Instead, the transition matrix of an indecomposable Markov chain (see 2.6.2) does verify it. To prove this assertion let

$$\mathbf{P} = \begin{pmatrix} \mathbf{P}_1 & \mathbf{0} \\ \mathbf{R} & \mathbf{T} \end{pmatrix}$$

denote the transition matrix of an indecomposable Markov chain, where $\mathbf{P}_1$ is a regular stochastic matrix. It is easily seen that

$$\mathbf{P}^n = \begin{pmatrix} \mathbf{P}_1^n & \mathbf{0} \\ \mathbf{R}_n & \mathbf{T}^n \end{pmatrix},$$

where $\mathbf{R}_{n+1} = \sum_{m=0}^{n} \mathbf{T}^m \mathbf{R} \mathbf{P}_1^{n-m} = \sum_{m=0}^{n} \mathbf{T}^{n-m} \mathbf{R} \mathbf{P}_1^m$, $n \geqslant 1$. Taking into account that $\mathbf{P}_1^n \to \mathbf{\Pi}_1 = \mathbf{e}\pi_1'$ (by Theorem 4.2) and $\mathbf{T}^n \to \mathbf{0}$ (by Corollary 2 to Theorem 2.7) as $n \to \infty$, the convergence being geometrical, we get $\mathbf{R}_{n+1} = \sum_{m=0}^{n} \mathbf{T}^{n-m} \mathbf{R} \mathbf{e}\pi_1' + \sum_{m=0}^{n} \mathbf{T}^{n-m} \mathbf{R}(\mathbf{P}_1^m - \mathbf{e}\pi_1') \to \left(\sum_{m \geqslant 0} \mathbf{T}^m\right) \mathbf{R}\mathbf{e}\pi_1' = \mathbf{N}\mathbf{R}\mathbf{e}\pi_1' = \mathbf{e}\pi_1'$. (By Exercise 3.5 we have $\mathbf{NRe} = \mathbf{e}$.) This means that for n_0 large enough $\mathbf{P}^{n_0}$ is a Markov matrix, therefore $\alpha(\mathbf{P}^{n_0}) > 0$.

Calling *mixing* a stochastic matrix $\mathbf{P}$ such that $\alpha(\mathbf{P}^{n_0}) > 0$ for some natural number n_0*), the above considerations allow us to state

Theorem 4.3. *If* $\mathbf{P}$ *is a mixing stochastic matrix then* $\mathbf{P}^n$ *converges to a stable stochastic matrix* $\mathbf{\Pi} = \mathbf{e}\pi'$, *where* $\pi = (\pi(i))_{i \in S}$ *is a probability vector whose nonnull entries correspond to the recurrent states of the Markov chain with transition matrix* $\mathbf{P}$. *We have*

$$|p(n, i, j) - \pi(j)| \leqslant (1 - \alpha(\mathbf{P}^{n_0}))^{n/n_0 - 1}$$

whatever the states i *and* j *and the natural number* n.

Clearly, the nonnull entries of π can be computed by applying equations (4.1) to the submatrix $\mathbf{P}_1$ of $\mathbf{P}$ (see above).

As in the case of a regular stochastic matrix, for a mixing stochastic matrix $\mathbf{P}$ the vector π is the only probability vector such that $\pi'\mathbf{P} = \pi'$ and the general solution of the equation $\mathbf{P}\rho = \rho$ is $\rho = c\mathbf{e}$, with c an arbitrary constant. (It then follows that the nonnull entries of π can also be computed using equations (4.2).) These assertions can be proved as follows. If $\mathbf{P}$ is an arbitrary probability vector, then $\lim_{n\to\infty} \mathbf{p}'\mathbf{P}^n = \mathbf{p}'\,\mathbf{\Pi} = \mathbf{p}'\mathbf{e}\pi' = \pi'$, since $\mathbf{p}'\mathbf{e} = 1$. Assume that $\mathbf{p}'\mathbf{P} = \mathbf{p}'$. It follows that $\mathbf{p}'\mathbf{P}^n = (\mathbf{p}'\mathbf{P})\,\mathbf{P}^{n-1} = \mathbf{p}'\mathbf{P}^{n-1} = \ldots = \mathbf{p}'$, and therefore $\mathbf{p}' = \lim_{n\to\infty} \mathbf{p}'\mathbf{P}^n = \pi'$. Next, $\mathbf{P}^n\rho = \mathbf{P}^{n-1}(\mathbf{P}\rho) = \mathbf{P}^{n-1}\rho = \ldots = \rho$. Letting $n \to \infty$ we get $\mathbf{\Pi}\rho = \rho$, i.e., all the entries of ρ equal $\pi'\rho$, therefore $\rho = c\mathbf{e}$.

Equations (4.3) are also true for a mixing stochastic matrix $\mathbf{P}$.

---

*) It follows from the above that a stochastic matrix is mixing if and only if the corresponding Markov chain is either regular or indecomposable.

4.1.4. The method of proof making use of the ergodicity coefficient α (that allowed us to state Theorem 4.3) is in fact based on an idea of A. A. Markov, who considered the special case where $p(i, j) > 0$, $i, j \in S$.

For proving Theorem 4.2 other methods were also devised.

A very interesting method recently extended to more general cases (see [129, 130, and 131]) is due to W. Doeblin [82]. It is known as the "two particle method" and starts by considering two *independent* copies $(X_1(n))_{n\geq 0}$ and $(X_2(n))_{n\geq 0}$ of the given chain which are observed simultaneously as a two dimensional Markov chain $\widetilde{X}(n) = (X_1(n), X_2(n))_{n\geq 0}$, with state space $S\times S = ((i, j) : i, j \in S)$ and transition probabilities $p((i, j), (k, l)) = p(i, k)\, p(j, l)$.

The first step in Doeblin's treatment consists of proving the fact that, under the regularity hypothesis, if the two copies start in any states i and j (a condition constantly preserved in what follows) then with probability 1 they will meet, that is they will reach the same state at some time. For a proof notice that the probability that $X_1(n_0) = X_2(n_0)$ equals

$$\sum_{k\in S} p(n_0, i, k)\, p(n_0, j, k) \geqslant \min_{i,j\in S} p(n_0, i, j) = \varepsilon > 0.$$

Similarly, the conditional probability that $X_1(2n_0) = X_2(2n_0)$ given that $X_1(n_0) \neq X_2(n_0)$ exceeds ε and, in general, the conditional probability that $X_1(nn_0) = X_2(nn_0)$ given that $X_1(mn_0) \neq X_2(mn_0)$, $1 \leqslant m < n$, exceeds ε. Consequently, the probability that the two copies meet by time nn_0 is at least $1-(1-\varepsilon)^n$, which approaches 1 as $n\to\infty$.

Now, denote by $r_{ij}(n)$ the probability that the two copies first meet after time n. The next step consists of proving the inequality

$$|p(n, i, k) - p(n, j, k)| \leqslant r_{ij}(n). \tag{4.6}$$

(which holds for an arbitrary Markov chain). To proceed, denote by $r_{ij}(m, l)$ the probability that the two copies first meet at time m in state l and by $r_{ij}^l(n)$ the probability that the first copy is in state l at time n and the two copies first meet after time n. By symmetry we have $r_{ij}(n) = r_{ji}(n)$, and $r_{ij}(m, l) = r_{ji}(m, l)$. Obviously, $r_{ij}(n) \geqslant r_{ij}^l(n)$. With this notation we have

$$p(n, i, k) = \sum_{m=1}^{n} \sum_{l\in S} r_{ij}(m, l)\, p(n-m, l, k) + r_{ij}^k(n),$$

$$p(n, j, k) = \sum_{m=1}^{n} \sum_{l\in S} r_{ij}(m, l)\, p(n-m, l, k) + r_{ji}^k(n),$$

Hence (4.6) follows at once.

We have already seen that for a regular Markov chain we have $r_{ij}(n) \leqslant (1-\varepsilon)^{n/n_0 - 1}$, so that

$$|p(n, i, k) - p(n, j, k)| \leqslant (1-\varepsilon)^{n/n_0 - 1}.$$

The existence of the limit $\lim_{n\to\infty} p(n, i, k)$ not depending on i may then be proved as in 4.1.3.

Another method for proving Theorem 4.2 is based on concepts and results from information theory. For details the reader should consult [323], pp. 597—601, [198], and [230].

Finally, a geometric method for studying general homogeneous Markov chains will be presented later on in 5.1.3.

4.2 THE STATIONARY DISTRIBUTION

4.2.1. The probability distribution π enjoys the remarkable property of being the only initial distribution making the Markov chain with transition matrix $\mathbf{P}$ a stationary sequence. More precisely, we shall prove

Theorem 4.4. *Let $(X(n))_{n\geqslant 0}$ be a Markov chain with mixing transition matrix* $\mathbf{P}$. *Whatever the states i_l, $0\leqslant l\leqslant m$, and the natural number m the probability*

$$\mathsf{P}_\pi(X(n+l) = i_l,\ 0\leqslant l\leqslant m)$$

does not depend on $n\geqslant 0$. The probability distribution π is the only initial distribution with the above property.

Proof. We have

$$\mathsf{P}_\pi(X(n+l) = i_l, 0\leqslant l\leqslant m)$$

$$= \sum_{i\in S} \mathsf{P}_\pi(X(0) = i)\ \mathsf{P}_\pi(X(n+l) = i_l,\ 0\leqslant l\leqslant m \mid X(0)=i)$$

$$= \sum_{i\in S} \pi(i) p(n, i, i_0)\ p(i_0, i_1) \dots p(i_{m-1}, i_m) = \pi(i_0)\ p(i_0, i_1) \dots p(i_{m-1}, i_m)$$

on account of the fact that $\pi'\mathbf{P} = \pi'$ implies $\pi'\mathbf{P}^n = \pi'$ for all natural numbers n, as has been previously shown. Therefore the above probability does not depend on n.

Conversely, if for an initial distribution $\mathbf{p}$ the probabilities considered do not depend on n, then, in particular, $\mathsf{P}_\mathbf{p}(X(0) = i) = \mathsf{P}_\mathbf{p}(X(1) = i)$, whence $p(i) = \sum_{j\in S} p(j)\ p(j, i)$, that is $\mathbf{p}' = \mathbf{p}'\mathbf{P}$. As has already been noticed, this equation implies that $\mathbf{p} = \pi$.

Theorem 4.4 enables us to call the limit distribution π *stationary* (or *invariant*).

For the case of a two state Markov chain the stationary distribution has already been computed in 1.11.2.

Another important special case is that of regular Markov chains for which the stationary distribution π is the uniform distribution over the state space S (i.e., all the components of π are equal). In this case the equation $\pi' = \pi'\mathbf{P}$ implies we must have $\sum_{i \in S} p(i, j) = 1$, $j \in S$, i.e., the transition matrix is doubly stochastic (see 1.11.1). Unicity of π as a solution of the equation $\mathbf{p}' = \mathbf{p}'\mathbf{P}$, where $\mathbf{p}$ is a probability vector, implies that, conversely, the stationary distribution for a regular doubly stochastic matrix is the uniform distribution over the state space.

A classical example of a Markov chain with doubly stochastic matrix occurs when modelling a certain technique for shuffling a deck of cards. The interested reader may consult [102], pp. 406 and 422.

4.2.2. Now, we shall compute the stationary distribution for Example 2 from 2.2. Assuming that $p_i > 0$, $q_i > 0$, $0 \leqslant i \leqslant l$, and that at least one of the numbers r_i, $0 \leqslant i \leqslant l$, is different from 0, the states of the chain constitute a unique aperiodic recurrent class. The equation $\pi' = \pi'\mathbf{P}$ leads to the system

$$\pi(0)r_0 + \pi(1)q_1 = \pi(0),$$

$$\pi(i-1)\, p_{i-1} + \pi(i)\, r_i + \pi(i+1)\, q_{i+1} = \pi(i), \quad 1 \leqslant i < l,$$

$$\pi(l-1)\, p_{l-1} + \pi(l)\, r_l = \pi(l).$$

Solving the above equations successively yields

$$\pi(1) = \frac{p_0}{q_1}\pi(0), \quad \pi(2) = \frac{p_0 p_1}{q_1 q_2}\pi(0),$$

and, in general,

$$\pi(i) = \frac{p_0 \cdots p_{i-1}}{q_1 \cdots q_i}\pi(0), \quad 1 \leqslant i \leqslant l,$$

whence

$$\pi(0) = \frac{1}{1 + \sum_{i=1}^{l} (p_0 \cdots p_{i-1})/(q_1 \cdots q_i)},$$

$$\pi(i) = \frac{(p_0 \cdots p_{i-1})/(q_1 \cdots q_i)}{1 + \sum_{i=1}^{l} (p_0 \cdots p_{i-1})/(q_1 \cdots q_i)}, \quad 1 \leqslant i \leqslant l.$$

In particular, in the case of the Bernoulli model for which

$$p_i = \left(1 - \frac{i}{l}\right)^2, \quad 0 \leqslant i \leqslant l-1,$$

$$q_i = \left(\frac{i}{l}\right)^2, \quad 1 \leqslant i \leqslant l,$$

we have

$$\frac{p_0 \dots p_{i-1}}{q_1 \dots q_i} = \left(\frac{l \dots (l-i+1)}{1 \dots i}\right)^2 = \binom{l}{i}^2, \qquad 1 \leqslant i \leqslant l.$$

Taking into account that $\sum_{i=0}^{l} \binom{l}{i}^2 = \binom{2l}{l}$, we get

$$\pi(i) = \binom{l}{i}^2 \Big/ \binom{2l}{l}, \qquad 0 \leqslant i \leqslant l. \ ^{*)}$$

Notice that $\pi(i)$ is the probability of getting i white and $l-i$ black balls in a drawing at random of l balls from an urn containing l white and l black balls.

4.2.3. Finding the stationary distribution for Example 8 from 2.2 is still easier. The equation $\boldsymbol{\pi}' = \boldsymbol{\pi}'\mathbf{P}$ leads to the system

$$\pi(0)q + \pi(m-1) = \pi(0), \qquad \pi(0)pp_{m-1} = \pi(1),$$

$$\pi(0)pp_{m-l} + \pi(l-1) = \pi(l), \qquad 2 \leqslant l \leqslant m-1.$$

The last $m-1$ equations yield successively

$$\pi(1) = pp_{m-1}\,\pi(0),$$

$$\pi(2) = p(p_{m-1} + p_{m-2})\,\pi(0),$$

$$\dots\dots\dots\dots\dots\dots\dots\dots$$

and, in general,

$$\pi(l) = p\left(\sum_{i=1}^{l} p_{m-i}\right)\pi(0), \qquad 1 \leqslant l \leqslant m-1.$$

Taking into account that $\sum_{i=0}^{m-1} \pi(i) = 1$ we get

$$\pi(0) = \frac{1}{1 + p\sum_{i=1}^{m-1} ip_i}.$$

4.2.4. We should not believe that it is always possible to find simple expressions for the components of the vector $\boldsymbol{\pi}$. Thus, for Example 6

---

*) This result was first obtained by A.A. Markov who studied the Bernoulli urn scheme, as well as a more general scheme involving several urns arranged in a circle, in a series of papers in the period 1912–1918 (see [260], pp. 439–463, 551–571, 589–595).

from 2.2 (we know that if $0 < p_0 < 1$ and $P_m > 0$ the associated Markov chain is regular) equation $\boldsymbol{\pi}' = \boldsymbol{\pi}'\mathbf{P}$ leads to the system

$$\begin{aligned} \pi(l) &= \pi(0)\, p_l + \sum_{j=1}^{l+1} \pi(j)\, p_{l-j+1}, \quad 0 \leqslant l \leqslant m-1, \\ \pi(m) &= \pi(0)\, P_m + \sum_{j=1}^{m} \pi(j)\, P_{m-j+1}, \end{aligned} \tag{4.7}$$

whose solution can be compactly written by means of the following trick. Set $p_l = 0$ for $l \geqslant m$ and consider the infinite system

$$\tilde{\pi}(l) = \tilde{\pi}(0)\, p_l + \sum_{j=1}^{l+1} \tilde{\pi}(j)\, p_{l-j+1}, \quad l \geqslant 0. \tag{4.8}$$

As the first m equations of system (4.7) are identical to the first m equations of system (4.8) we should have $\pi(l) = c\tilde{\pi}(l)$, $0 \leqslant l \leqslant m-1$, c being a constant. [Each $\tilde{\pi}(l)$, $0 \leqslant l \leqslant m-1$, is the same function of $\tilde{\pi}(0)$ as $\pi(l)$ is of $\pi(0)$.] For $|z| \leqslant 1$ let us consider the generating functions $P(z) = \sum_{k=0}^{m-1} p_k z^k$ and $\Pi(z) = \sum_{l \geqslant 0} \tilde{\pi}(l) z^l$. Multiplying the lth equation of system (4.8) by z^l and summing over $l \geqslant 0$ yields

$$\begin{aligned} \Pi(z) &= \tilde{\pi}(0)\, P(z) + \frac{1}{z} \sum_{l \geqslant 0} \sum_{j=1}^{l+1} \tilde{\pi}(j)\, z^j p_{l-j+1} z^{l-j+1} = \\ &= \tilde{\pi}(0)\, P(z) + \frac{1}{z} \sum_{j \geqslant 1} \tilde{\pi}(j) z^j \sum_{r=l-j+1 \geqslant 0} p_r z^r \\ &= \tilde{\pi}(0)\, P(z) + \frac{1}{z} (\Pi(z) - \tilde{\pi}(0))\, P(z), \end{aligned}$$

whence

$$\Pi(z) = \frac{\tilde{\pi}(0)\,(1-z)\,P(z)}{P(z) - z}.$$

It follows that

$$\pi(l) = \frac{\pi(0)}{l\,!} \frac{d^l}{dz^l} \left[\frac{(1-z)\,P(z)}{P(z)-z} \right] \Bigg|_{z=0} = c_l \pi(0), \quad 1 \leqslant l \leqslant m-1.$$

Then, using the last equation of system (4.7),

$$\pi(m) = \frac{1}{p_0} \left[\pi(0)\, P_m + \sum_{j=1}^{m-1} \pi(j)\, P_{m-j+1} \right] = c_m \pi(0).$$

Finally

$$\pi(0) = \frac{1}{1 + \sum_{l=1}^{m} c_l}.$$

4.3 THE FUNDAMENTAL MATRIX

4.3.1. The part played by the matrix $\mathbf{N} = (\mathbf{I} - \mathbf{T})^{-1}$ in the theory of absorbing Markov chains is assumed in the theory of regular Markov chains by the matrix $\mathbf{Z} = [\mathbf{I} - (\mathbf{P} - \mathbf{\Pi})]^{-1} = (z(i, j))$.

Theorem 4.5. *Let* $\mathbf{P}$ *be a regular stochastic matrix and* $\mathbf{\Pi}$ *the limiting matrix* $\lim_{n\to\infty} \mathbf{P}^n$. *Then the matrix* $\mathbf{Z} = [\mathbf{I} - (\mathbf{P} - \mathbf{\Pi})]^{-1}$ *exists and equals* $\mathbf{I} + \sum_{n\geqslant 1} (\mathbf{P}^n - \mathbf{\Pi})$.

Proof. We show that $(\mathbf{P} - \mathbf{\Pi})^n = \mathbf{P}^n - \mathbf{\Pi}$ and the theorem will then follow from Theorem 1.7.

Indeed, by equations (4.3) we have

$$(\mathbf{P} - \mathbf{\Pi})^n = \sum_{l=0}^{n} \binom{n}{l} (-1)^{n-l}\, \mathbf{P}^l\, \mathbf{\Pi}^{n-l}$$

$$= \mathbf{P}^n + \sum_{l=0}^{n-1} \binom{n}{l} (-1)^{n-l}\, \mathbf{\Pi} = \mathbf{P}^n - \mathbf{\Pi}.$$

The matrix $\mathbf{Z} = [\mathbf{I} - (\mathbf{P} - \mathbf{\Pi})]^{-1}$ is called the *fundamental matrix* of the regular Markov chain with transition matrix $\mathbf{P}$.

Proposition 4.6. *Let* $\mathbf{P}$ *be a regular stochastic matrix. We have*

1) $\mathbf{PZ} = \mathbf{ZP}$; 3) $\pi'\mathbf{Z} = \pi'$;

2) $\mathbf{Ze} = \mathbf{e}$; 4) $\mathbf{I} - \mathbf{Z} = \mathbf{\Pi} - \mathbf{PZ}$.

Proof. The first three points are easily justified and the proof is left to the reader.

To prove 4) we multiply $\mathbf{Z} = \mathbf{I} + \sum_{n\geqslant 1} (\mathbf{P}^n - \mathbf{\Pi})$ by $\mathbf{I} - \mathbf{P}$. We get

$$(\mathbf{I} - \mathbf{P})\, \mathbf{Z} = (\mathbf{I} - \mathbf{P}) + (\mathbf{P} - \mathbf{\Pi}) = \mathbf{I} - \mathbf{\Pi},$$

whence 4) follows at once.

Noticc that by virtue of 3) the sum of the entries of any row of the matrix $\mathbf{Z}$ equals 1. But $\mathbf{Z}$ is *not* a stochastic matrix as some of its entries may be negative.

Remark. On account of the results from 4.1.3. concerning the extension of some of the properties of regular Markov chains to Markov chains with mixing transition matrix, it is easily seen that Theorem 4.5 and Proposition 4.6 also hold for mixing matrices.

4.3.2. We can compute in terms of the fundamental matrix the moments of the first passage time to various states of a regular Markov main. It is recalled that τ_j, the first passage time to state j, is defined as

$$\tau_j = \min(n \geqslant 1 : X(n) = j).$$

In particular, we shall find that the mean value $\mathsf{E}_i(\tau_i)$, which is called the *mean recurrence time* of state i, can be very simply expressed in terms of the stationary distribution $\boldsymbol{\pi}$.

Theorem 4.7 ([195], p. 79). *In a regular Markov chain the mean values* $m(i, j) = \mathsf{E}_i(\tau_j)$ *are finite whatever the states i and j. In particular,* $m(i, i) = 1/\pi(i)$, $i \in S$. *The matrix* $\mathbf{M} = (m(i, j))_{i,j \in S}$ *is given by* $\mathbf{M} = (\mathbf{I} - \mathbf{Z} + \mathbf{E}\mathbf{Z}_{dg})\, \boldsymbol{\Pi}_{dg}^{-1}$.

Proof. Assuming the existence of the mean values $m(i, j)$ we first prove that $\mathbf{M}$ satisfies the equation

$$\mathbf{M} = \mathbf{P}(\mathbf{M} - \mathbf{M}_{dg}) + \mathbf{E}. \tag{4.9}$$

We have

$$m(i, j) = \mathsf{E}_i(\tau_j) = 1 \cdot \mathsf{P}_i(\tau_j = 1) + \sum_{n \geqslant 2} n \mathsf{P}_i(\tau_j = n) = \mathsf{P}_i(\tau_j = 1)$$

$$+ \sum_{n \geqslant 2} n \sum_{k \in S} \mathsf{P}_i(X(n) = j,\ X(l) \neq j,\ 1 \leqslant l < n \mid X(1) = k)\, \mathsf{P}_i(X(1) = k).$$

But

$$\mathsf{P}_i(X(n) = j,\ X(l) \neq j,\ 1 \leqslant l < n \mid X(1) = k)$$

$$= \begin{cases} (1 - \delta(k, j))\, \mathsf{P}_k(X(n-1) = j,\ X(l) \neq j,\ 1 \leqslant l < n-1), & \text{if } n > 2, \\ (1 - \delta(k, j))\, \mathsf{P}_k(X(1) = j), & \text{if } n = 2, \end{cases}$$

$$= (1 - \delta(k, j))\, \mathsf{P}_k(\tau_j = n - 1).$$

Therefore

$$m(i, j) = \mathsf{P}_i(\tau_j = 1) + \sum_{n \geqslant 2} (n - 1 + 1) \sum_{k \neq j} p(i, k)\, \mathsf{P}_k(\tau_j = n - 1)$$

$$= \mathsf{P}_i(\tau_j = 1) + \sum_{n \geqslant 2} \sum_{k \neq j} p(i, k)\, \mathsf{P}_k(\tau_j = n - 1) + \sum_{k \neq j} p(i, k)\, m(k, j)$$

$$= \sum_{k \neq j} p(i, k)\, m(k, j) + 1,$$

since

$$\mathsf{P}_i(\tau_j = 1) + \sum_{n\geqslant 2}\sum_{k\neq j} p(i, k)\,\mathsf{P}_k(\tau_j = n-1) = \mathsf{P}_i(\tau_i = 1)$$

$$+\sum_{n\geqslant 2}\sum_{k\neq j} \mathsf{P}_i(X(1) = k)\,\mathsf{P}_i(X(n) = j,\ X(l) \neq j,\ 1 \leqslant l < n \mid X(1)=k)$$

$$= \mathsf{P}_i(\tau_j = 1) + \sum_{n\geqslant 2} \mathsf{P}_i(\tau_j = n) = f(i, j) = 1$$

by Theorem 2.9. This proves (4.9).

Let us now show that the mean values $m(i, j)$ are finite. Assume that $i \neq j$ and consider a new Markov chain with transition probabilities

$$\tilde{p}(i', j') = \begin{cases} p(i', j'), & \text{if } i' \neq j,\ j' \in S, \\ \delta(i', j'), & \text{if } i' = j,\ j' \in S. \end{cases}$$

Obviously, the new chain is an absorbing one with a single absorbing state j and transient states $i' \in S$, $i' \neq j$. The mean first passage time $m(i, j)$ in the original chain is nothing but the mean time to absorption in the new chain given it starts in i. As by Theorem 3.2 the latter is finite we conclude that $m(i, j) < \infty$. The fact that $m(i, i)$ is also finite follows from the equality $m(i, i) = \sum_{k\neq i} p(i, k)\, m(i, k) + 1$ (see above).

To prove that $m(i, i) = 1/\pi(i)$ we multiply equation (4.9) by $\boldsymbol{\pi} = (\pi(i))_{i\in S}$. We get

$$\boldsymbol{\pi}'\mathbf{M} = \boldsymbol{\pi}'\mathbf{P}(\mathbf{M} - \mathbf{M}_{dg}) + \boldsymbol{\pi}'\mathbf{E} = \boldsymbol{\pi}'(\mathbf{M} - \mathbf{M}_{dg}) + \boldsymbol{\pi}'\mathbf{E},$$

whence $\boldsymbol{\pi}'\mathbf{M}_{dg} = \boldsymbol{\pi}'\mathbf{E} = \mathbf{e}'$, that is $\pi(i)\, m(i, i) = 1$.

It remains to compute the matrix $\mathbf{M}$. We first show that equation (4.9) in which the unknown is $\mathbf{M}$ has a unique solution. Assume for a contradiction that there exist two solutions $\mathbf{M}^{(1)}$ and $\mathbf{M}^{(2)}$ of (4.9). Then, as we have just seen, $m^{(1)}(i, i) = m^{(2)}(i, i) = 1/\pi(i)$, that is $\mathbf{M}^{(1)}_{dg} = \mathbf{M}^{(2)}_{dg}$. It follows that

$$\mathbf{M}^{(1)} - \mathbf{M}^{(2)} = \mathbf{P}(\mathbf{M}^{(1)} - \mathbf{M}^{(2)}).$$

This means that any column of the matrix $\mathbf{M}^{(1)} - \mathbf{M}^{(2)}$ is a vector, say $\boldsymbol{\rho}$, satisfying $\mathbf{P}\boldsymbol{\rho} = \boldsymbol{\rho}$, hence on account of a previous remark (see p. 140) all of its components are identical. Since the diagonal entries of the matrix $\mathbf{M}^{(1)} - \mathbf{M}^{(2)}$ are null, all the columns of this matrix should equal $\mathbf{0}$, therefore $\mathbf{M}^{(1)} = \mathbf{M}^{(2)}$.

On account of the uniqueness of the solution of equation (4.9) all that remains is to prove that $\mathbf{M} = (\mathbf{I} - \mathbf{Z} + \mathbf{E}\mathbf{Z}_{dg})\boldsymbol{\Pi}_{dg}^{-1}$ satisfies the equation. We have

$$\mathbf{M} - \boldsymbol{\Pi}_{dg}^{-1} = (-\mathbf{Z} + \mathbf{E}\mathbf{Z}_{dg})\,\boldsymbol{\Pi}_{dg}^{-1},$$

$$\mathbf{P}(\mathbf{M} - \boldsymbol{\Pi}_{dg}^{-1}) = (-\,\mathbf{P}\mathbf{Z} + \mathbf{E}\mathbf{Z}_{dg})\,\boldsymbol{\Pi}_{dg}^{-1},$$

whence, taking into account that $\mathbf{EZ}_{\mathrm{dg}}\,\mathbf{\Pi}_{\mathrm{dg}}^{-1} = \mathbf{M} - \mathbf{\Pi}_{\mathrm{dg}}^{-1} + \mathbf{Z}\mathbf{\Pi}_{\mathrm{dg}}^{-1}$ and that, by Proposition 4.6, $\mathbf{I} - \mathbf{Z} = \mathbf{\Pi} - \mathbf{PZ}$

$$\mathbf{P}(\mathbf{M} - \mathbf{\Pi}_{\mathrm{dg}}^{-1}) = \mathbf{M} - \mathbf{\Pi}\mathbf{\Pi}_{\mathrm{dg}}^{-1} = \mathbf{M} - \mathbf{E}.$$

Since $\mathbf{M}_{\mathrm{dg}} = \mathbf{\Pi}_{\mathrm{dg}}^{-1}$ Theorem 4.7 is proved.

Corollary. $\mathsf{E}_\pi(\tau_j) = z(j, j)/\pi(j), \quad j \in S.$

Proof. We have

$$\mathsf{E}_\pi(\tau_j) = \sum_{i \in S} \pi(i)\,\mathsf{E}_i(\tau_j) = \sum_{i \in S} \pi(i)\, m(i, j),$$

whence $(\mathsf{E}_\pi(\tau_j))_{j \in S} = \boldsymbol{\pi}'\mathbf{M}$. But

$$\boldsymbol{\pi}'\mathbf{M} = \boldsymbol{\pi}'(\mathbf{I} - \mathbf{Z} + \mathbf{EZ}_{\mathrm{dg}})\,\mathbf{\Pi}_{\mathrm{dg}}^{-1} = (\boldsymbol{\pi}' - \boldsymbol{\pi}' + \mathbf{e}'\mathbf{Z}_{\mathrm{dg}})\,\mathbf{\Pi}_{\mathrm{dg}}^{-1}$$

$$= \mathbf{e}'\mathbf{Z}_{\mathrm{dg}}\,\mathbf{\Pi}_{\mathrm{dg}}^{-1} = (z(j, j)/\pi(j))_{j \in S}.$$

Similar arguments ([195], p. 83) show that the mean values $w(i, j) = \mathsf{E}_i(\tau_j^2)$ exist whatever the states i and j. In particular, $w(i, i) = 2z(i, i)/\pi^2(i) - 1/\pi(i)$. The matrix $\mathbf{W} = (w(i, j))_{i, j \in S}$ is given by

$$\mathbf{W} = \mathbf{M}(2\mathbf{Z}_{\mathrm{dg}}\,\mathbf{\Pi}_{\mathrm{dg}}^{-1} - \mathbf{I}) + 2(\mathbf{ZM} - \mathbf{E}(\mathbf{ZM})_{\mathrm{dg}}).$$

More generally, the moments $\mathsf{E}_i(\tau_j^r)$ exist for any $r \geqslant 1$. This is an immediate consequence of Theorem 3.2 applied in conjunction with the trick used to prove the existence of the mean values $\mathsf{E}_i(\tau_j)$, $i, j \in S$.

4.3.3. In his researches concerning irreversibility in thermodynamics the Polish physicist Marjan Smoluchowski used another definition of the recurrence time of a state i in an evolutionary physical system (see [182]). In the context of Markov chains, the *Smoluchowskian recurrence time* θ_i of a nonabsorbing state i is the number of steps between an exit from state i and the next return to state i. With this convention, consecutive occurrences of a state do not mean recurrence. We have

$$\mathsf{E}(\theta_i) = \frac{\sum_{j \neq i} p(i, j)\,\mathsf{E}_j(\tau_i)}{\sum_{j \neq i} p(i, j)} = \frac{(m(i, i) - 1)}{(1 - p(i, i))} = \frac{1 - \pi(i)}{\pi(i)\,(1 - p(i, i))}.$$

The advantage of Smoluchowski's definition is manifest in situations where the random variables $X(n)$ of the chain are derived from a (continuous time) Markov process $(Y(t))_{t \geqslant 0}$ (see Chapter 8). In many applications $X(n) = Y(na)$, $a > 0$, the system being thus observed at times $0, a, 2a, \ldots$. In this case $\mathsf{E}(\theta_i)$ equals $(1 - \pi(i))a/\pi(i)(1 - p(a, i, i))$. The limit of the ratio $a/(1 - p(a, i, i))$ as $a \to 0$ exists and is different from 0. Therefore it is often possible to define the limiting mean recurrence time as observations are taken continuously. With the usual definition of mean recurrence time, the limiting mean recurrence time is always 0 $(=\lim_{a\to 0} a/\pi(i))$. The reason is quite clear: for small a the probability of a state being followed by itself is very close to 1 so that too much weight is attached to what, in fact, is only a false recurrence.

4.3.4. We now propose to compute the mean values $m(i, j)$ for the random walk with elastic barriers for which $p_0 = r_l = p$, $r_0 = q_l = q$ (see Example 1 from 2.2).

Notice first that by virtue of 4.2.2 the stationary distribution $\pi = (\pi(i))_{0\leqslant i\leqslant l}$, is given by

$$\pi(i) = \begin{cases} a^i(a-1)/(a^{l+1}-1), \text{ if } a = p/q \neq 1, \\ 1/(l+1), \text{ if } p = q = 1/2, \quad 0 \leqslant i \leqslant l. \end{cases}$$

Next, let us find the fundamental matrix $\mathbf{Z}$. This can be done in a direct manner by inverting the matrix $\mathbf{I} - \mathbf{P} + \mathbf{\Pi}$ but we prefer to use a trick whose applicability is more general (cf. [195], p. 121). Let us remark that if state 0 is made absorbing the resulting Markov chain is nothing but the random walk with absorbing barrier at 0 and elastic barrier at l. The fundamental matrix $\mathbf{N} = (\mathbf{I} - \mathbf{T})^{-1}$ of the latter was computed in 3.1.2. Let us denote by $\mathbf{N}^*$ the matrix constructed from $\mathbf{N}$ by inserting a first row and first column of zeros. We shall prove that $\mathbf{Z} = \mathbf{\Pi} + (\mathbf{I} - \mathbf{\Pi})\mathbf{N}^* (\mathbf{I} - \mathbf{\Pi})$. Indeed, setting $\widetilde{\boldsymbol{\pi}}' = (\pi(1), \ldots, \pi(l))$, $\widetilde{\mathbf{p}}' = (p, 0, \ldots, 0)$ (l components) we can write

$$\mathbf{P} = \begin{pmatrix} q & \widetilde{\mathbf{p}}' \\ \mathbf{e} - \mathbf{Te} & \mathbf{T} \end{pmatrix}, \quad \mathbf{\Pi} = \begin{pmatrix} \pi(0) & \widetilde{\boldsymbol{\pi}}' \\ \pi(0)\mathbf{e} & \mathbf{e}\widetilde{\boldsymbol{\pi}}' \end{pmatrix}, \quad \mathbf{N}^* = \begin{pmatrix} 0 & 0 \\ 0 & \mathbf{N} \end{pmatrix}.$$

whence

$$\mathbf{I} - \mathbf{P} = \begin{pmatrix} 1-q & -\widetilde{\mathbf{p}}' \\ \mathbf{Te} - \mathbf{e} & \mathbf{I} - \mathbf{T} \end{pmatrix}, \quad \mathbf{I} - \mathbf{\Pi} = \begin{pmatrix} 1-\pi(0) & -\widetilde{\boldsymbol{\pi}}' \\ -\pi(0)\,\mathbf{e} & \mathbf{I} - \mathbf{e}\widetilde{\boldsymbol{\pi}}' \end{pmatrix},$$

$$(\mathbf{I} - \mathbf{P})\,\mathbf{N}^* = \begin{pmatrix} 0 & -\widetilde{\mathbf{p}}'\mathbf{N} \\ 0 & \mathbf{I} \end{pmatrix},$$

$$(\mathbf{I} - \mathbf{P})\,\mathbf{N}^*(\mathbf{I} - \mathbf{\Pi}) = \begin{pmatrix} \pi(0)\,\widetilde{\mathbf{p}}'\mathbf{Ne} & -\widetilde{\mathbf{p}}'\mathbf{N}(\mathbf{I} - \mathbf{e}\widetilde{\boldsymbol{\pi}}') \\ -\pi(0)\,\mathbf{e} & \mathbf{I} - \mathbf{e}\widetilde{\boldsymbol{\pi}}' \end{pmatrix}.$$

Now, the equation $\boldsymbol{\pi}' = \boldsymbol{\pi}'\mathbf{P}$ implies the equalities

$$\pi(0)\,q + \widetilde{\boldsymbol{\pi}}'(\mathbf{e} - \mathbf{Te}) = \pi(0), \quad \pi(0)\,\widetilde{\mathbf{p}} + \widetilde{\boldsymbol{\pi}}'\mathbf{T} = \widetilde{\boldsymbol{\pi}}',$$

whence

$$\frac{\widetilde{\boldsymbol{\pi}}'}{\pi(0)} = \widetilde{\mathbf{p}}'(\mathbf{I} - \mathbf{T})^{-1} = \widetilde{\mathbf{p}}'\mathbf{N}. \tag{4.10}$$

On account of (4.10) we get

$$(\mathbf{I} - \mathbf{P})\,\mathbf{N}^*(\mathbf{I} - \mathbf{\Pi}) = \begin{pmatrix} 1 - \pi(0) & -\tilde{\pi}' \\ -\pi(0)\,\mathbf{e} & \mathbf{I} - \mathbf{e}\tilde{\pi}' \end{pmatrix} = \mathbf{I} - \mathbf{\Pi}.$$

By making use of equations (4.3) we can write $(\mathbf{I} - \mathbf{P} + \mathbf{\Pi})\,[\mathbf{\Pi} + (\mathbf{I} - \mathbf{\Pi})\,\mathbf{N}^*(\mathbf{I} - \mathbf{\Pi})] = \mathbf{\Pi} + (\mathbf{I} - \mathbf{P})\,\mathbf{N}^*(\mathbf{I} - \mathbf{\Pi}) = \mathbf{\Pi} + \mathbf{I} - \mathbf{\Pi} = \mathbf{I}$. Hence $\mathbf{Z} = (\mathbf{I} - \mathbf{P} + \mathbf{\Pi})^{-1} = \mathbf{\Pi} + (\mathbf{I} - \mathbf{\Pi})\,\mathbf{N}^*(\mathbf{I} - \mathbf{\Pi})$.

The problem we raised is now easily answered. It follows from the above equation and Theorem 4.7 that

$$m(i, j) = \frac{1}{\pi(j)}\,(n(j, j) - n(i, j) + \delta(i, j)) + \sum_{k=1}^{l} (n(i, k) - n(j, k)),$$

for $1 \leqslant i,\ j \leqslant l$, and

$$m(i, 0) = \sum_{k=1}^{l} n(i, k), \quad m(0, i) = \frac{1}{\pi(i)}\,n(i, i) - \sum_{k=1}^{l} n(i, k),$$

for $1 \leqslant i \leqslant l$ and, obviously, $m(0, 0) = 1/\pi(0)$. Substituting for the values $n(i, j)$ computed in 3.1.2 we get

$$m(i, j) = \begin{cases} a^{-i}(a^{l+1} - 1)/(a - 1), & \text{if } i = j, \\[2mm] \dfrac{1}{2p - 1}\left(\dfrac{a^{n-j+1} - a^{n-i+1}}{a - 1} - i + j\right), & \text{if } i > j,\ 0 \leqslant i,\ j \leqslant l, \\[2mm] \dfrac{1}{2p - 1}\left(j - i - \dfrac{a^j - a^i}{a^{i+j}(a - 1)}\right), & \text{if } i < j, \end{cases}$$

for $p \neq 1/2$ and

$$m(i, j) = \begin{cases} l + 1, & \text{if } i = j, \\ j(2l - i + 1)\,i - (2l - j + 1)\,j, & \text{if } i > j, \quad 0 \leqslant i,\ j \leqslant l, \\ (j + 1) - i(i + 1) & \text{if } i < j, \end{cases}$$

for $p = 1/2$.

We close by noticing that equation (4.10) has a simple and interesting probabilistic meaning. It says that the mean value of the number of visits to state $j \neq 0$ between two occurrences of state 0 equals $\pi(j)/\pi(0)$ (see 3.2.10). At the same time we want to stress that the trick we used is generally applicable to any regular Markov chain and any state of it. This allows us, for instance, to assert that for a regular Markov chain the mean value of the number of visits to state $j \neq i$ between two occurrences of state i equals $\pi(j)/\pi(i)$.

4.3.5. Let f be a real valued function defined on the state space S of a Markov chain with mixing transition matrix. We shall prove that $\lim_{n\to\infty} \frac{1}{n} \operatorname{var}_{\mathbf{p}} \left(\sum_{l=0}^{n-1} f(X(l)) \right)$ exists, does not depend on $\mathbf{p}$ and can be expressed in terms of the fundamental matrix $\mathbf{Z}$.

Theorem 4.8. *For a Markov chain with mixing transition matrix we have*

$$\lim_{n\to\infty} \frac{1}{n} \operatorname{var}_{\mathbf{p}} \left(\sum_{l=0}^{n-1} f(X(l)) \right) = \sum_{i,j\in S} f(i)\, c(i, j)\, f(j),$$

whatever the initial distribution $\mathbf{p}$ *where*

$$c(i, j) = \pi(i)\, z(i, j) + \pi(j)\, z(j, i) - \pi(i)\, \delta(i, j) - \pi(i)\, \pi(j),\ i, j \in S.$$

Proof. Let us first consider the case $\mathbf{p} = \boldsymbol{\pi}$. Set

$$\sigma_n^2(f) = \mathsf{E}_\pi(f^2(X(0))) - \mathsf{E}_\pi^2(f(X(0)))$$

$$+ 2\sum_{l=1}^{n-1} \frac{n-l}{n} [\mathsf{E}_\pi(f(X(0))f(X(l))) - \mathsf{E}_\pi^2 f(X)0)))], \qquad n > 1.$$

We have

$$\sigma_n^2(f) = \sum_{i\in S} \pi(i) f^2(i) - \sum_{i,j\in S} \pi(i)\, \pi(j)\, f(i)\, f(j)$$

$$+ 2\sum_{l=1}^{n-1} \frac{n-l}{n} \left(\sum_{i,j\in S} \pi(i)\, p(l, i, j) f(i) f(j) - \sum_{i,j\in S} \pi(i)\, \pi(j)\, f(i)\, f(j) \right)$$

$$= \sum_{i,j\in S} \left[\pi(i) \left(\delta(i, j) + \sum_{l=1}^{n-1} \frac{n-l}{n} (p(l, i, j) - \pi(j)) \right) \right.$$

$$\left. + \pi(j) \left(\delta(i, j) \sum_{l=1}^{n-1} \frac{n-l}{n} (p(l, j, i) - \pi(i)) \right) - \pi(i)\, \delta(i, j) - \pi(i)\, \pi(j) \right] f(i) f(j).$$

The convergence of the series $z(i, j) = \delta(i, j) + \sum_{l\geq 1} (p(l, i, j) - \pi(j))$ implies that

$$z(i, j) = \delta(i, j) + \lim_{n\to\infty} \frac{\sum_{m=2}^{n} \sum_{l=1}^{m-1} (p(l, i, j) - \pi(j))}{n}$$

$$= \delta(i, j) + \lim_{n\to\infty} \sum_{l=1}^{n-1} \frac{n-l}{n} (p(l, i, j) - \pi(j))$$

(convergence in the Cesàro sense).

It then follows that

$$\lim_{n\to\infty} \sigma_n^2(f) = \sigma^2(f) = \sum_{i,j\in S} f(i)\, c(i,j)\, f(j). \tag{4.11}$$

Let us remark that by Theorem 4.4 the probabilities $\mathsf{P}_\pi(X(m) = i)$ and $\mathsf{P}_\pi(X(m) = i,\ X(m+l) = j)$ do not depend on m, thus the mean values $\mathsf{E}_\pi(f(X(m)))$, $\mathsf{E}_\pi(f^2(X(m)))$ and $\mathsf{E}_\pi(f(X(m))\, f(X(m+l)))$ also do not depend on m [therefore they equal $\mathsf{E}_\pi(f(X(0)))$, $\mathsf{E}_\pi(f^2(X(0)))$ and $\mathsf{E}_\pi(f(X(0))\, f(X(l)))$, respectively]. Consequently,

$$\operatorname{var}_\pi\left(\sum_{l=0}^{n-1} f(X(l))\right) = \mathsf{E}_\pi\left(\sum_{l=0}^{n-1} f(X(l))\right)^2 - \mathsf{E}_\pi^2\left(\sum_{l=0}^{n-1} f(X(l))\right)$$

$$= n\mathsf{E}_\pi(f^2(X(0))) + 2\sum_{l=1}^{n-1}(n-l)\mathsf{E}_\pi(f(X(0))f(X(l))) - (n\mathsf{E}_\pi(f(X(0)))^2$$

$$= n\Bigg[\mathsf{E}_\pi(f^2(X(0))) - \mathsf{E}_\pi^2(f(X(0)))$$

$$+ 2\sum_{l=1}^{n-1}\frac{n-l}{n}[\mathsf{E}_\pi(f(X(0))f(X(l))) - \mathsf{E}_\pi^2(f(X(0)))]\Bigg] = n\sigma_n^2(f),$$

so that, by equation (4.11), the theorem is proved in the case where $\mathbf{p} = \boldsymbol{\pi}$. The case $\mathbf{p} \neq \boldsymbol{\pi}$ obtains from the previous one by noticing that whatever the state $i \in S$ we have

$$\lim_{n\to\infty} \frac{\sum_{l=1}^{n-1}[\mathsf{E}_i(f^2(X(l))) - \mathsf{E}_\pi(f^2(X(0)))]}{n} = 0,$$

$$\lim_{n\to\infty} \frac{\sum_{l=1}^{n-1}[\mathsf{E}_i^2(f(X(l))) - \mathsf{E}_\pi^2(f(X(0)))]}{n} = 0, \tag{4.12}$$

$$\lim_{n\to\infty}\left(\frac{\sum_{0\le r<s\le n-1}[\mathsf{E}_i(f(X(r))f(X(s))) - \mathsf{E}_i(f(X(r)))\mathsf{E}_i(f(X(s)))]}{n}\right.$$

$$\left. - \frac{\sum_{l=1}^{n-1}\frac{n-l}{n}[\mathsf{E}_\pi f(X(0)\, f(X(l))) - \mathsf{E}_\pi^2 f(f(X(0)))]}{n}\right) = 0.$$

For example

$$\frac{\sum_{l=0}^{n-1}[\mathsf{E}_i f^2(X(l))) - \mathsf{E}_\pi(f^2(X(0)))]}{n} = \sum_{j\in S}\frac{\sum_{l=0}^{n-1}(p(l,i,j) - \pi(j))}{n} f^2(j),$$

and, as $\lim_{l\to\infty} p(l, i, j) = \pi(j)$, convergence in the Cesàro sense holds, i.e.,

$$\lim_{n\to\infty} \frac{\sum_{l=0}^{n-1} (p(l, i, j) - \pi(j))}{n} = 0.$$

Hence the first equation (4.12) is verified. On the other hand, denoting A_n, B_n, and C_n the expressions under 'lim' in (4.12), it is easily seen that

$$\frac{\text{var}_i\left(\sum_{l=0}^{n-1} f(X(l))\right) - n\sigma_n^2(f)}{n} = A_n - B_n + 2C_n.$$

This shows that Theorem 4.8 is true for the initial distribution concentrated in an arbitrary state $i \in S$. On account of the fact that $\text{var}_p = \sum_{i\in S} p(i)\ \text{var}_i$, the theorem is proved for any initial distribution **p**.

The importance of Theorem 4.8 lies in the fact that the quantity $\sigma^2(f)$ is involved in two fundamental limit theorems (the central limit theorem and the law of the iterated logarithm) for Markov chains with mixing transition matrix*). More precisely, it can be proved that Theorems 1.5 and 1.6 apply to the random variables $X_n = f(X(n))$, $n \geqslant 1$, with

$$m = \mathsf{E}_\pi(f(X(0))) = \sum_{i\in S} \pi(i) f(i), \quad \sigma = \sigma(f).\ ^{**)}$$

For the probability P occurring there one may take P_p whatever the initial distribution **p** of the Markov chain. Theorem 1.4 (the law of large numbers) also applies but it does not involve $\sigma(f)$.

An important special case obtains if we consider a fixed recurrent state k and take

$$f(i) = \begin{cases} 1, & \text{if } i = k, \\ 0, & \text{if } i \neq k, \end{cases}$$

a choice yielding $m = \pi(k)$, $\sigma(f) = \sqrt{c(k, k)}$. In this case the sum $\sum_{l=0}^{n-1} f(X(l))$ is nothing but the number of occurrences of state k in the first n steps, a quantity we shall denote by $\nu_n(k)$. Then the law of large

*) These results are in fact special cases of a theory applicable to much more general dependent random variables. The interested reader should consult the monograph [167]. At the same time, in the case of a regular Markov chain the results can be obtained by reducing the problem to the case of independent random variables. In this context Theorem 2.11 plays an essential part (see [52], I. § 14).

**) Necessary and sufficient conditions for having $\sigma > 0$ (or, equivalently, $\sigma = 0$) were given by Mihoc [270] and Doeblin [81].

numbers says that the (relative) frequency $\nu_n(k)/n$ of occurrence of state k in the first n steps converges to $\pi(k)$ as $n \to \infty$ as stated in Theorem 1.4. The central limit theorem and the law of the iterated logarithm bring further precision. If $c(k, k) > 0$ then

$$\lim_{n\to\infty} \mathsf{P}_{\mathbf{p}} \left(\frac{\nu_n(k)/n - \pi(k)}{\sqrt{c(k, k)/n}} < x \right) = \frac{1}{\sqrt{2\pi}} \int_{-\infty}^{x} e^{-u^2/2}\, du$$

whatever the initial distribution $\mathbf{p}$ and the real number x and the limits superior and inferior of the ratio.

$$\frac{\nu_n(k)/n - \pi(k)}{\sqrt{2c(k, k)(\ln \ln n)/n}}$$

are equal to 1 and -1, respectively, as stated in Theorem 1.6.

4.4 CYCLIC MARKOV CHAINS

4.4.1. We remember that a Markov chain whose states constitute a unique periodic recurrent class was called *cyclic* (see 2.6.1).

When applied to the transition matrix of a cyclic Markov chain of period d Theorem 1.10 tells us that the d roots of order d of unity are eigenvalues of this matrix. The remaining eigenvalues are smaller that 1 in absolute value. Conversely, if an irreducible stochastic matrix has $d > 1$ eigenvalues of modulus 1, then this matrix is the transition matrix of a cyclic Markov chain of period d. The same theorem shows that, suitably defined, the left eigenvector corresponding to the eigenvalue $\lambda_1 = 1$ provides a stationary distribution for the cyclic chain under consideration, just as for regular Markov chains. In what follows we shall proceed to a direct investigation of cyclic Markov chains.

4.4.2. Many properties of regular Markov chains extend to cyclic ones. Let us consider a cyclic Markov chain of period d whose state space S has cyclic subclasses $\mathfrak{C}_0, \mathfrak{C}_1, \ldots, \mathfrak{C}_{d-1}$. As it was noticed in 2.4.4 then dth power of its transition matrix $\mathbf{P}$ assumes the form

$$\mathbf{P}^d = \begin{array}{c} \mathfrak{C}_0 \\ \mathfrak{C}_1 \\ \cdot \\ \cdot \\ \cdot \\ \mathfrak{C}_{d-1} \end{array} \overset{\begin{array}{cccc} \mathfrak{C}_0 & \mathfrak{C}_1 & \cdots & \mathfrak{C}_{d-1} \end{array}}{\begin{pmatrix} \times & & & \\ & \times & & \mathbf{0} \\ & & \ddots & \\ & \mathbf{0} & & \\ & & & \times \end{pmatrix}}.$$

Hence it may be considered as the transition matrix of a recurrent Markov chain with d aperiodic classes.

By Theorem 4.2 the matrix $\mathbf{P}^{nd}$ converges as $n \to \infty$ to a matrix $\mathbf{A}$ of the form

$$\mathbf{A} = \begin{array}{c} \\ \mathcal{C}_0 \\ \mathcal{C}_1 \\ \vdots \\ \mathcal{C}_{d-1} \end{array} \begin{array}{c} \begin{array}{cccc} \mathcal{C}_0 & \mathcal{C}_1 & \cdots & \mathcal{C}_{d-1} \end{array} \\ \begin{pmatrix} \mathbf{\Pi}_0 & & & \mathbf{0} \\ & \mathbf{\Pi}_1 & & \\ & & \ddots & \\ \mathbf{0} & & & \mathbf{\Pi}_{d-1} \end{pmatrix} \end{array},$$

where $\mathbf{\Pi}_r = \mathbf{e}\pi'_r$, $0 \leqslant r \leqslant d-1$, is a positive stable (i.e., with identical rows) stochastic matrix. Obviously, $\mathbf{P}^d\mathbf{A} = \mathbf{A}\mathbf{P}^d = \mathbf{A}$. Therefore, $\mathbf{P}^{nd+r}$, $0 \leqslant r \leqslant d-1$, converges as $n \to \infty$ to the matrix $\mathbf{P}^r\mathbf{A} = \mathbf{A}\mathbf{P}^r$. This means that the sequence $(\mathbf{P}^n)_{n\geqslant 0}$ has d convergent subsequences and this fact implies the convergence of the arithmetic mean $(\mathbf{I} + \mathbf{P} + \ldots + \mathbf{P}^{n-1})/n$ to the matrix $\mathbf{\Pi} = d^{-1}\sum_{r=0}^{d-1}\mathbf{P}^r\mathbf{A} = d^{-1}\sum_{r=0}^{d-1}\mathbf{A}\mathbf{P}^r$. It is easily verified that $\mathbf{\Pi}$ is a stable stochastic matrix, more precisely $\mathbf{\Pi} = \mathbf{e}\pi'$, where

$$\begin{array}{c} \qquad\quad \mathcal{C}_0 \quad \mathcal{C}_1 \ldots \mathcal{C}_{d-1} \\ \pi' = d^{-1}(\pi'_0, \pi'_0, \ldots, \pi'_{d-1}). \end{array}$$

Also $\mathbf{P\Pi} = \mathbf{\Pi P} = \mathbf{\Pi} = \mathbf{\Pi}^2$, hence $\pi'\mathbf{P} = \pi'$. The reader will be able to prove without any difficulty that the unique solution of the equation $\mathbf{p}'\mathbf{P} = \mathbf{p}'$ is $\mathbf{p} = \pi$ (see the proof for the case of Markov chains with mixing transition matrix). This fact allows us to conclude that Theorem 4.4 is true for cyclic chains, too. On the other hand, it is easily seen that the difference between the matrices $(\mathbf{I} + \mathbf{P} + \ldots + \mathbf{P}^{n-1})/n$ and $\mathbf{\Pi}$ is of the order of $1/n$. Consequently, the exponential rate of convergence present in regular Markov chains (Theorem 4.2) no longer holds in cyclic Markov chains.

Finally, let us remark that, as in the proof of Theorem 4.5 we can prove that $(\mathbf{P} - \mathbf{\Pi})^n = \mathbf{P}^n - \mathbf{\Pi}$, $n \geqslant 1$.

4.4.3. We are now able to introduce the fundamental matrix for a cyclic Markov chain. Let us consider the identity

$$\left(\mathbf{I} + \sum_{l=1}^{n}\frac{n-l}{n}(\mathbf{P}^l - \mathbf{\Pi})\right)(\mathbf{I} - (\mathbf{P} - \mathbf{\Pi}))$$

$$= \left(1 + \frac{1}{n}\right)\mathbf{I} - \frac{n+1}{n}\left(\frac{\mathbf{I} + \mathbf{P} + \ldots + \mathbf{P}^n}{n+1} - \mathbf{\Pi}\right) - \frac{\mathbf{\Pi}}{n}.$$

The right side converges to $\mathbf{I}$ as $n \to \infty$. It follows that the inverse $\mathbf{Z} = (\mathbf{I} - (\mathbf{P} - \mathbf{\Pi}))^{-1}$ exists and

$$\mathbf{Z} = \mathbf{I} + \lim_{n\to\infty}\sum_{l=1}^{n}\frac{n-l}{n}(\mathbf{P}^l - \mathbf{\Pi}).$$

The matrix $\mathbf{Z}$ is called the *fundamental matrix* of the cyclic Markov chain considered.

The reader will be able to prove without difficulty that this matrix verifies all the equations in the statement of Proposition 4.6.

4.4.4. The generalization of the basic properties of the fundamental matrix to cyclic Markov chains in such a manner that the period d does not appear explicitly makes it possible to assert that most of the results in 4.2 are valid for cyclic chains, too. In particular, this is true of Theorem 4.7 and the method developed in 4.3.4. In the proof of Theorem 4.8 only convergence in the Cesàro sense of the probabilities $p(n, i, j)$ to the limit values $\pi(j)$, $j \in S$, was needed. Therefore this theorem applies also to cyclic Markov chains. This fact eventually ensures the possibility of extending to cyclic Markov chains the fundamental limit theorems (see [52], I.14).

4.4.5. The results given in 4.4.2 allow us to answer a problem raised in 3.2.6, namely, which are the left eigenvectors corresponding to the eigenvalue $\lambda = 1$ in the case of an absorbing Markov chain with r recurrent classes $\mathcal{C}^1, \ldots, \mathcal{C}^r$. It is now obvious that if the stationary distributions corresponding to these classes are $\boldsymbol{\pi}^1, \ldots, \boldsymbol{\pi}^r$, then the vectors $\mathbf{u}_l = (u_l(i))_{i \in S}$, $1 \leqslant l \leqslant r$, where

$$u_l(i) = \begin{cases} \pi^l(i), & \text{if} \quad i \in \mathcal{C}^l, \\ 0, & \text{if} \quad i \notin \mathcal{C}^l, \end{cases}$$

constitute a basis of the eigenspace corresponding to the eigenvalue $\lambda = 1$.

4.5 REVERSED MARKOV CHAINS

4.5.1. We showed in 2.1.4 that the Markov property is temporally reversible, i.e., it still holds for the chain observed in reverse order. This new chain is called the *reversed* Markov chain. We are now able to go on with studying this matter.

First, it is easily seen that the transition probabilities of the reversed Markov chain, namely

$$\mathsf{P}_p(X(n-1)=j \mid X(n)=i) = \frac{\mathsf{P}_p(X(n-1) = j) \; \mathsf{P}_p(X(n) = i \mid X(x-1) = j)}{\mathsf{P}_p(X(n) = i)}$$

do not depend on n if and only if the probabilities $\mathsf{P}_p(X(n) = i)$, $i \in S$, do not depend on n. For a Markov chain having transient states this condition obviously cannot be satisfied. On the other hand, by Theorem

4.4 for an ergodic*) Markov chain the condition is satisfied if and only if $\mathbf{p} = \boldsymbol{\pi}$, the stationary distribution of the chain. In this case the transition probabilities $\hat{p}(i, j)$ of the reversed Markov chain assume the form

$$\hat{p}(i, j) = \frac{\pi(j)}{\pi(i)} p(j, i), \quad i, j \in S.$$

Clearly, $\sum_{j \in S} \hat{p}(i, j) = 1$, $i \in S$, i.e., $\hat{\mathbf{P}} = (\hat{p}(i, j))_{i,j \in S}$ is a stochastic matrix. We have $\hat{\mathbf{P}} = \mathbf{\Pi}_{dg}^{-1}\mathbf{P}'\mathbf{\Pi}_{dg}$. This equation is fundamental for proving formulae connecting various characteristics of the original Markov chain (with transition matrix $\mathbf{P}$) with the corresponding ones of the reversed Markov chain (with transition matrix $\hat{\mathbf{P}}$).

Theorem 4.9. *The original and the reversed Markov chains have the same stationary distribution. The following equalities hold*

$$\hat{\mathbf{Z}} = \mathbf{\Pi}_{dg}^{-1}\mathbf{Z}'\mathbf{\Pi}_{dg}, \quad \hat{\mathbf{C}} = \mathbf{C},^{**}$$

$$\hat{\mathbf{M}} - \mathbf{M} = \mathbf{Z}\mathbf{\Pi}_{dg}^{-1} - (\mathbf{Z}\mathbf{\Pi}_{dg}^{-1})'.$$

Proof. On account of the fact that $\boldsymbol{\pi}'\mathbf{P} = \boldsymbol{\pi}'$ we have

$$\boldsymbol{\pi}'\hat{\mathbf{P}} = \boldsymbol{\pi}'\mathbf{\Pi}_{dg}^{-1}\mathbf{P}'\mathbf{\Pi}_{dg} = \mathbf{e}'\mathbf{P}'\mathbf{\Pi}_{dg} = (\mathbf{P}\mathbf{e})'\mathbf{\Pi}_{dg} = \mathbf{e}'\mathbf{\Pi}_{dg} = \boldsymbol{\pi}'.$$

It follows that $\hat{\mathbf{\Pi}} = \mathbf{\Pi}$ and then

$$\begin{aligned}\hat{\mathbf{Z}} &= (\mathbf{I} - \hat{\mathbf{P}} + \hat{\mathbf{\Pi}})^{-1} = (\mathbf{I} - \mathbf{\Pi}_{dg}^{-1}\mathbf{P}'\mathbf{\Pi}_{dg} + \mathbf{\Pi}_{dg}^{-1}\mathbf{\Pi}'\mathbf{\Pi}_{dg})^{-1} \\ &= \mathbf{\Pi}_{dg}^{-1}(\mathbf{I} - \mathbf{P}' + \mathbf{\Pi}')^{-1}\,\mathbf{\Pi}_{dg} = \mathbf{\Pi}_{dg}^{-1}\mathbf{Z}'\mathbf{\Pi}_{dg}.\end{aligned}$$

The proof of the remaining two equalities is left to the reader.

4.5.2. An important special case is that where $\hat{\mathbf{P}} = \mathbf{P}$. An ergodic Markov chain with this property is called *reversible.* It follows from the equality $\mathbf{\Pi}_{dg}^{-1}\mathbf{P}'\mathbf{\Pi}_{dg} = \mathbf{P}$ that $\mathbf{P}'\mathbf{\Pi}_{dg} = \mathbf{\Pi}_{dg}\mathbf{P}$, hence $(\mathbf{\Pi}_{dg}\mathbf{P})' = \mathbf{\Pi}_{dg}\mathbf{P}$. Therefore *an ergodic Markov chain is reversible if and only if* $\mathbf{\Pi}_{dg}\mathbf{P}$ *is a symmetric matrix.* This amounts to the equations

$$\pi(i)\, p(i, j) = \pi(j)\, p(j, i), \quad i, j \in S.$$

It is easily verified that any ergodic two state Markov chain is reversible. Next, any ergodic Markov chain with symmetric transition

---

*) The case of a general recurrent Markov chain reduces to this case of a single recurrent class. See Theorem 5.3.

**) The entries of the matrix $\mathbf{C} = (c(i, j))_{i,j \in S}$ have been defined in the statement of Theorem 4.8.

matrix $[p(i, j) = p(j, i),\ i, j \in S]$ is also reversible since its stationary distribution is the uniform distribution over S (see 4.2.1 — a symmetric stochastic matrix is a special case of a doubly stochastic matrix).

It follows at once by using results from 4.2.2 that the random walk from Example 2 from 2.2 for which*) $p_i > 0$, $q_i > 0$, $0 \leqslant i \leqslant l$, is reversible. In particular, this means that the Markov chains associated with the Bernoulli and Ehrenfest models are reversible. On the other hand the Markov chain from Example 8 from 2.2 is not reversible (see 4.2.3).

4.6 THE EHRENFEST MODEL

4.6.1. The Ehrenfest model was introduced in Example 2 from 2.2. It leads to a cyclic Markov chain of period 2 with state space $0, 1, \ldots, 2l$**) and whose positive transition probabilities are

$$p(i,\ i-1) = \frac{i}{2l}, \qquad 1 \leqslant i \leqslant 2l,$$

$$p(i,\ i+1) = 1 - \frac{i}{2l}, \qquad 0 \leqslant i < 2l.$$

In 6.3.1 a more general Markov chain will be considered. It follows from the results there that the eigenvalues of the transition matrix of the Markov chain associated with the Ehrenfest model are $\lambda_r = 1 - r/l$, $0 \leqslant r \leqslant 2l$. In accordance with the general theory, among these values there appear $1 (= \lambda_0)$ and $-1 (= \lambda_{2l})$. The left (right) eigenvectors $\mathbf{u}_r = (u_r(i))_{0 \leqslant i \leqslant 2l}$ $(\mathbf{v}_r = (v_r(i))_{0 \leqslant i \leqslant 2l})$ corresponding to the eigenvalue λ_r are given by

$$u_r(i) = 2^{r-2l} \sum_{h=\max(0, r+i-2l)}^{\min(r, i)} (-1)^{r-h} \binom{r}{h} \binom{2l-r}{i-h},$$

$$v_r(i) = 2^{-r} \sum_{h=\max(0, r+i-2l)}^{\min(r, i)} (-1)^{r-h} \binom{i}{h} \binom{2l-i}{r-h}, \qquad 0 \leqslant i, r \leqslant 2l.$$

It is easily seen that $v_r(i) = 2^{2l-i-r} (-1)^{r-i} u_i(r)$, $0 \leqslant i$, $r \leqslant 2l$.

In particular, $\mathbf{u}_0$ is the unique stationary distribution of the chain. Its components $u_0(i) = \binom{2l}{i} 2^{-2l}$, $0 \leqslant i \leqslant 2l$, are precisely the terms of the binomial expansion $(1/2 + 1/2)^{2l}$. In other words, the stationary

---

*) It was assumed in 4.2.2 that at least one of the numbers r_i, $0 \leqslant i \leqslant l$, was non-null. This condition ensures aperiodicity. Now we know the results there are valid for cyclic chains, too, therefore for the case where all the r_i are null.

**) The reader's attention should be drawn the fact that the even number $2l$ has no particular significance in the mathematical treatment. All the results are true for an arbitrary natural number, even or odd.

distribution is the binominal distribution with parameters $2l$ and $1/2$. Therefore, whatever the initial number of balls in the first urn, after a long time the probability of finding a certain number of balls in it is approximately the same as if the $2l$ balls had been distributed at random between the two urns (thus each ball having probability $1/2$ of being in the first urn).

The spectral representation (1.12) allows us to write the nth power of the transition matrix of the Markov chain associated with the Ehrenfest model as

$$\mathbf{P}^n = \sum_{r=0}^{2l} \lambda_r^n \mathbf{v}_r \mathbf{u}_r', \tag{4.13}$$

whence

$$\mathbf{\Pi} = \lim_{n\to\infty} \frac{\mathbf{I} + \mathbf{P} + \ldots + \mathbf{P}^{n-1}}{n} = \mathbf{v}_0 \mathbf{u}_0'.$$

Then

$$\mathbf{Z} = \mathbf{I} + \lim_{n\to\infty} \sum_{l=1}^{n} \frac{n-l}{n} (\mathbf{P}^l - \mathbf{\Pi}) = \mathbf{I} + \sum_{r=1}^{2l} \frac{\lambda_r}{1-\lambda_r} \mathbf{v}_r \mathbf{u}_r'.$$

Now we can compute the mean value $m(i, j)$ of the first passage time from state i to state j, $0 \leqslant i, j \leqslant 2l$ (Theorem 4.7). The reader will check without any difficulty that

$$m(i, j) = \left(1 + \sum_{r=1}^{2l} \frac{\lambda_r}{1-\lambda_r} (v_r(j) - v_r(i))\, u_r(j)\right) \frac{1}{u_0(j)}, \quad 0 \leqslant i, j \leqslant 2l.$$

In particular,

$$m(i, i) = 2^{2l} \Big/ \binom{2l}{i}, \quad 0 \leqslant i \leqslant 2l.$$

Another way of obtaining the values $m(i, j)$ will be described in 5.4.4.

4.6.2. The Ehrenfest model is a simple and convenient model of the exchange of heat between two isolated bodies of unequal temperatures. (The exchange of heat is conceived not as an orderly process, as in classical thermodynamics, but as a random process, as in the kinetic theory of matter.) An argument supporting the above assertion is the fact that it allows us to derive Newton's law of cooling.

The temperatures of the two bodies are measured by the numbers of balls in the two urns. If we choose as origin the value l, then the temperature of the first body at time n is given by the mean value $\mathsf{E}_i(X(n) - l) = \mathsf{E}_i(X(n)) - l$, assuming the initial number of balls in the first urn was i. We have

$$\mathsf{E}_i(X(n)) = \sum_{j=0}^{2l} j p(n, i, j).$$

We might compute this mean value using equation (4.13), but a simpler and more elegant method is to use the generating function

$$g_i(n, z) = \sum_{j=0}^{2l} p(n, i, j)\, z^j$$

of $X(n)$. We have successively

$$g_i(n+1, z) = \sum_{j=0}^{2l} p(n+1, i, j)\, z^j = p(n+1, i, 0)$$

$$+ \sum_{j=1}^{2l-1} \left(p(n, i, j-1) \left(1 - \frac{j-1}{2l}\right) + p(n, i, j+1) \frac{j+1}{2l} \right) z^j + p(n+1, i, 2l)\, z^{2l}$$

$$= p(n, i, 1) \frac{1}{2l} + z\,(g_i(n, z) - p(n, i, 2l-1)\, z^{2l-1} - p(n, i, 2l)\, z^{2l})$$

$$- \frac{z^2}{2l} (g_i'(n, z) - (2l-1)\, p(n, i, 2l-1)\, z^{2l-2} - 2l p(n, i, 2l)\, z^{2l-1})$$

$$+ \frac{1}{2l} (g_i'(n, z) - p(n, i, 1)) + p(n, i, 2l-1) \left(1 - \frac{2l-1}{2l}\right) z^{2l}$$

$$= z g_i(n, z) - \frac{z^2 - 1}{2l} g_i'(n, z),$$

where $g_i'(n, z) = d/dz(g_i(n, z))$. On account of the fact that $g_i'(n, 1) = \mathsf{E}_i(X(n))$ (see 1.4.6), differentiating the above recurrence relation yields

$$\mathsf{E}_i(X(n+1)) = 1 + \left(1 - \frac{1}{l}\right) \mathsf{E}_i(X(n)),$$

whence

$$\mathsf{E}_i(X(n+1) - l) = \left(1 - \frac{1}{l}\right) (\mathsf{E}_i(X(n) - l)).$$

Noticing that $\mathsf{E}_i(X(0) - l) = i - l$ we finally get

$$\mathsf{E}_i(X(n) - l) = (i - l) \left(1 - \frac{1}{l}\right)^n,$$

which in the limit as $l \to \infty$, $1/l\Delta \to \gamma > 0$, $n\Delta = t$, $i - l = i'$(constant) approaches $i' e^{-\gamma t}$ (Newton's law of cooling).

4.6.3. The Ehrenfest model may also be used to discuss irreversibility in thermodynamics (see [181], pp. 385—387). The particular context is that of a famous paradox, which at the turn of the century was on the point of foiling L. Bolzmann's attempt to explain thermodynamics on the basis of a kinetic theory of matter. In classical thermodynamics the exchange of heat between two bodies of unequal temperatures is irreversible. On the other hand, if the bodies are treated as a dynamical system the celebrated "Wiederkehrsatz" of H. Poincaré asserts that "almost every" state of the system will be achieved arbitrarily closely. Thus the irreversibility postulated in classical thermodynamics and the "recurrence" properties of dynamical systems seem to be irreconcilable. Bolzmann explained the paradox by the fact that the "Poincaré cycles" (time intervals after which states "nearly recur" for the first time) are enormous when compared with time intervals involved in ordinary experiments so that predictions based on classical thermodynamics can be accepted without reservation. Though correct in principle, the explanation was set forth in an unconvincing manner and the controversy raged on. The situation was clarified through the efforts of Ehrenfest and Smoluchowski who showed that irreversibility should be interpreted in a statistical manner. Their explanation is easily discussed by appealing to the Ehrenfest model. We know that the states of the associated Markov chain are recurrent or, in other words, *any state is bound to recur with probability* 1. This is to be considered as the statistical analogue of Poincaré's "Wiederkehrsatz". Next, the mean recurrence time of state i is $m(i, i) = 2^{2l} \Big/ \binom{2l}{i}$, $0 \leqslant i \leqslant 2l$. This is the statistical analogue of a "Poincaré cycle" telling us, roughly speaking, how long, on the average, one will have to wait for state i to recur. If i and $2l - i$ differ considerably, then $m(i, i)$ is fantastically large. For instance, if $i = 0$ and $l = 5000$ then $m(i, i) = 2^{10000}$ steps (of the order of 10^{3000} years if one step is performed in one second!) and the thermodynamic process will appear to be irreversible. On the other hand, if i and $2l - i$ are nearly equal, then $m(i, i)$ is small. For instance, if $i = l = 5000$ then $m(i, i) = 2^{10000} \Big/ \binom{10000}{5000} \approx 50\sqrt{2\pi} \approx 125$ steps (by using Stirling's formula), and so it makes no sense to speak of irreversibility.

4.6.4. We noted in 4.5.2 that the Markov chain associated with the Ehrenfest model is reversible*). A few remarks are in order. It is sometimes maintained that a process modeled by this chain (as the thermodynamic one) has a "direction" by virtue of the strong tendency to equilibrium (convergence to the stationary distribution). There are experimental simulations of an Ehrenfest process showing this tendency very clearly, especially in the case where the initial state differs considerably from l (see [183]). However, if the observation of the process begins after it has been going for a long time, one may consider that it started in equilibrium. Reversibility of the Markov chain then makes the process appear the same when looked at in the backward direction

*) The concepts of reversibility of a Markov chain and reversibility of the thermodynamic process should not be confused. The latter in fact amounts to the recurrence of the states of the process.

as it appears in the forward direction. This may be also put as follows. If a physicist is faced with the record of a long sequence of outcomes then he would be unable to tell whether they were ordered in increasing or in decreasing time.

EXERCISES

4.1 [306] A stochastic matrix $\mathbf{P}$ of order r is regular if and only if $\mathbf{P}^{(r-1)^2+1} > 0$.

4.2 [170] Some power of a stochastic matrix $\mathbf{P}$ of order r is a Markov matrix if and only if $\mathbf{P}^{(r-1)(r-2)+1}$ is a Markov matrix.

4.3 A stochastic matrix $\mathbf{P}$ is mixing if and only if $\mathbf{P}^m$ is a Markov matrix for some natural number m.

4.4 [251] A stochastic matrix $\mathbf{P}$ of order r is mixing if and only if $\alpha(\mathbf{P}^{r^2/2-r+1}) > 0$ or $\alpha(\mathbf{P}^{r^2/2-r+3/2}) > 0$ according as r is even or odd.

4.5 Deduce from Theorems 4.2 and 4.3 that

$$|\lambda_2| \leqslant (1 - \alpha(\mathbf{P}^{n_0}))^{1/n_0}.$$

4.6 Show that a two state Markov chain has a mixing transition matrix if and only if $|\lambda_2| = |1 - p(1, 2) - p(2, 1)| < 1$. (Hint: see 1.11.2.)

4.7 Show that the fundamental matrix of a two state Markov chain with mixing transition matrix is given by

$$\mathbf{Z} = \frac{1}{p(1,2)+p(2,1)} \begin{pmatrix} p(2,1) + \dfrac{p(1,2)}{p(1,2)+p(2,1)} & p(1,2) - \dfrac{p(1,2)}{p(1,2)+p(2,1)} \\ p(2,1) - \dfrac{p(2,1)}{p(1,2)+p(2,1)} & p(1,2) + \dfrac{p(2,1)}{p(1,2)+p(2,1)} \end{pmatrix}.$$

4.8 Prove ([195], p. 81) that the matrix $\overline{\mathbf{M}} = \mathbf{M} - \mathbf{M}_{dg}$ (see 4.3.2) is nonsingular. Show then that $\pi = (\mathbf{e}'\mathbf{Z}_{dg}\mathbf{e} - 1)(\overline{\mathbf{M}}^{-1}\mathbf{e})$ and $\mathbf{P} = \mathbf{I} - (\mathbf{M}_{dg} - \mathbf{E})\overline{\mathbf{M}}^{-1}$. Conclude that $\mathbf{P}$ is completely determined by $\overline{\mathbf{M}}$.

4.9 Show that the matrices $\mathbf{M}$ and $\mathbf{C}$ (see Theorems 4.7 and 4.8) for a regular two state Markov chain are given by

$$\mathbf{M} = \begin{pmatrix} \dfrac{p(1,2)+p(2,1)}{p(2,1)} & \dfrac{1}{p(1,2)} \\ \dfrac{1}{p(2,1)} & \dfrac{p(1,2)+p(2,1)}{p(1,2)} \end{pmatrix}$$

and

$$\mathbf{C} = \frac{p(1,2)\,p(2,1)\,[2 - p(1,2) - p(2,1)]}{[p(1,2)+p(2,1)]^3} \begin{pmatrix} 1 & -1 \\ -1 & 1 \end{pmatrix}.$$

4.10 *First passage time to a set of states.* Consider an ergodic Markov chain $(X(n))_{n\geq 0}$ with stationary distribution $\boldsymbol{\pi} = (\pi(i))_{i\in S}$. Let τ_A denote the first passage time to a set of states $A \subset S$, i.e.,

$$\tau_A = \min(n \geq 1 : X(n) \in A).$$

Put $m(i, A) = \mathsf{E}_i(\tau_A)$. Prove [222] that

i) $m(i, A) < \infty$ whatever $i \in S$, $A \subset S$;

ii) $m(i, A) = 1 + \sum_{j\notin A} p(i, j)\, m(j, A)$ whatever $i \in S$, $A \subset S$;

iii) $\sum_{i\in A}\sum_{j\notin A} \pi(i)\, p(i, j)\, m(j, A) = \sum_{j\notin A} \pi(j)$ whatever $A \subset S$;

iv) $\sum_{i\in A} \pi(i)\, m(i, A) = 1$ whatever $A \subset S$. (H i n t s : i) $\tau_A = \min_{i\in A} \tau_i$; ii) adapt the proof for the case $A = (j)$ — see Theorem 4.7; iii) use the equality $\boldsymbol{\pi}' = \boldsymbol{\pi}'\mathbf{P}$; iv) use ii) and iii).)

4.11 *Continuation.* Consider the probability distributions $\mathbf{p}_1$ and $\mathbf{p}_2$ over A defined by

$$p_1(i) = \mathsf{P}_\pi(X(n) = i \mid X(n) \in A) = \frac{\pi(i)}{\sum_{i\in A} \pi(i)}, \quad i \in A,$$

$$p_2(i) = \mathsf{P}_\pi(X(n+1) = i \mid X(n) \notin A,\ X(n+1) \in A) =$$

$$= \frac{\sum_{j\notin A} \pi(j)\, p(j, i)}{\sum_{i\in A}\sum_{j\notin A} \pi(j)\, p(j, i)}, \quad i \in A.$$

Prove [222] that whatever $A \subset S$

j) $\sum_{i\in A} p_1(i)\, m(i, A) = 1 \Big/ \sum_{i\in A} \pi(i)$;

jj) $\sum_{i\in A} p_2(i)\, m(i, S - A) = \left(1 - \frac{\sum_{i\in A}\sum_{j\in A} \pi(i)\, p(i,j)}{\sum_{i\in A} \pi(i)}\right)^{-1}.$

Show that the above quantities may be interpreted as the mean recurrence time of A and the mean sojourn time in A under P_π, respectively. (Compare with the equations $m(i, i) = 1/\pi(i)$ and $\mathsf{E}_i(\rho(i)) = (1 - p(i, i))^{-1}$ — see 3.3.2). (H i n t: use iv) and iii) above.)

4.12 *Smoluchowskian recurrence time of a set.* Define the Smoluchowskian recurrence time θ_A of a set A (under P_π) as the number of steps between an exit from A (from any state of A) and the next return to A (to any state of A). Show that $\mathsf{E}_\pi(\theta_A^s) = \mathsf{E}_{\mathbf{p}}(\tau_A^s)$, $s \geqslant 1$, where $\mathbf{p}$ is the probability distribution over $S-A$ defined by

$$p(i) = \frac{\sum_{j \in A} \pi(j)\, p(j, i)}{\sum_{j \in A} \pi(j) \sum_{k \notin A} p(j, k),}, \quad i \in S - A.$$

Deduce that

$$\mathsf{E}_\pi(\theta_A) = \frac{1 - \sum_{i \in A} \pi(i)}{\sum_{i \in A} \pi(i) \sum_{j \notin A} p(i, j)}.$$

Prove [250] that

$$\mathsf{E}_\pi(\theta_A^s) = \frac{(-1)^k\,(k!a_s - 1 + \sum_{j \in A} \pi(j))}{\sum_{j \in A} \pi(j) \sum_{k \notin A} p(j, k),}, \quad s \geqslant 1,$$

where the a_s, $s \geqslant 1$, are constants depending on A.

4.13 *Generalizing Theorem* 4.8. Let f and g be two real valued functions defined on the state space S of a Markov chain with mixing transition matrix.

Prove that

$$\lim_{n\to\infty} \frac{1}{n} \operatorname{cov}_{\mathbf{p}}\left(\sum_{l=0}^{n-1} f(X(l)),\ \sum_{l=0}^{n-1} g(X(l))\right) = \sum_{i, j \in S} f(i)\, c(i, j)\, g(j).$$

(H i n t: use instead of $\sigma_n^2(f)$ the quantity

$$\sigma_n^2(f, g) = \mathsf{E}_\pi(f(X(0))\, g(X(0))) - \mathsf{E}_\pi(f(X(0)))\, \mathsf{E}_\pi(g(0)))$$

$$+ \sum_{l=1}^{n-1} \frac{n-l}{n} [\mathsf{E}_\pi(f(X(0))\, g(X(l))) - \mathsf{E}_\pi(f(X(0)))\, \mathsf{E}_\pi(g(X(0)))]$$

$$+ \sum_{l=1}^{n-1} \frac{n-l}{n} [\mathsf{E}_\pi(g(X(0))\, f(X(l))) - \mathsf{E}_\pi(f(X(0)))\, \mathsf{E}_\pi(g(X(0)))].)$$

4.14 Consider a Markov chain $(X(n))_{n \geqslant 0}$ with mixing transition matrix and put

$$p(n, i, (j_1, \ldots, j_m)) = \mathsf{P}_i(X(n) = j_1, \ldots, X(n + m - 1) = j_m),$$

$$\pi(j_1, \ldots, j_m) = \begin{cases} \pi(j_1), \text{ if } m = 1, \\ \pi(j_1)\, p(j_1, j_2) \ldots p(j_{m-1}, j_m), \text{ if } m \geqslant 2, \end{cases}$$

whatever $i, j_1, \ldots, j_m \in S$, $m \geqslant 1$, $n \geqslant 0$.

Prove that

$$\sum_{(j_1, \ldots, j_m) \in A^{(m)}} |(p(n, i, (j_1, \ldots, j_m)) - \pi(j_1, \ldots, j_m))|$$
$$\leqslant \sup_{A \subset S} \left| \sum_{j \in A} (p(n, i, j) - \pi(j)) \right|$$

whatever $i \in S$, $m \geqslant 1$, $n \geqslant 0$, and the set $A^{(m)}$ of m-tuples of states from S. Deduce that there are constants $a \geqslant 0$, $0 < b < 1$ such that whatever $m \geqslant 1$, $n \geqslant 0$, the initial distribution $\mathbf{p}$, and the random events E_1 and E_2 defined in terms of the random variables $X(0), \ldots, X(m)$ and $X(m+n)$, $X(m+n+1), \ldots$, respectively, one has

$$|\mathsf{P}_{\mathbf{p}}(E_1 \cap E_2) - \mathsf{P}_{\mathbf{p}}(E_1)\, \mathsf{P}_{\mathbf{p}}(E_2)| \leqslant ab^n\, \mathsf{P}_{\mathbf{p}}(E_1).$$

(This is precisely the kind of dependence alluded to in the first footnote on p. 140.)

4.15 Prove that a cyclic Markov chain of period greater than 2 cannot be reversible.

4.16 Prove that the reversed Markov chain is of the same type as the initial one (regular or cyclic).

4.17 Prove that the eigenvalues of the transition matrix $\mathbf{P}$ of a reversible Markov chain are real and that the matrix is diagonalizable. (Hint: consider the symmetric matrix

$$\mathbf{A} = (\delta(i, j) \sqrt{\pi(i)})\, \mathbf{P}(\delta(i, j)/\sqrt{\pi(j)}) = (p(i, j) \sqrt{\pi(i)}/\sqrt{\pi(j)}).$$

Show that $\mathbf{A}$ and $\mathbf{P}$ have the same eigenvalues. Then use 1.9.4 and 1.9.6.)

4.18 *Kolmogorov's conditions for reversibility* [215]. Prove that an ergodic Markov chain is reversible if and only if

$$p(i_1, i_2) \ldots p(i_{n-1}, i_n)\, p(i_n, i_1) = p(i_1, i_n)\, p(i_n, i_{n-1}) \ldots p(i_2, i_1)$$

whatever the states $i_1, \ldots, i_n$ and the natural number $n \geqslant 2$. (In words: given an arbitrary starting state i_1 the probability of any sequence of steps ultimately returning to i_1 has the same probability whether this cycle is traced in one sense or the other.)

4.19 Prove [400] that a Markov chain with positive transition matrix $\mathbf{P}$ is reversible if and only if for a fixed state i

$$p(i, j)\, p(j, k)\, p(k, i) = p(i, k)\, p(k, j)\, p(j, i)$$

whatever the states j and k. (Hint: the vector $\mathbf{u} = (p(i, j)/p(j, i))_{j \in S}$ satisfies $\mathbf{u}'\mathbf{P} = \mathbf{u}'$, hence $\mathbf{u}$ is proportional to $\boldsymbol{\pi}$.)

Note. The minimal sufficient condition of the type above has not yet been determined in the case where some of the entries of $\mathbf{P}$ are null.

CHAPTER 5

General Properties of Markov Chains

The aim of this chapter is to take up various topics in Markov chains regardless of their type.

5.1 ASYMPTOTIC BEHAVIOUR OF TRANSITION PROBABILITIES

5.1.1. The results proved in the previous chapters allow us to solve the problem of the asymptotic behaviour as $n \to \infty$ of the n step transition probabilities $p(n, i, j)$ of an arbitrary Markov chain.

By Corollary 2 of Theorem 2.7 if j is a transient state then

$$\lim_{n\to\infty} p(n, i, j) = 0 \tag{5.1}$$

whatever the state $i \in S$.

Theorem 4.7 and the considerations in 4.4.2 lead to the equality

$$\lim_{n\to\infty} p(nd_i, i, i) = \frac{d_i}{m(i, i)} \tag{5.2}$$

whatever the recurrent state i of period d_i. Obviously, $p(n, i, i) = 0$ if $n \not\equiv 0 \pmod{d_i}$.

Let us define $f_r(i, j)$ for any pair of states $i, j \in S$ by

$$f_r(i, j) = \sum_{m\geqslant 0} f(md_j + r, i, j), \quad 1 \leqslant r \leqslant d_j,$$

where d_j is the period of state j. It is clear that

$$f_r(i, j) = \mathsf{P}_i(X(n) = j \text{ for at least one value } n \equiv r \pmod{d_j}))$$

and

$$\sum_{r=1}^{d_j} f_r(i, j) = f(i, j), \; i, j \in S.$$

The solution of the problem raised above is given by

Theorem 5.1. *Whatever states $i, j \in S$ we have*

$$\lim_{n\to\infty} p(nd_j + r, i, j) = f_r(i, j)\, d_j/m(j, j), \quad 1 \leqslant r \leqslant d_j.$$

Proof. For a transient state j we have $m(j, j) = \infty$ and the theorem holds by equation (5.1). It remains to consider the case of a recurrent state j. As $p(n, j, j) = 0$ if $n \not\equiv 0 \pmod{d_j}$, equation (2.3) yields

$$0 \leqslant p(nd_j + r, i, j) - \sum_{m=0}^{n'} f(md_j + r, i, j)\, p((n-m)\, d_j, j, j)$$

$$\leqslant \sum_{m=n'+1}^{n} f(md_j + r, i, j)$$

for any $n > n'$. On account of (5.2) letting first $n \to \infty$ then $n' \to \infty$ the theorem follows.

Corollary. *The Césaro limit*

$$\lim_{n\to\infty} \frac{1}{n} \sum_{m=1}^{n} p(m, i, j) = \pi(i, j)$$

exists and equals $f(i, j)/m(j, j)$ *whatever the states* $i, j \in S$. *The matrix* $\mathbf{\Pi} = (\pi(i, j))_{i,j \in S}$ *is stochastic and satisfies the equations*

$$\mathbf{\Pi P} = \mathbf{P\Pi} = \mathbf{\Pi} = \mathbf{\Pi}^2.$$

Proof. The existence of the limit $\pi(i, j)$ follows at once from the fact that by Theorem 5.1 the sequence $(p(m, i, j))_{m\geqslant 1}$ has d_j convergent subsequences. Then $\pi(i, j)$ equals

$$\frac{1}{d_j} \sum_{r=1}^{d_j} f_r(i, j)\, d_j/m(j, j) = f(i, j)/m(j, j), \quad i, j \in S.$$

Furthermore, we have

$$\mathbf{\Pi} = \lim_{n\to\infty} \frac{1}{n} \sum_{m=1}^{n} \mathbf{P}^m,$$

whence

$$\mathbf{\Pi P} = \mathbf{P\Pi} = \lim_{n\to\infty} \frac{1}{n} \sum_{m=2}^{n+1} \mathbf{P}^m = \mathbf{\Pi}.$$

It follows that for all n

$$\mathbf{\Pi} = \mathbf{\Pi P}^n = \mathbf{P}^n \mathbf{\Pi} = \mathbf{\Pi}\left(\frac{1}{n} \sum_{m=1}^{n} \mathbf{P}^m\right) = \left(\frac{1}{n} \sum_{m=1}^{n} \mathbf{P}^m\right)\mathbf{\Pi},$$

whence, letting $n \to \infty$, we get $\mathbf{\Pi} = \mathbf{\Pi}^2$.

Remark. If states i and j belong to the same recurrent class, then $f(i, j) = 1$ (see Theorem 2.9), and therefore $\pi(i, j)$ does not depend on i. This is just a well known property of ergodic chains.

It is possible to represent the matrix $\mathbf{\Pi}$ as a polynomial in the matrix $\mathbf{P}$. Let us consider the characteristic polynomial $f(\lambda) = \det\,(\lambda\mathbf{I} - \mathbf{P})$. If we denote by $h(\lambda)$ the greatest common divisor of the polynomials $f(\lambda)$ and $f'(\lambda)$, then the roots of the polynomial $g(\lambda)=f(\lambda)/h(\lambda)$ are simple and coincide with the distinct roots of $f(\lambda)$. Set

$$f(\lambda) = a(\lambda - 1)^r\,\varphi(\lambda),$$

$$g(\lambda) = b(\lambda - 1)\,\psi(\lambda),$$

where the constants a and b are chosen so that $\varphi(1) = \psi(1) = 1$. (Thus we assume that there are r recurrent classes, each contributing once the eigenvalue $\lambda = 1$.)

Theorem 5.2. [389] $\mathbf{\Pi} = \varphi(\mathbf{P}) = \psi(\mathbf{P})$.

The *proof* follows from the Jordan canonical form of $\mathbf{P}$ (see 1.9.7).

In particular, the above theorem shows that for a two state Markov chain (see 1.11.2) we have

$$\mathbf{\Pi} = \begin{cases} \dfrac{1}{1-\lambda_2}(\mathbf{P} - \lambda_2\mathbf{I}), & \text{if } \lambda_2 \neq 1, \\ \mathbf{I}, & \text{if } \lambda_2 = 1. \end{cases}$$

5.1.2. A probability distribution $\mathbf{p}$ is said to be a *stationary* distribution for a Markov chain with transition matrix $\mathbf{P}$ if and only if $\mathbf{p}'\mathbf{P} = \mathbf{p}'$. A careful examination of the proof of Theorem 4.4 leads us to conclude that a Markov chain is a stationary sequence if and only if its initial probability distribution is stationary. We are now able to find all the stationary distributions of an arbitrary Markov chain.

Theorem 5.3. *A probability distribution is stationary if and only if it is a convex linear combination of the probability distributions* $(\pi(i, j))_{j\in S}$, $i \in S - T$.

Proof. Let $\mathbf{p}$ be a stationary distribution. Then the equality $\mathbf{p}'\mathbf{P} = \mathbf{p}'$ implies that $\mathbf{p}'\mathbf{P}^n = \mathbf{p}'$ whatever $n \geqslant 1$. Hence

$$\mathbf{p}'\left(\sum_{m=1}^{n} \mathbf{P}^m\right)/n = \mathbf{p}'$$

and, by the corollary to Theorem 5.1, letting $n \to \infty$ we get $\mathbf{p}'\mathbf{\Pi} = \mathbf{p}'$, that is

$$p(j) = \sum_{i\in S} p(i)\,\pi(i, j), \quad j \in S. \tag{5.3}$$

Since $\pi(i, j) = 0$ if $j \in T$, we have $p(j) = 0$, $j \in T$. Therefore equations (5.3) tell us that $\mathbf{p}$ is a convex linear combination of the probability distributions $(\pi(i, j))_{j\in S}$, $i \in S - T$ (with coefficients $p(i)$, $i \in S - T$).

Conversely, if $\mathbf{p}$ is a probability distribution over S such that $p(j) = 0$, $j \in T$, then by the corollary to Theorem 5.1 we have $\mathbf{p}'\mathbf{\Pi}\mathbf{P} = \mathbf{p}'\mathbf{\Pi}$, that is $(\mathbf{p}'\mathbf{\Pi})'$, which is a convex linear combination of the probability distributions $(\pi(i, j))_{j \in S}$, $i \in S - T$, is a stationary distribution.

5.1.3. The asymptotic behaviour of the n step transition probabilities of a Markov chain can be studied by means of a geometric method [314 and 406] which is briefly this: the transition matrix $\mathbf{P}$ is identified with a linear operator f on the simplex Σ of all the probability distributions over the states of the chain. The intersection $\mathfrak{K}$ of all the images $f(\Sigma)$, $f^2(\Sigma) = f(f(\Sigma))$, $f^3(\Sigma)$, ... is a simplex whose vertices are permuted by f. The position of $\mathfrak{K}$ in Σ and this permutation determine the behaviour of the probabilities $p(n, i, j)$ as $n \to \infty$ and also locate the vertices of the simplex of the stationary distributions of the chain. Let us remark that the images $f(\Sigma)$, $f^2(\Sigma)$, ... represent the totality of probability distribution over the state space of the chain at times 1, 2, ... Intuitively, the intersection $\mathfrak{K}$ of these images, should represent the totality of possible "ultimate" probability distributions and, therefore, some connection between $\mathfrak{K}$ and the asymptotic behaviour of the n step transition probabilities is to be expected.

Let us recall that by *convex polytope* in a Euclidean space we mean the convex hull of finitely many points $\mathbf{x}_1, \ldots, \mathbf{x}_r$ of the space*). The point $\mathbf{x}_l$ is a *vertex* of the polytope if and only if the convex hull of the points $\mathbf{x}_h$, $1 \leq h \leq r$, $h \neq l$ does not contain it. A convex polytope is a *simplex* if and only if none of its vertices is in the flat**) determined by the other ones. A linear function f mapping a simplex Σ into itself is called a *linear operator* on Σ. The convex hull of any subset of the vertices of a simplex is called a subsimplex of Σ. The subsimplices of a simplex are themselves simplices. If A is a non empty subset of a simplex Σ, the subsimplex of Σ with fewest vertices containing A is called the *carrier* of A in Σ and denoted by carr (A).

We state without proof the following result.

Theorem 5.4 [341]. *If f is a linear operator on a simplex Σ then*

1) *The intersection $\mathfrak{K}$ of the iterates $f^n(\Sigma)$, $n \geq 1$, is a simplex.*

2) *The vertices of $\mathfrak{K}$ are permuted by f and hence fall into r disjoint classes $E_1, \ldots, E_r$ on each of which f is a cyclic permutation.*

3) *If we consider an arbitrary vertex $\mathbf{k}_{l0} \in E_l$ and define $\mathbf{k}_{ls} = f^s(\mathbf{k}_{l0})$, $1 \leq l \leq r$, $s \geq 0$ (it is obvious that $\mathbf{k}_{ls} = \mathbf{k}_{ls'}$ when $s \equiv s' \pmod{d_l}$, where*

---

*) The *convex hull* of the points $\mathbf{x}_1, \ldots, \mathbf{x}_r$ is the set of all points $c_1\mathbf{x}_1 + \ldots + c_r\mathbf{x}_r$ with $c_1, \ldots, c_r \geq 0$, $\sum_{i=1}^{r} c_i = 1$.

**) The *flat* determined by the points $\mathbf{x}_1, \ldots, \mathbf{x}_r$ is the set of all points $c_1\mathbf{x}_1 + \ldots + c_r\mathbf{x}_r$ with $\sum_{i=1}^{r} c_i = 1$ ($c_1, \ldots, c_r$ can take both positive and negative values).

d_l is the number of elements in E_l), then the sets E_{ls} = the carrier of $\mathbf{k}_{ls}$ in Σ, $1 \leqslant l \leqslant r$, $0 \leqslant s < d_l$, are disjoint, $f^t(E_{ls}) \subseteq E_{ls'}$ and

$$\bigcap_{n \equiv t (\mathrm{mod}\ d_l)} f^n(E_{ls}) = \mathbf{k}_{ls'}$$

when $s' = t + s \pmod{d_l}$.

4) *If $\mathfrak{K}_l$ denotes the subsimplex of $\mathfrak{K}$ whose vertices are the elements of E_l, $1 \leqslant l \leqslant r$, then the set of all f-fixed points in Σ is a simplex whose vertices are the barycentres of the $\mathfrak{K}_l$.*

The remarks below show the correspondence between the geometric and probabilistic concepts.

Each state i, $1 \leqslant i \leqslant m$, of the Markov chain with transition matrix $\mathbf{P} = (p(i, j))_{1 \leqslant i, j \leqslant m}$ corresponds to the m-dimensional vector ξ_i whose only nonnull component, 1, is its ith component. Let us define

$$f(\mathbf{x}) = \left(\sum_{i=1}^{m} x(i)\, p(i, j)\right)_{1 \leqslant j \leqslant m}$$

for any $\mathbf{x} \in \Sigma$ = (convex hull of the vectors ξ_i, $1 \leqslant i \leqslant m$) $= ((x(i))_{1 \leqslant i \leqslant m} : x(i) \geqslant 0, \sum_{i=1}^{m} x(i) = 1)$. Clearly, Σ is a simplex and f a linear operator on Σ.

a) *$i \to j$ if and only if $\xi_j \in \bigcup_{n \geqslant 1}$ carr $(f^n(\xi_i))$.*

Proof. This is a direct consequence of the definitions of the relation "$\to$" and the carrier of a subset of Σ.

b₁) *If $R(\mathfrak{K}) = (i : \xi_i \in$ carr $(\mathfrak{K}))$ then whatever the state j there exists a state $i \in R(\mathfrak{K})$ such that $j \to i$.*

Proof. Let $D =$ carr $(\bigcup_{n \geqslant 1} f^n(\xi_j))$. Then $f(D) \subseteq D$ since the image by a linear operator of the carrier of a subset of a simplex is contained in the carrier of the image of the subset. Hence $\bigcap_{n \geqslant 1} f^n(D)$ is a non empty subset of both $\mathfrak{K}$ and D. There exists, therefore, a vertex ξ_i of $R(\mathfrak{K})$ which is also a vertex of D. As the vertices of D are also those of $\bigcup_{n \geqslant 1}$ carr $(f^n(\xi_i))$ by a) we get that $j \to i \in R(\mathfrak{K})$.

b₂) *If $i \in R(\mathfrak{K})$ and $i \to j$, then $j \in R(\mathfrak{K})$ and $j \to i$.*

Proof. The hypothesis made implies that ξ_i is a vertex of some E_{ls} and ξ_j is a vertex of carr $(f^t(\xi_i))$ for some $t \geqslant 1$ by a). Thus $f^t(\xi_i) \in f^t(E_{ls})$. But $f^t(E_{ls}) \subseteq E_{ls'}$ when $s' \equiv s + t \pmod{d_l}$ by Theorem 5.4 point 3). Hence ξ_j is a vertex of $E_{ls'}$ and consequently $j \in R(\mathfrak{K})$. Furthermore, carr $(f^n(\xi_j)) = E_{ls}$ for a sufficiently large $n \equiv s - s' \pmod{d_l}$ by the same Theorem 5.4 point 3). But $\xi_i \in E_{ls}$ and hence $j \to i$ by a).

b) *State i is recurrent if and only if ξ_i is a vertex of* carr $(\mathfrak{K})$.

Proof. The states of $R(\mathfrak{K})$ are recurrent by b_2) and the properties of recurrence (see 2.5.6). On the other hand, if $j \notin R(\mathfrak{K})$ then $j \to i \in R(\mathfrak{K})$ by b_1) and $i \not\to j$ by b_2). Consequently, j is transient. Hence $R(\mathfrak{K})$ contains all the recurrent states.

c) *The recurrent classes are*

$$\mathfrak{C}^l = \left(i : \xi_i \in \bigcup_{s=0}^{d_l-1} E_{ls}\right)$$

with cyclic subclasses $\mathfrak{C}^l_s = (i : \xi_i \in E_{ls})$, $1 \leqslant l \leqslant r$, $0 \leqslant s \leqslant d_l - 1$.

Proof. The sets $\mathfrak{C}^l_s$ are pairwise disjoint and non empty by Theorem 5.4 point 3). If $i \in \mathfrak{C}^l_s$ and state i leads to state j in one step, then $\xi_i \in E_{ls}$ and $\xi_j \in$ carr $(f(\xi_i))$. As $f(E_{ls}) \subseteq E_{l,\,s+1}$ we have $f(\xi_i) \in E_{l,\,s+1}$, hence $\xi_j \in E_{l,\,s+1}$, that is $j \in \mathfrak{C}^l_{s+1}$.

d) *If state j is transient then* $\lim\limits_{n\to\infty} p(n, i, j) = 0$ *whatever the state i.*

Proof. Let $\mathbf{x}^{(n)}$ denote the point of carr $(\mathfrak{K})$ closest to $f^n(\xi_i)$. The distance between $f^n(\xi_i)$ and $\mathbf{x}^{(n)}$ converges to zero as $n \to \infty$ by the very definition of $\mathfrak{K}$. But on account of b) we have $x^{(n)}(j) = 0$ whatever the transient state j. Therefore $\lim\limits_{n\to\infty} p(n, i, j) = 0$ since the jth component of $f^n(\xi_i)$ is $p(n, i, j)$.

e) *If $i \in \mathfrak{C}^l_s$ then*

$$\lim_{n\to\infty} p(nd_l + t, i, j) = k_{ls'}(j)$$

whenever $s' = s + t \pmod{d_l}$ *whatever* $0 \leqslant t \leqslant d_l - 1$ *and state* j.

Proof. This is an immediate consequence of (5.4).

It is worth noting that $k_{ls}(j)$, the jth component of $\mathbf{k}_{ls}$, equals 0 if and only if $j \in \mathfrak{C}^l_s$ because of the definitions of carr (E_{ls}) and $\mathfrak{C}^l_s$.

f) Theorem 5.4 point 4) is a translation into a geometric language of Theorem 5.3.

5.2 THE TAIL σ-ALGEBRA

5.2.1. Now we take up another aspect of the asymptotic behaviour of Markov chains.

In 2.1.5 we defined the concept of a random event posterior to time n in a Markov chain. The set of all these random events form a σ-algebra, $\mathcal{X}^n$, the smallest σ-algebra that contains all the random events of the form

$$\{X(n) = i_n, \ldots, X(n+m) = i_{n+m}\},\ i_n, \ldots, i_{n+m} \in S,\ m \geqslant 0.$$

The intersection of the σ-algebras $\mathcal{X}^n$, $n \geq 0$, which we shall denote by $\mathcal{T}$, is again a σ-algebra called the *tail* σ-algebra of the Markov chain considered. It may be thought of as the set of all the random events which are posterior to any time $n \geq 0$.

5.2.2. In general, given a probability space $(\Omega, \mathcal{X}, \mathsf{P})$, a random event $A \in \mathcal{X}$ is called a P-*atom* if and only if $\mathsf{P}(A) > 0$ and whatever the random event $B \subset A$ either $\mathsf{P}(B) = 0$ or $\mathsf{P}(B) = \mathsf{P}(A)$. Notice that a P-atom is determined modulo null P-probability random events (i.e., random events E such that $\mathsf{P}(E) = 0$). In other words one cannot distinguish between A and $A \Delta E = (A \cap E^c) \cup (A^c \cap E)$. It is not difficult to prove that the set of the P-atoms of $\mathcal{X}$ is either finite or countably infinite and that Ω can be partitioned as

$$\Omega = N \cup \Big(\bigcup_{r \in I} A_r\Big), \tag{5.5}$$

where the random event N may be absent and the index set I may be empty, finite or countably infinite. If present, N is a P-completely nonatomic random event (i.e., for any positive number $b \leq \mathsf{P}(N)$ there exists a random event $B \subset N$ such that $\mathsf{P}(B) = b$) and, if I is not empty, the A_i, $i \in I$, are P-atoms. The decomposition (5.5) is unique modulo null P-probability random events. The σ-algebra $\mathcal{X}$ is said to be P-*atomic* if and only if $N = \emptyset$. A P-atomic σ-algebra is said to be *finite* if and only if the set of its atoms is finite. In particular, it is said to be *trivial* if and only if there is only one atom (identical, of course, to the sure random event Ω).

5.2.3. It will be proved in 7.6.1 that the tail σ-algebra of a (homogeneous or nonhomogeneous) finite Markov chain is finite and that the number of its atoms does not exceed the number of states of the chain. In the homogeneous case a precise description of the atoms can be given. Namely we have

Theorem 5.5 [33]. *Consider a recurrent Markov chain with r recurrent classes $\mathcal{C}^l$ with cyclically moving subclasses $\mathcal{C}^l_0, \ldots, \mathcal{C}^l_{d_l-1}$, $1 \leq l \leq r$. Then the* P-*atoms of the tail σ-algebra $\mathcal{T}$ are the random events $A_{l,s} = \{X(0) \in \mathcal{C}^l_s\}$, $0 \leq s \leq d_l - 1$, $1 \leq l \leq r$, for which $\Sigma_{i \in \mathcal{C}^l_s} p(i) > 0$.*

Proof. The basic fact used in the proof is as follows: Whatever the random event A determined by the random variables of the Markov chain considered, the random variable f^n_A with values $f^n_A(\omega) = \mathsf{P}(A \mid X(n) = i)$ for $\omega \in \{X(n) = i\}$, $i \in S$, converges as $n \to \infty$ to the indicator χ_A of the random event A,

$$\chi_A(\omega) = \begin{cases} 1, & \text{if } \omega \in A, \\ 0, & \text{if } \omega \notin A, \end{cases}$$

for all $\omega \in \Omega - E$, where E is a random event such that $\mathsf{P}(E) = 0$*).

*) This is a special case of the following more general theorem: let $(X(n))_{n \geq 0}$ be a sequence of discrete random variables and A an arbitrary random event determined by them. Then the random variable f^n_A with values $f^n_A(\omega) = \mathsf{P}(A \mid X(n) = i_n, \ldots, X(0) = i_0)$ for $\omega \in \{X(n) = i_n, \ldots, X(0) = i_0\}$ converges to χ_A as $n \to \infty$. The advanced reader will recognize here a special case of the martingale convergence theorem.

We shall treat just the case where $\Sigma_{i\in\mathcal{C}_s^l} p(i) > 0$ for all the values $0 \leqslant s \leqslant d_l - 1$, $1 \leqslant l \leqslant r$. The reader will be able to adjust the proof in order to make it work in the case where some of the above quantities are null.

Let d denote the smallest common multiple of the periods d_l, $1 \leqslant l \leqslant r$. Then, for any sufficiently large n we have $\mathsf{P}_p(X(nd) = i) > 0$ whatever the state i (see 4.4.2). This allows us to write

$$\mathsf{P}_p(A \mid X(nd) = i) = \sum_{j\in S} \mathsf{P}_p(A \cap (X((m+n)d) = j) \mid X(nd) = i)$$

$$= \sum_{j\in S} p(md, i, j)\, \mathsf{P}_p(A \mid X((m+n)\, d) = j) \tag{5.6}$$

for all n large enough ($\geqslant n_0$), $m \geqslant 0$, $A \in \mathcal{T}$, $i \in S$.

Next, consider the sequences $(\mathsf{P}_p(A \mid X((m+n)\, d) = j))_{m\geqslant 0}$, $j \in S$. Since they are bounded one can find an increasing sequence $(n_s)_{s\geqslant 1}$ of natural numbers such that the limits

$$\lim_{s\to\infty} \mathsf{P}_p(A \mid X(n_s d) = j),\ j \in S,$$

exist.

Now, notice that the matrix $\mathbf{P}^d$ may be thought of as the transition matrix of a Markov chain with $d_1 + \dots + d_r$ recurrent aperiodic classes. Consequently, by Theorem 4.2 the matrix $\mathbf{P}^{nd}$ approaches as $n \to \infty$ a matrix of the form

$$\begin{pmatrix} \Pi_{1,0} & & & & & & \\ & \ddots & & & & \mathbf{0} & \\ & & \Pi_{1,d_1-1} & & & & \\ & & & \ddots & & & \\ & & & & \Pi_{r,0} & & \\ & \mathbf{0} & & & & \ddots & \\ & & & & & & \Pi_{r,d_r-1} \end{pmatrix},$$

where on the main diagonal there appear positive stable stochastic matrices. Replacing m by $n_s - n$ in (5.6) and letting $s \to \infty$ we deduce that the conditional probability $\mathsf{P}_p(A \mid X(nd) = i)$ does not depend on $n \geqslant n_0$ and that the vector $(\mathsf{P}_p(A \mid X(nd) = i))_{i\in S}$ is a linear combination, with coefficients depending on A and $\mathbf{P}$, of $d_1 + \dots + d_r$ linearly independent vectors whose components are constant.

Furthermore, if A is a P_p-atom of $\mathcal{T}$ then there exists $i \in S$ such that $\mathsf{P}_p(A \mid X(nd) = i) > 0$ whatever $n \geqslant n_0$ [otherwise it would follow that $\mathsf{P}_p(A) = \sum_{i\in S} \mathsf{P}_p(X(nd) = i)\, \mathsf{P}_p(A \mid X(nd) = i) = 0$] and the result stated at the beginning of the proof leads us to conclude that $\mathsf{P}_p(A \mid X(nd) = i)$

$= 1$, $n \geqslant n_0$ (the fact that i is a recurrent state is essential). We are now able to prove that if $A_1, \ldots, A_q$ are distinct P_p-atoms of $\mathcal{T}$, then the vectors $\mathbf{v}_m = (\mathsf{P}_p(A_m \mid X(n_0 d) = i))_{i \in S}$ $1 \leqslant m \leqslant q$, are linearly independent. Indeed, if $\sum_{m=1}^{q} c_m \mathbf{v}_m = \mathbf{0}$, on account of the fact that $\mathsf{P}_p(A_1 \mid X(n_0 d) = i) = 1$ for a certain $i \in S$ [implying $\mathsf{P}_p(A_2 \mid X(n_0 d) = i) = \ldots = \mathsf{P}_p(A_q \mid X(n_0 d) = i) = 0$], we conclude that $c_1 = 0$, and in a similar way, $c_2 = \ldots = c_q = 0$.

It follows from the above that there can be at most $d_1 + \ldots + d_r$ atoms. Since the $d_1 + \ldots + d_r$ disjoint random events $A_{l,s}$, $0 \leqslant s \leqslant d_l - 1$, $1 \leqslant l \leqslant r$, belong to $\mathcal{T}$ [we have $A_{l,s} = \{X(nd_l) \in \mathcal{C}_s^l$ for any $n \geqslant 0\}$] they are, clearly, the P_p-atoms of $\mathcal{T}$.

Thus the proof is complete.

Corollary. *The tail σ-algebra of a regular Markov chain is trivial under any initial distribution.*

Let us now consider the case of an absorbing Markov chain. We shall prove

Theorem 5.7 [164]. *Consider an absorbing Markov chain with r recurrent classes $\mathcal{C}^l$ with cyclically moving subclasses $\mathcal{C}_0^l, \ldots, \mathcal{C}_{d_l-1}^l$, $1 \leqslant l \leqslant r$. Then the P_p-atoms of the tail σ-algebra $\mathcal{T}$ are the random events*

$$B_{l,s} = \lim_{n \to \infty} \{X(nd) \in \mathcal{C}_s^l\}^{*)}$$

for which

$$\sum_{i \in S} \sum_{j \in \mathcal{C}_s^l} p(i)\, f_{d_l}(i, j) > 0, \quad 0 \leqslant s \leqslant d_l - 1, \quad 1 \leqslant l \leqslant r.$$

Proof. Let ν denote the time to absorption into the set of recurrent states, i.e., $\nu = \min(n : X(n) \in \mathcal{C}^1 \cup \ldots \cup \mathcal{C}^r)$. Clearly, ν is a stopping time for the Markov chain considered. By the strong Markov property (Theorem 2.1), the sequence $(X(n + \nu))_{n \geqslant 0}$ is a homogeneous Markov chain with state space $\mathcal{C}^1 \cup \ldots \cup \mathcal{C}^r$ and whose transition probabilities coincide with the corresponding ones for the initial chain $(X(n))_{n \geqslant 0}$. It is easily seen (try to prove it !) that the P_p-atoms of the latter are precisely the $\mathsf{P}_{\bar{p}}$-atoms of the recurrent Markov chain $(X(n + \nu))_{n \geqslant 0}$, where $\bar{\mathbf{p}}$ is the probability distribution over $\mathcal{C}^1 \cup \ldots \cup \mathcal{C}^r$ defined by

$$\bar{p}(i) = \mathsf{P}_p(X(\nu) = i), \quad i \in \mathcal{C}^1 \cup \ldots \cup \mathcal{C}^r.$$

Therefore, Theorem 5.7 follows from Theorem 5.5.

Corollary. *The tail σ-algebra of an indecomposable Markov chain is trivial under any initial distribution.*

---

*) The limit of a sequence of random events has been defined in 1.2.5.

The study of the tail σ-algebra $\mathcal{T}$ thus establishes a little expected connection between the random events posterior to any time $n \geqslant 0$ and the grouping of the states into classes and subclasses. Though we did not get essentially new information, the results obtained offer suggestions for the investigation of the nonhomogeneous case (see 7.6.1).

5.3 LIMIT THEOREMS FOR PARTIAL SUMS

5.3.1. Another important aspect of the asymptotic properties of Markov chains concerns the behaviour of the sum $T_n = \sum_{l=0}^{n-1} f(X(l))$ as $n \to \infty$. Here f is a real valued function defined on the state space S. We have already discussed this problem in the case of regular or indecomposable Markov chains (see 4.3.5). We shall now make some remarks for the general case.

Let $\nu_n(i)$ denote the number of occurrences (equivalently, the *occupation time*) of state i in the first n steps. Clearly, T_n can be expressed in terms of the $\nu_n(i)$ as $T_n = \sum_{i \in S} \nu_n(i) f(i)$. Consequently, the asymptotic behaviour of T_n may be derived from that of the random vector $(\nu_n(i))_{i \in S}$. The result we state next is fundamental.

Proposition 5.8. *Let* $\mathbf{D}(\mathbf{t})$, $\mathbf{t} = (t_j)_{j \in S}$, *denote the diagonal matrix with diagonal entries* $\exp(\mathrm{i}t_j)$, $j \in S$. *Then*

$$\left(\mathsf{E}_i\left[\exp\left(\mathrm{i}\sum_{j \in S} t_j \nu_n(j)\right)\right]\right)_{i \in S} = (\mathbf{D}(\mathbf{t})\mathbf{e})'(\mathbf{PD}(\mathbf{t}))^{n-1}\mathbf{e}. \tag{5.7}$$

Proof. It is easy to prove in a similar way to 3.2.2 that the probability $\mathsf{P}_i(\nu_n(j) = r_{n,j},\ j \in S)$ is the coefficient of $\prod_{j \in S} z_j^{r_{n,j}}$ in the product

$$(\delta(i,j)\, z_j)'_{j \in S}\, (\mathbf{PD}(\mathbf{z}))^{\sum_{j \in S} r_{n,j} - 1}\, \mathbf{e},$$

where $\mathbf{D}(\mathbf{z}) = (\delta(i,j) z_j)_{i,j \in S}$, $|z_j| \leqslant 1$, $j \in S$. Therefore, on account of the fact that $\sum_{j \in S} \nu_n(j) = n$, we have

$$\mathsf{E}_i\left[\exp\left(\mathrm{i}\sum_{j \in S} t_j \nu_n(j)\right)\right] = \sum_{\substack{r_{n,j} \geqslant 0 \\ \sum_{j \in S} r_{n,j} = n}} \prod_{j \in S} (\exp(\mathrm{i}t_j))^{r_{n,j}}\, \mathsf{P}_i(\nu_n(j) = r_{n,j},\ j \in S) =$$

$$= (\delta(i,j) \exp(\mathrm{i}t_j))'_{j \in S}\, (\mathbf{PD}(\mathbf{t}))^{n-1}\mathbf{e},$$

so that (5.7) follows at once.

Proposition 5.8 shows that the solution of the problem depends on the asymptotic behaviour of the nth power of the matrix $\mathbf{PD}(\mathbf{t})$ (which for $\mathbf{t} = \mathbf{0}$ reduces to $\mathbf{P}$). The Perron formula (see 1.9.6) may then be used to solve the problem. This is the way followed by O. Onicescu and G. Mihoc [296, 297, and 299] to whom fundamental results are due (among these we note the central limit theorem for ergodic Markov chains). The discussion of this topic is beyond the scope of this book. A modern treatment with the corresponding references may be found in the impressive memoir [137] (see also [4] and [143], Ch. IX)). In fact, in [137] one considers the more general case where there is given a sequence of Markov chains, the chain of index n having transition matrix $\mathbf{P}_n$, $n \geqslant 1$, and the sum T_n being associated with it. Without any loss of generality one may assume that the limit $\lim_{n\to\infty} \mathbf{P}_n = \mathbf{P}$ exists and is, therefore, a stochastic matrix. In this setting the problem amounts to the asymptotic behaviour of the matrix $(\mathbf{P}_n\mathbf{D}(\mathbf{t}))^n$ as $n \to \infty$.

5.3.2. For the reader to form an idea of the diversity of the limit distributions of the partial sums T_n we shall present the complete list of the possible situations for two state Markov chains. This study was made by R.L. Dobrušin [76], using a direct method. Assume, as we did at the end of the preceding subparagraph, that we are given a sequence of two state Markov chains, the chain of index n having transition matrix

$$\mathbf{P}_n = \begin{pmatrix} 1-a_n & a_n \\ b_n & 1-b_n \end{pmatrix}, \quad n \geqslant 1,$$

and that

$$\mathbf{P} = \lim_{n\to\infty} \mathbf{P}_n = \begin{pmatrix} 1-a & a \\ b & 1-b \end{pmatrix}.$$

Denote by $\nu_n(i)$, $i = 1, 2$, the number of occurrences of state i in the first n steps for the chain of index n, $n \geqslant 1$. Clearly, $\nu_n(2) = n - \nu_n(1)$, and the asymptotic behaviour of $T_n = \nu_n(1)\ f(1) + \nu_n(2)f(2) = (f(1) - f(2))\nu_n(1) + nf(2)$ can be derived from that of $\nu_n(1)$. We distinguish the following four cases.

1. $\mathbf{P}$ is the transition matrix of a regular Markov chain. This happens if and only if $0 < ab < 1$.

In this case for all $-\infty < x < \infty$

$$\lim_{n\to\infty} \mathsf{P}_1 \left(\frac{\nu_n(1) - m_n}{d_n} < x \right) = \frac{1}{\sqrt{2\pi}} \int_{-\infty}^{x} e^{-u^2/2}\, du, \tag{5.8}$$

where

$$m_n = nb_n/(a_n + b_n),\ d_n^2 = na_nb_n(2 - a_n - b_n)/(a_n + b_n)^3.$$

2. $\mathbf{P}$ is the transition matrix of a cyclic Markov chain of period 2, that is

$$\mathbf{P} = \begin{pmatrix} 0 & 1 \\ 1 & 0 \end{pmatrix}.$$

In this case, if

$$\lim_{n\to\infty} n(\mathbf{P}_n - \mathbf{P}) = \begin{pmatrix} u & -u \\ -v & v \end{pmatrix}, \quad u,\ v < \infty,$$

then for all $-\infty < x < \infty$

$$\begin{aligned} \lim_{n\to\infty} \mathsf{P}_1(\nu_{2n}(1) - n < x) &= \sum_{k\in E_x} r_k, \\ \lim_{n\to\infty} \mathsf{P}_1(\nu_{2n+1}(1) - n - 1 < x &= \sum_{k\in E'_x} r_k, \end{aligned} \tag{5.9}$$

where E_x and E'_x denote the sets of integers k such that the integer part of $(k+1)/2$, or $k/2$, respectively, is less than x and

$$r_k = \left(\frac{u}{v}\right)^{\frac{k}{2}} e^{-\frac{u+v}{2}} I_k(\sqrt{uv}), \text{ if } uv \neq 0;$$

$$r_k = \frac{u^k e^{-u}}{k!}, \quad k \geqslant 0 \text{ and } r_k = 0,\ k < 0, \text{ if } v = 0;$$

$$r_{-k} = \frac{v^k e^{-v}}{k!}, \quad k \geqslant 0 \text{ and } r_k = 0,\ k > 0, \text{ if } u = 0.$$

The function I_k is the modified Bessel function of order k defined by the series

$$I_k(w) = \sum_{l\geqslant 0} \frac{1}{l!(k+l)!}\left(\frac{w}{2}\right)^{k+2l}.$$

The limit distributions (5.9) coincide with the distributions of the random variables $[(\eta_1 - \eta_2 + 1)/2]$ and $[(\eta_1 - \eta_2)/2]$*), respectively, where η_1 and η_2 are independent Poisson random variables with parameters $u/2$ and $v/2$.

If $\lim_{n\to\infty} n(\mathbf{P}_n - \mathbf{P})$ is a matrix with at least one infinite entry, then (5.8) holds.

3. **P** is the transition matrix of a Markov chain with one transient state.

If this is state 1, then

$$\mathbf{P} = \begin{pmatrix} 1-a & a \\ 0 & 1 \end{pmatrix},$$

where $0 < a \leqslant 1$.

In this case, if $\lim_{n\to\infty} nb_n = v < \infty$, then

$$\lim_{n\to\infty} \mathsf{P}_1(\nu_n(1) < x) = ae^{-v} \sum_{1\leqslant k<x} L_{k-1}\left(-\frac{av}{1-a}\right)(1-a)^{k-1}, \quad x>1,$$

* The notation $[x]$ designates the integer part of the real number x.

where L_k is the Laguerre polynominal of order k, $k \geqslant 0$, defined by the generating function

$$\frac{1}{1-z} e^{-\frac{wz}{1-z}} = \sum_{k \geqslant 0} L_k(w) z^k.$$

This limit distribution (first obtained in [219]) coincides with the distribution of the sum $\eta_1 + \ldots + \eta_\mu$ of a random number μ of terms of a sequence $(\eta_n)_{n \geqslant 1}$ of independent and identically distributed random variables such that η_1 has a geometric distribution with parameter $1 - a$, $\mu - 1$ has a Poisson distribution with parameter v, and μ and $(\eta_n)_{n \geqslant 1}$ are independent.

If $\lim_{n \to \infty} nb_n = \infty$, then (5.8) holds.

If the transient state is state 2, that is if

$$\mathbf{P} = \begin{pmatrix} 1 & 0 \\ b & 1-b \end{pmatrix},$$

where $0 < b \leqslant 1$, then, under the assumption that $\lim_{n \to \infty} na_n = u < \infty$, we have

$$\lim_{n \to \infty} \mathsf{P}_1(\nu_n(1) - n < x)$$

$$= e^{-u} \sum_{\substack{k > -x \\ k \geqslant 0}} \left[L_k\left(-\frac{bu}{1-b}\right) + L_{k-1}\left(-\frac{bu}{1-b}\right)\right] (1-b)^k, \tag{5.10}$$

with the convention $L_{-1}(w) = 0$; under the assumption that $\lim_{n \to \infty} na_n = \infty$, (5.8) again holds. The limit distribution (5.10) (first obtained in [219], too) coincides with the distribution of the random variable $-(\eta_1 + \ldots + \eta_\mu)$, where μ and $(\eta_n)_{n \geqslant 1}$ enjoy the same independence properties as in the case where state 1 is transient, with the only difference that η_1 has a geometric distribution with parameter $1 - b$ and μ has a Poisson distribution with parameter u.

4. $\mathbf{P}$ is the unit matrix of order 2.

In this case, if $\lim_{n \to \infty} n(\mathbf{P}_n - \mathbf{P})$ is a matrix with infinite entries, then (5.8) holds. If $\lim_{n \to \infty} nb_n = v < \infty$ and $\lim_{n \to \infty} na_n = \infty$, then

$$\lim_{n \to \infty} \mathsf{P}_1(a_n \nu_n(1) < x) = \begin{cases} 0, & \text{if } x \leqslant 0, \\ \int_0^x e^{-(v+t)} I_0(2\sqrt{vt})\, dt, & \text{if } x > 0. \end{cases}$$

This limit distribution coincides with the distribution of the sum $\eta_1 + \ldots + \eta_\mu$ of a random number μ of terms of a sequence $(\eta_n)_{n \geqslant 1}$ of independent and identically distributed random variables such that η_1 has an exponential distribution with parameter 1, $\mu - 1$ has a Poisson distribution with parameter v, and μ and $(\eta_n)_{n\geqslant 1}$ are independent.

If $\lim\limits_{n\to\infty} na_n = u < \infty$ and $\lim\limits_{n\to\infty} nb_n = \infty$, then*)

$$\lim_{n\to\infty} \mathsf{P}_1(b_n(\nu_n(1)-n)<x)=\begin{cases} 1-e^{-u}-\displaystyle\int_0^{-x} e^{-(u+t)}\sqrt{\frac{u}{t}}\, I_1(2\sqrt{ut})\,dt, & \text{if } x<0,\\ 1-e^{-u}, \text{ if } x=0, & \\ 1, \text{ if } x>0. & \end{cases}$$

This limit distribution coincides with the distribution of the random variable $-(\eta_1 + \dots + \eta_\mu)$, where μ and $(\eta_n)_{n\geqslant 1}$ enjoy the same properties as in the preceding situation, with the only difference that μ has a Poisson distribution with parameter u.

If $\lim\limits_{n\to\infty} na_n = u < \infty$ and $\lim\limits_{n\to\infty} bn_n = v < \infty$, then**)

$$\lim_{n\to\infty} \mathsf{P}_1\left(\frac{\nu_n(1)}{n} < x\right)=\begin{cases} 0, \quad \text{if } x\leqslant 0,\\ \displaystyle\int_0^x e^{(-u+v)t-v}\Big[uI_0(2\sqrt{uvt(1-t)}) + \\ \quad + \sqrt{\frac{uvt}{1-t}}\, I_1(2\sqrt{uvt(1-t)})\Big]\,dt, \text{ if } 0<x<1,\\ 1-e^{-u}, \text{ if } x=1\\ 1, \qquad \text{if } x>1. \end{cases}$$

There exist no other limit distributions different from those described above. This means that for any other convergence conditions of $\mathbf{P}_n$ to $\mathbf{P}$ or any other normings of the random variable $\nu_n(1)$, either a limit distribution does not exist or it is degenerate (i.e., corresponds to a constant random variable).

5.4 GROUPED MARKOV CHAINS

5.4.1. When the number of states of a Markov chain is very large we may be interested in reducing its study to that of a Markov chain with a smaller number of states. A natural procedure for achieving this aim is the grouping (lumping) of states. More precisely, let $(X(n))_{n\geqslant 0}$ be a Markov chain with state space S, transition matrix $\mathbf{P}$ and (variable) initial distribution $\mathbf{p}$. Consider a decomposition $S = S_1 \cup \dots \cup S_q$ of

---

*) This corrects the original statement in [76].

**) For an interpretation of this limit distribution see Exercise 8.4.

the state space in pairwise disjoint sets. To simplify the writing we shall denote the sets $S_1, \ldots, S_q$ by $\hat{1}, \ldots, \hat{q}$, respectively. Let us define a new sequence of random variables $(Y(n))_{n\geqslant 0}$ by

$$Y(n) = \hat{k} \text{ if and only if } X(n) \in S_k, \quad n \geqslant 0, \quad 1 \leqslant k \leqslant q.$$

The problem we want to solve is to find conditions for $(Y(n))_{n\geqslant 0}$ to be a homogeneous Markov chain under $\mathsf{P}_{\mathbf{p}}$, whatever the choice of $\mathbf{p}$, with transition probabilities independent of $\mathbf{p}$. If this happens, then $(X(n))_{n\geqslant 0}$ is said to be *groupable* with respect to the partition $S = S_1 \cup \ldots \cup S_q$ and $(Y(n))_{n\geqslant 0}$ is called a *grouped* Markov chain *).

Put $p(i, A) = \sum_{j \in A} p(i, j)$ whatever $A \subset S$ and $i \in S$.

Proposition 5.9. *A necessary and sufficient condition for a Markov chain to be groupable with respect to a partition* $S = S_1 \cup \ldots \cup S_q$ *is that the probabilities* $p(i, S_l)$ *have the same value* $\hat{p}(\hat{k}, \hat{l})$ *for all states* $i \in S_k$, *whatever the pair of subsets* S_k, S_l, $1 \leqslant k, l \leqslant q$. *The matrix* $\widehat{\mathbf{P}} = (\hat{p}(\hat{k}, \hat{l}))$ *is the transition matrix of the grouped Markov chain.*

Proof. If the chain is groupable, then

$$\mathsf{P}_{\mathbf{p}}(Y(1) = \hat{l} \mid Y(0) = \hat{k})$$

does not depend on the initial distribution $\mathbf{p}$, whenever $\mathsf{P}_{\mathbf{p}}(Y(0) = \hat{k}) > 0$. In particular, this should happen for all the initial distributions concentrated in states $i \in S_k$. Therefore the condition is necessary. Let us prove it is also sufficient. We have indeed

$$\mathsf{P}_{\mathbf{p}}(Y(n+1) = \hat{l} \mid Y(n) = \hat{k},\ Y(n-1) = \hat{i}_{n-1}, \ldots, Y(0) = \hat{i}_0)$$

$$= \frac{\sum_{i \in S_k} \mathsf{P}_{\mathbf{p}}(Y(n+1) = \hat{l},\ X(n) = i,\ Y(n-1) = \hat{i}_{n-1}, \ldots, Y(0) = \hat{i}_0)}{\mathsf{P}_{\mathbf{p}}(Y(n) = \hat{k},\ Y(n-1) = \hat{i}_{n-1}, \ldots, Y(0) = \hat{i}_0)}$$

$$= \frac{\hat{p}(\hat{k}, \hat{l}) \sum_{i \in S_k} \mathsf{P}_{\mathbf{p}}(X(n) = i,\ Y(n-1) = \hat{i}_{n-1}, \ldots, Y(0) = \hat{i}_0)}{\mathsf{P}_{\mathbf{p}}(Y(n) = \hat{k},\ Y(n-1) = \hat{i}_{n-1}, \ldots, Y(0) = \hat{i}_0)} = \hat{p}(\hat{k}, \hat{l})$$

[by the Markov property —see (2.2)].

---

*) Notice that if we define the function f on S by $f(i) = \hat{k}$ for all states $i \in S_k$, $1 \leqslant k \leqslant q$, then $Y(n) = f(X(n))$, $n \geqslant 0$. Therefore the grouped chain is nothing but a function of the orginal chain. The problem is then to find conditions for a function of a Markov chain to be a new Markov chain with transition probabilities independent of the initial distribution of the former. Our presentation leans on [195], § 6.3. For more general cases the reader may consult [37, 45, 325, 327, 328, and 329].

5.4.2. Consider now a Markov chain with r states that is groupable with respect to a partition $S = S_1 \cup \ldots \cup S_q$. Denote by $\mathbf{B}$ a $q \times r$ matrix such that its kth row is a probability vector whose nonnull components are those corresponding to the states in S_k and by $\mathbf{C}$ an $r \times q$ matrix such that the nonnull components of its kth column are equal to 1 and correspond to the states in S_k, too, $1 \leqslant k \leqslant q$. It is easy to check that the transition matrix of the grouped chain is given by $\hat{\mathbf{P}} = \mathbf{BPC}$.

In the following it will be convenient to assume that the states are numbered so that those in S_k precede those in S_{k+1}, $1 \leqslant k < q$.

We first prove a matrix variant of Proposition 5.9.

Theorem 5.10. *A necessary and sufficient condition for a Markov chain to be groupable with respect to a partition* $S = S_1 \cup \ldots \cup S_q$ *is that the equality*

$$\mathbf{CBPC} = \mathbf{PC}. \tag{5.11}$$

holds.

Proof. The convention concerning state numbering implies that the matrix $\mathbf{CB}$ has the form

$$\mathbf{CB} = \begin{pmatrix} \mathbf{\Pi}_1 & & & \\ & \cdot & & 0 \\ & & \cdot & \\ & 0 & & \cdot \\ & & & \mathbf{\Pi}_q \end{pmatrix},$$

where $\mathbf{\Pi}_1, \ldots, \mathbf{\Pi}_q$ are positive stable stochastic matrices. Equation (5.11) amounts to the fact that the columns of the matrix $\mathbf{PC}$ are right eigenvectors of the matrix $\mathbf{CB}$ corresponding to the eigenvalue $\lambda = 1$. By a remark made in 4.1.2 (see the footnote on p. 124) this happens if and only if the components of any column of $\mathbf{PC}$ corresponding to the states of S_k, $1 \leqslant k \leqslant q$, are equal. But this is just the grouping condition from Proposition 5.9.

An immediate consequence of (5.11) is that

$$\hat{\mathbf{P}}^2 = \mathbf{BPCBPC} = \mathbf{BP}^2\mathbf{C}$$

and, in general,

$$\hat{\mathbf{P}}^n = \mathbf{BP}^n\mathbf{C}, \quad n \geqslant 1. \tag{5.12}$$

5.4.3. Now we shall deal with the question of how the fundamental matrices are modified when passing to the grouped chain. We first consider a groupable absorbing Markov chain with transition matrix of the form

$$\mathbf{P} = \left(\begin{array}{c|c} \mathbf{P}_1 & \mathbf{0} \\ \hline \mathbf{R} & \mathbf{T} \end{array}\right)$$

and restrict ourselves to the case where only states of the same kind are gouped. This restriction allow us to represent the matrices **B** and **C** as

$$\mathbf{B} = \left(\begin{array}{c|c} \mathbf{B}_1 & \mathbf{0} \\ \hline \mathbf{0} & \mathbf{B}_2 \end{array}\right), \quad \mathbf{C} = \left(\begin{array}{c|c} \mathbf{C}_1 & \mathbf{0} \\ \hline \mathbf{0} & \mathbf{C}_2 \end{array}\right).$$

Clearly, we have $\mathbf{B}_1\mathbf{C}_1 = \mathbf{I}$, $\mathbf{B}_2\mathbf{C}_2 = \mathbf{I}$. The necessary and sufficient condition for groupability (5.11) leads us to the equalities

$$\mathbf{C}_1\mathbf{B}_1\mathbf{P}_1\mathbf{C}_1 = \mathbf{P}_1\mathbf{C}_1, \quad \mathbf{C}_2\mathbf{B}_2\mathbf{R}\mathbf{C}_1 = \mathbf{R}\mathbf{C}_1, \quad \mathbf{C}_2\mathbf{B}_2\mathbf{T}\mathbf{C}_2 = \mathbf{T}\mathbf{C}_2. \qquad (5.13)$$

(The first equality is trivially verified if $\mathbf{P}_1 = \mathbf{I}$, that is, if all the recurrent states are absorbing.)

Then, the transition matrix of the grouped chain is

$$\hat{\mathbf{P}} = \mathbf{BPC} = \left(\begin{array}{c|c} \mathbf{B}_1\mathbf{P}_1\mathbf{C}_1 & \mathbf{0} \\ \hline \mathbf{B}_2\mathbf{R}\mathbf{C}_1 & \mathbf{B}_2\mathbf{T}\mathbf{C}_2 \end{array}\right),$$

whence

$$\hat{\mathbf{R}} = \mathbf{B}_2\mathbf{R}\mathbf{C}_1, \quad \hat{\mathbf{T}} = \mathbf{B}_2\mathbf{T}\mathbf{C}_2.$$

The third equation (5.13) shows that $\mathbf{T}_2 = \mathbf{B}_2\mathbf{T}\mathbf{C}_2\mathbf{B}_2\mathbf{T}\mathbf{C}_2 = \mathbf{B}_2\mathbf{T}^2\mathbf{C}_2$ and, in general, $\mathbf{T}^n = \mathbf{B}_2\mathbf{T}^n\mathbf{C}_2$, $n \geqslant 1$. Therefore

$$\begin{aligned} \hat{\mathbf{N}} &= \mathbf{I} + \hat{\mathbf{T}} + \hat{\mathbf{T}}^2 + \ldots = \mathbf{B}_2\mathbf{I}\mathbf{C}_2 + \mathbf{B}_2\mathbf{T}\mathbf{C}_2 + \mathbf{B}_2\mathbf{T}^2\mathbf{C}_2 + \ldots \\ &= \mathbf{B}_2(\mathbf{I} + \mathbf{T} + \mathbf{T}^2 + \ldots)\mathbf{C}_2 = \mathbf{B}_2\mathbf{N}\mathbf{C}_2. \end{aligned}$$

It follows in particular that the mean time to absorption for the grouped chain is given by

$$(\mathsf{E}_{\hat{i}}(\nu))_{\hat{i} \in \hat{T}} = \mathbf{B}_2\mathbf{N}\mathbf{C}_2\mathbf{e} = \mathbf{B}_2\mathbf{N}\mathbf{e} = \mathbf{B}_2(\mathsf{E}_i(\nu))_{i \in T}.$$

An important consequence of this result is the fact that *for a groupable absorbing Markov chain the mean time to absorption has the same value for all starting states in the same subset* $S_k \subset T$. Indeed on account of the very construction of the matrix **B** the last equality allows us to assert that the sum $\sum_{i \in S_k} p(i)\, \mathsf{E}_i(\nu)$ has the same value for all $p(i) > 0$, $i \in S_k$, such that $\sum_{i \in S_k} p(i) = 1$. But it is easy to show that this is possible if and only if $\mathsf{E}_i(\nu)$ does not depend on $i \in S_k$.

Now let us turn to the case of a groupable ergodic Markov chain. Equation (5.12) allows us to write

$$\lim_{n\to\infty} \frac{\hat{\mathbf{P}} + \dots + \hat{\mathbf{P}}^n}{n} = \mathbf{B}\left(\lim_{n\to\infty} \frac{\mathbf{P} + \dots + \mathbf{P}^n}{n}\right)\mathbf{C} = \mathbf{B\Pi C}.$$

Hence the grouped chain is also ergodic and $\hat{\mathbf{\Pi}} = \mathbf{B\Pi C}$. In other words the stationary distribution of the grouped chain is given by

$$\hat{\pi}(\hat{k}) = \sum_{i \in S_k} \pi(i), \quad 1 \leqslant k \leqslant q.$$

Next we have $\hat{\mathbf{Z}} = \mathbf{BZC}$ (by using the infinite series representation of the fundamental matrix — see Theorem 4.5).

It seems that there is no simple connection between $\mathbf{M}$ and $\hat{\mathbf{M}}$. But we can assert that *for a groupable ergodic Markov chain the mean first passage time from a state in S_k to the set S_l has the same value for all the states in S_k*. The proof is obtained by making absorbing all the states in S_l and applying the result we have just established for groupable absorbing Markov chains. In particular, if S_l consists of a single state, then $\hat{m}(\hat{i}, \hat{l})$ may be found from $\mathbf{M}$ whatever $\hat{i}$, $1 \leqslant i \leqslant q$.

5.4.4. It was shown in [195], § 7.3 that the Markov chain associated with the Ehrenfest model can be considered as a grouped Markov chain. Let us define the *microscopic* state of the system consisting of the two urns of the Ehrenfest model (see Example 2 from 2.2) as the vector $\boldsymbol{\alpha} = (\alpha(i))_{1 \leqslant i \leqslant 2l}$ in which the component $\alpha(i)$ equals 1 or 0 according as the ball with number i is in the first or in the second urn, $1 \leqslant i \leqslant 2l$. In all there are 2^{2l} microscopic states. The evolution of the microscopic state of the Ehrenfest model is described by a Markov chain, too. If the system is in a microscopic state $\boldsymbol{\alpha}$, then the choice of a ball and its transfer to the other urn brings the system into a microscopic state $\boldsymbol{\beta}$ that is found from $\boldsymbol{\alpha}$ by changing only one of its components. From any microscopic state it is possible to move in one step to one of $2l$ microscopic states, the transitions each occurring with probability $1/2l$ independently of the states previously observed. It is possible to reach any state $\boldsymbol{\beta}$ from a given state $\boldsymbol{\alpha}$ but it is possible to return to $\boldsymbol{\alpha}$ only in an even number of steps and, in particular, in two steps. The *(microscopic)* chain is therefore ergodic of period 2. Next, if it is possible to reach $\boldsymbol{\beta}$ from $\boldsymbol{\alpha}$ in one step, then the reverse transition is also possible in one step (the probability being $1/2l$ in both cases). It follows that the transition matrix is symmetric. Hence (see 4.5.2) the microscopic chain is reversible and its stationary distribution is the uniform distribution over the 2^{2l} microscopic states (i.e., all the components are equal to 2^{-2l}).

For instance, for the case $l = 2$ there are 16 microscopic states: $\boldsymbol{\alpha}_1 = (0000)$, $\boldsymbol{\alpha}_2 = (0001)$, $\boldsymbol{\alpha}_3 = (0010)$, $\boldsymbol{\alpha}_4 = (0100)$, $\boldsymbol{\alpha}_5 = (1000)$, $\boldsymbol{\alpha}_6 = (0011)$, $\boldsymbol{\alpha}_7 = (0101)$, $\boldsymbol{\alpha}_8 = (0110)$, $\boldsymbol{\alpha}_9 = (1001)$, $\boldsymbol{\alpha}_{10} = (1010)$, $\boldsymbol{\alpha}_{11} = (1100)$, $\boldsymbol{\alpha}_{12} = (0111)$, $\boldsymbol{\alpha}_{13} = (1011)$, $\boldsymbol{\alpha}_{14} = (1101)$, $\boldsymbol{\alpha}_{15} = (1110)$, $\boldsymbol{\alpha}_{16} = (1111)$, and the transition matrix is

$$\begin{pmatrix}
0 & 1/4 & 1/4 & 1/4 & 1/4 & 0 & 0 & 0 & 0 & 0 & 0 & 0 & 0 & 0 & 0 & 0 \\
1/4 & 0 & 0 & 0 & 0 & 1/4 & 1/4 & 0 & 1/4 & 0 & 0 & 0 & 0 & 0 & 0 & 0 \\
1/4 & 0 & 0 & 0 & 0 & 1/4 & 0 & 1/4 & 0 & 1/4 & 0 & 0 & 0 & 0 & 0 & 0 \\
1/4 & 0 & 0 & 0 & 0 & 0 & 1/4 & 1/4 & 0 & 0 & 1/4 & 0 & 0 & 0 & 0 & 0 \\
1/4 & 0 & 0 & 0 & 0 & 0 & 0 & 0 & 1/4 & 1/4 & 1/4 & 0 & 0 & 0 & 0 & 0 \\
0 & 1/4 & 1/4 & 0 & 0 & 0 & 0 & 0 & 0 & 0 & 0 & 1/4 & 1/4 & 0 & 0 & 0 \\
0 & 1/4 & 0 & 1/4 & 0 & 0 & 0 & 0 & 0 & 0 & 0 & 1/4 & 0 & 1/4 & 0 & 0 \\
0 & 0 & 1/4 & 1/4 & 0 & 0 & 0 & 0 & 0 & 0 & 0 & 1/4 & 0 & 0 & 1/4 & 0 \\
0 & 1/4 & 0 & 0 & 1/4 & 0 & 0 & 0 & 0 & 0 & 0 & 0 & 1/4 & 1/4 & 0 & 0 \\
0 & 0 & 1/4 & 0 & 1/4 & 0 & 0 & 0 & 0 & 0 & 0 & 0 & 1/4 & 0 & 1/4 & 0 \\
0 & 0 & 0 & 1/4 & 1/4 & 0 & 0 & 0 & 0 & 0 & 0 & 0 & 0 & 1/4 & 1/4 & 0 \\
0 & 0 & 0 & 0 & 0 & 1/4 & 1/4 & 1/4 & 0 & 0 & 0 & 0 & 0 & 0 & 0 & 1/4 \\
0 & 0 & 0 & 0 & 0 & 1/4 & 0 & 0 & 1/4 & 1/4 & 0 & 0 & 0 & 0 & 0 & 1/4 \\
0 & 0 & 0 & 0 & 0 & 0 & 1/4 & 0 & 1/4 & 0 & 1/4 & 0 & 0 & 0 & 0 & 1/4 \\
0 & 0 & 0 & 0 & 0 & 0 & 0 & 1/4 & 0 & 1/4 & 1/4 & 0 & 0 & 0 & 0 & 1/4 \\
0 & 0 & 0 & 0 & 0 & 0 & 0 & 0 & 0 & 0 & 0 & 1/4 & 1/4 & 1/4 & 1/4 & 0
\end{pmatrix}$$

The microscopic Markov chain can be thought as a random walk on the corner points of the $2l$-dimensional unit cube. (The microscopic states are precisely the corner points of the $2l$-dimensional unit cube of $2l$-dimensional Euclidean space.) The states that are accessible from a given state $\boldsymbol{\alpha}$ are the corner points connected to $\boldsymbol{\alpha}$ by an edge of the cube. There are $2l$ such corner points and the probability of moving to any one is $1/(2l)$. Let us define the distance $\Delta(\boldsymbol{\alpha}, \boldsymbol{\beta})$ between two microscopic states $\boldsymbol{\alpha} = (\alpha(i))_{1 \leqslant i \leqslant 2l}$ and $\boldsymbol{\beta} = (\beta(i))_{1 \leqslant i \leqslant 2l}$ as the smallest number of steps needed to reach $\boldsymbol{\beta}$ from $\boldsymbol{\alpha}$ $\left[\text{clearly, we have } \Delta(\boldsymbol{\alpha}, \boldsymbol{\beta}) = \sum_{i=1}^{2l} |\alpha(i) - \beta(i)|\right]$. Then the above interpretation of the microscopic chain allows us to conclude that the mean value of the first passage time from a given state to another given state depends only on the distance between these states. Let $m_d(l) = m_d$ denote the mean value of the first passage time from a microscopic state $\boldsymbol{\beta}$ to a microscopic state $\boldsymbol{\alpha} \neq \boldsymbol{\beta}$ such that $\Delta(\boldsymbol{\alpha}, \boldsymbol{\beta}) = d$, $0 < d \leqslant 2l$.*) Notice

*) By Theorem 4.7 the mean recurrence time of any microscopic state equals 2^{2l}, the reciprocal of the common value of the components of the stationary distribution.

that the equality $\Delta(\boldsymbol{\alpha}, \boldsymbol{\beta}) = d$ amounts to the fact that $\boldsymbol{\alpha}$ and $\boldsymbol{\beta}$ have exactly d different components. Therefore there are d states $\boldsymbol{\alpha}'$ such that $\Delta(\boldsymbol{\alpha}, \boldsymbol{\alpha}') = 1$, $\Delta(\boldsymbol{\alpha}', \boldsymbol{\beta}) = d-1$, and, similarly, there are $2l-d$ states $\boldsymbol{\alpha}''$ such that $\Delta(\boldsymbol{\alpha}, \boldsymbol{\alpha}'') = 1$, $\Delta(\boldsymbol{\alpha}'', \boldsymbol{\beta}) = d+1$. This leads to the equation

$$m_d = 1 + \frac{d}{2l} m_{d-1} + \frac{2l-d}{2l} m_{d+1}, \quad 0 < d \leqslant 2l,$$

with the convention that $m_0 = m_{2l+1} = 0$. Putting

$$A(k, l) = \sum_{i=0}^{k} \binom{2l}{i} \Big/ \binom{2l-1}{k}, \quad 0 \leqslant k \leqslant 2l-1,$$

then it is easily found that

$$m_d = \sum_{s=1}^{d} A(2l-s, l), \quad 0 < d \leqslant 2l.$$

Now, let us prove that the microscopic Markov chain associated with the Ehrenfest model is groupable with respect to the partition $S = S_0 \cup \ldots \cup S_{2l}$, where S_i is the set of all the microscopic states having i components equal to 1 (this amounts to the presence of i balls in the first urn), $0 \leqslant i \leqslant 2l$. The set S_i has $\binom{2l}{i}$ elements, $0 \leqslant i \leqslant 2l$. From any microscopic state in S_i, the chain moves either to a microscopic state in S_{i-1} or to a microscopic state in S_{i+1}, the corresponding probabilities being $i/2l$ and $1-i/2l$, respectively. These probabilities are the same for all the microscopic states in S_i. Hence the condition of grouping is satisfied and the grouped chain is precisely the Markov chain associated with the Ehrenfest model (this may be called the *macroscopic* chain).

We shall use the above results to obtain alternative formulae for the mean first passage times $m(i, j)$, $0 \leqslant i, j \leqslant 2l$ for the macroscopic chain (compare with 4.6.1). Notice that for this chain

$$m(i+1, 0) = m(i+1, i) + m(i, 0), \quad 0 < i < 2l,$$

since 0 is accessible from $i+1$ only passing through i. Similarly,

$$m(i-1, 2l) = m(i-1, i) + m(i, 2l), \quad 0 < i < 2l.$$

In general

$$m(i, j) = \sum_{k=i}^{j-1} m(k, k+1), \quad i < j,$$

$$m(i, j) = \sum_{k=j}^{i-1} m(k+1, k), \quad i > j.$$

Therefore it is sufficient to determine only the values $m(i, 0)$, $0 < i \leqslant 2l$, and $m(i, 2l)$, $0 \leqslant i < 2l$. Noticing that states 0 and $2l$ originate from the microscopic states $(0, \ldots, 0)$ and $(1, \ldots, 1)$, respectively, and taking into account the property stated at the end of the preceding subparagraph we conclude that $m(i, 0) = m_i$, $0 < i \leqslant 2l$, $m(i, 2l) = m_{2l-i}$, $0 \leqslant i < 2l$. Hence

$$m(i, j) = \sum_{k=i}^{j-1} A(i, l),\ i<j,$$

$$m(i, j) = \sum_{k=j}^{i-1} A(2l-k-1,\ l),\ i>j.$$

5.4.5. We close this paragraph with a remark concerning the grouping procedure.

In 5.4.1 we gave conditions under which a function of a Markov chain is a new Markov chain whose transition probabilities are independent of the initial distribution of the original chain. If we drop out this last requirement, i.e., we require only that the sequence $(f(X(n))_{n \geqslant 0}$ be a Markov chain under $\mathsf{P}_{\mathbf{p}}$, where $\mathbf{p}$ is a *fixed* initial distribution, then we come to the concept of *weak* groupability. Necessary and sufficient conditions for a Markov chain to be weakly groupable with respect to a partition $S_1 \cup \ldots \cup S_q$ and an initial distribution $\mathbf{p}$ are known. The interested reader may consult [45] and [195], § 6.4.

5.5 EXPANDED MARKOV CHAINS

5.5.1. By grouping the states of a Markov chain under the conditions we have stated one may obtain a new Markov chain with a smaller number of states. At the expense of getting less information about the original chain we are faced with a more manageable Markov chain.

It is also possible to go in the opposite direction, i.e., to expand the state space. Now, at the expense of dealing with a less manageable Markov chain we can get more detailed information about the original chain.

Let $(X(n))_{n \geqslant 0}$ be a Markov chain with state space S and transition matrix $\mathbf{P}$. We shall consider the sequence of random variables $(X(0), X(1))$, $(X(1), X(2))$, ... It is easy to prove that this is (always) a Markov chain (note the difference with respect to the case of grouping of states) with state space $\bar{S}$ = all the pairs (i, j), $i, j \in S$, such that $p(i, j) > 0$. [The exclusion of a pair (i, j) such that $p(i, j) = 0$ is due to the fact that for such a pair we have $\mathsf{P}_{\mathbf{p}}(X(n) = i,\ X(n+1) = j) = 0$ whatever $n \geqslant 0$ and the initial distribution $\mathbf{p}$ of the original chain.]

Put $Y(n) = (X(n),\ X(n+1))$, $n \geqslant 0$. The sequence $(Y(n))_{n \geqslant 0}$ is said to be an *expanded* Markov chain. Its transition probabilities are given by

$$\begin{aligned} \bar{p}((i,j),(k,l)) &= \mathsf{P}(Y(m+1) = (k,l) \mid Y(m) = (i,j)) \\ &= \mathsf{P}(X(m+2) = l,\ X(m+1) = k \mid X(m+1) = j,\ X(m) = i) \\ &= \delta(j,k)\, p(j,l),\ (i,j),\ (k,l) \in \bar{S}. \end{aligned}$$

More generally, the n-step transition probabilities are given by

$$\bar{p}(n, (i, j), (k, l)) = \mathsf{P}(Y(m+n) = (k, l) \mid Y(m) = (i, j))$$

$$= \mathsf{P}(X(m+n+1) = l,\ X(m+n) = k \mid X(m+1) = j,\ X(m) = i)$$

$$= p(n-1, j, k)\, p(k, l),\quad n \geqslant 1,\ (i, j),\ (k, l) \in \bar{S}. \tag{5.14}$$

Hence, if the original Markov chain is ergodic, the expanded chain is also ergodic and the periods of the two Markov chains are equal. Similarly, an absorbing Markov chain leads to an expanded chain which is absorbing, too. A state $(i, j) \in \bar{S}$ is absorbing in the expanded chain if and only if $i = j$ and state i is absorbing in the original chain.

It is interesting to note that the original chain can be obtained from the expanded chain by grouping its states, the corresponding partition being $\bar{S} = \bigcup_{i \in S} \bar{S}_i$, where $\bar{S}_i = ((k, i) \in \bar{S} : k \in S)$. The proof of this assertion is left to the reader. It follows (see 5.4.3) the implications from the preceding paragraph are valid in both directions.

5.2.5. The stationary distribution $(\bar{\pi}((i, j)))_{(i,j) \in \bar{S}}$ of an ergodic expanded Markov chain is found from the stationary distribution $(\pi(i))_{i \in S}$ of the original chain by $\bar{\pi}((i, j)) = \pi(i)\, p(i, j)$, $(i, j) \in \bar{S}$. Indeed, we have

$$\bar{\pi}((i, j)) > 0,\quad (i, j) \in \bar{S},$$

$$\sum_{(i,j) \in \bar{S}} \bar{\pi}((i, j)) = \sum_{i, j \in S} \pi(i)\, p(i, j) = \sum_{i \in S} \pi(i) = 1$$

and

$$\sum_{(i,j) \in \bar{S}} \bar{\pi}((i, j))\bar{p}((i, j), (k, l)) = \sum_{i, j \in S} \pi(i) p(i, j) \delta(j, k) p(j, l)$$

$$= \sum_{j \in S} \pi(j) \delta(j, k) p(j, l) = \pi(k) p(k, l) = \bar{\pi}((k, l))$$

whatever the state $(k, l) \in \bar{S}$.

5.5.3. Equation (5.14) allows us to compute immediately the fundamental matrix for an absorbing or ergodic expanded Markov chain. Thus we have

$$\bar{n}((i, j), (k, l)) = \delta((i, j), (k, l)) + \sum_{n \geqslant 1} \bar{p}(n, (i, j), (k, l))$$

$$= \delta((i, j), (k, l)) + p(k, l) \sum_{n \geqslant 1} p(n-1, j, k)$$

$$= \delta((i, j), (k, l)) + p(k, l) n(j, k)$$

and

$$\bar{z}((i, j), (k, l)) = \delta((i, j), (k, l)) + \sum_{n \geq 1} [\bar{p}(n, (i, j), (k, l)) - \bar{\pi}((k, l))]$$

$$= \delta((i, j), (k, l)) + \sum_{n \geq 1} [p(n - 1, j, k,) \; p(k, l) - \pi(k) \; p(k, l)]$$

$$= \delta((i, j)), (k, l)) + p(k, l) \; (z(j, k) - \pi(k))$$

whatever the states (i, j), $(k, l) \in \bar{S}$.

5.5.4. One of the most important applications of the expanded Markov chain is as follows. Note first that $\nu_n((i, j))$, the number of occurrences of state $(i, j) \in \bar{S}$ in the expanded chain in the first n steps, is equal to the number $\nu_{n+1}(i, j)$ of values $0 \leq m \leq n-1$ for which $X(m) = i$ and $X(m+1) = j$, $n \geq 1$. In many applications the transition matrix **P** is unknown and its entries should be estimated from data. For an ergodic Markov chain an estimate of the transition probability $p(i, j)$ is

$$\nu_{n+1}(i, \; j)/\nu_n(i)$$

$\left[\nu_n(i) = \sum_{j \in S} \nu_{n+1}(i, \; j)\right.$ has been defined on p. $\left.140\right]$ Indeed, by the law of large numbers, the ratios $\nu_n(i, j)/n$ and $\nu_n(i)/n$ converge to $\bar{\pi}((i, j)) = \pi(i) \; p(i,j)$ and $\pi(i)$, respectively, as $n \to \infty$, as stated in Theorem 1.4.

For the general problem of estimating the transition matrix **P** the reader may consult the papers [29, 42, 365, 399, and 408] and the monographs [235 and 276].

5.6 EXTENDING THE CONCEPT OF A HOMOGENEOUS FINITE MARKOV CHAIN

5.6.1. The concept of a homogeneous finite Markov chain can be extended by dropping out various of the underlying assumptions and replacing them by more general ones. We shall discuss here the following situations: 1) The state space S is no longer a finite set but a countably infinite one; 2) The very concept of Markovian dependence is extended by postulating dependence either on a segment of fixed length into the past or on the whole past. In Chapter 8 we shall take up the situation for which the time parameter no longer assumes the discrete values 0, 1, 2, ... but varies continuously on the real positive semiaxis $[0, \infty)$.

5.6.2. A moment's reflexion shows that the existence theorem and all the results concerning the Markov property or the strong Markov property are still valid for the case where the state space S is countably infinite. The recurrent-transient dichotomy as well as the grouping of states into classes also hold. But further on there occur phenomena which are impossible in finite Markov chains. Thus, it may happen that a denumerable Markov chain has no recurrent state. This leads to the fact that the concepts of closed class and recurrent class are no longer identical. Next, there are denumerable Markov chains for which the absorption probability in the set of recurrent states is less than 1 (it may be even 0). Finally, there are recurrent denumerable Markov chains which do not possess a stationary distribution.

The standard reference for denumerable Markov chains is the monograph [52]. A concise treatment may be found in the first volume of the treatise [166].

To close these brief considerations we should point out that the Perron-Frobenius theory developed for nonnegative square matrices of finite order cannot be extended in all its generality to nonnegative matrices having countably many rows and columns. The interested reader may consult the monograph [335] and the memoir [388].

5.6.3. A more general dependence concept is that of multiple Markovian dependence of order $r \geqslant 2$. The Markov chains we have dealt with so far may be thought of as having multiplicity 1, hence they may be called *simple*. Without any loss of generality we shall consider the case $r = 2$. A sequence of random variables $(X(n))_{n \geqslant -1}$ with values in a (finite or countably infinite) set S is said to be a *(homogeneous) multiple Markov chain of order* 2 *(a double Markov chain,* for short) if and only if

$$\mathsf{P}(X(n+1) = i_{n+1} \mid X(n) = i_n, \ldots, X(-1) = i_{-1})$$
$$= \mathsf{P}(X(n+1) = i_{n+1} \mid X(n) = i_n, X(n-1) = i_{n-1})$$

for all $i_{-1}, \ldots, i_{n+1} \in S$, and $n \geqslant 0$, the value of the last conditional probability depending on i_{n-1} and i_n but not on n.

It is easily seen that if $(X(n))_{n \geqslant -1}$ is a homogeneous double Markov chain then the sequence

$$(X(-1), X(0)), (X(0), (X(1)), \ldots, (X(n-1), X(n)), \ldots \tag{5.15}$$

is a homogeneous (simple) Markov chain with state space $S^2 = ((i, j)\colon i, j \in S)$.

In this way with any multiple Markov chain one can associate a simple Markov chain. This is the origin of the rooted prejudice that the study of multiple Markov chains reduces to that of simple ones. The following considerations (see [163]) shows that things are different. If a state (i, j) is recurrent in the Markov chain (5.15), then its components i and j will also be recurrent in the original double Markov chain. But (if S is countably infinite) a state i may be recurrent in the double Markov chain without being a component of a recurrent state (i, j) in the simple Markov chain (5.15).

Important material on the asymptotic properties of multiple Markov chains are to be found in the work by G. Mihoc [269, 270, 273, and 274].

A more general dependence concept than that of multiple Markovian dependence was introduced in 1935 by O. Onicescu and G. Mihoc. A sequence of random variables $(X(n))_{n\geqslant 0}$ with values in a (finite or countably infinite) set S is said to be a *chain with complete connections* (or an *O-M chain*) if and only if the conditional probability

$$\mathsf{P}(X((n+1) = i_{n+1} \mid X(n) = i_n, \ldots, X(0) = i_0)$$

is a given function $\varphi_{i_n i_{n+1}}$ of the conditional probabilities

$$\mathsf{P}(X(n) = i \mid X(n-1) = i_{n-1}, \ldots, X(0) = i_0), \quad i \in S,$$

for all $i_0, \ldots, i_{n+1} \in S$, and $n \geqslant 0$. The case of a simple Markov chain obtain when the functions φ_{ij}, $i, j \in S$, are constant: $\varphi_{ij}(\cdot) = p(i, j)$. (How should one define the φ_{ij} to get multiple Markov chains of order r?)

The existence of O-M chains (and, implicitly, that of multiple Markov chains) can be proved in a similar manner to that of simple Markov chains (see 2.1.2).

For an extensive study of O-M chains and their generalizations the reader is referred to the monograph [167].

EXERCISES

5.1 Prove that

i) The limit matrix $\mathbf{\Pi}$ is stable if and only if there is only one recurrent class (either periodic or aperiodic);

ii) The limit matrix $\mathbf{\Pi}$ is positive if and only if the corresponding Markov chain is ergodic (so that $\mathbf{\Pi}$ should be stable);

iii) The limit matrix $\mathbf{\Pi}$ is stable and there is actual convergence of $\mathbf{P}^n$ to $\mathbf{\Pi}$ as $n \to \infty$ if and only if $\mathbf{P}$ is mixing.

5.2 Prove that if all the recurrent states are aperiodic then $\lim_{n\to\infty} \mathbf{P}^n = \mathbf{\Pi}$ and that the convergence is geometrically fast.

(Hint: use Theorem 5.1. Next, a) If i and j are in different recurrent classes then $p(n, i, j) = 0$; b) If i and j are in the same recurrent class then by Theorem 4.2 $p(n, i, j)$ approaches $\pi(i, j)$ $(= \pi(j, j))$ geometrically fast; c) If i is transient and j recurrent then $p(m+n, i, j) = \sum_{k\in S} p(m, i, k)\, p(n, k, j) = \sum_{k\in \mathcal{C}(j)} p(m, i, k)\, p(n, k, j) + \sum_{k\in T} p(m, i, k)\, p(n, k, j)$ and on account of b) (applied to $p(n, k, j)$) and Exercise 2.8 geometrically fast convergence still follows.)

5.3 *Taboo probabilities.* Given a subset $H \subset S$ define for all $i, j \in S$ and natural numbers n the "taboo" transition probability (see [52], I. § 9)

$$_Hp(1, i, j) = p(i, j),$$

$$_Hp(n, i, j) = \mathsf{P}_i(X(m) \notin H,\ 1 \leqslant m < n,\ X(n) = j), \quad n \geqslant 2.$$

Clearly, ${}_j p(n, i, j) = f(n, i, j)$, ${}_i p(n, i, j) = g(n, i, j)$, $i, j \in S$, $n \geqslant 1$ (see Chapter 2). Consider a Markov chain obtained from the original one by making absorbing all the states in H, i.e., the Markov chain with transition probabilities given by

$$\tilde{p}(i, j) = \begin{cases} p(i, j), & \text{if } i \notin H,\ j \in S \\ \delta(i, j), & \text{if } i \in H,\ j \in S. \end{cases}$$

Show that

$${}_H p(n, i, j) = \sum_{k \notin H} p(i, k)\, \tilde{p}(n-1, k, j)$$

whatever $i \in S$, $j \notin H$, $n \geqslant 1$. In particular,

$${}_H p(n, i, j) = \tilde{p}(n, i, j)$$

whatever $i, j \notin H$, $n \geqslant 1$. Deduce that

$${}_H p(m+n, i, j) = \sum_{k \notin H} {}_H p(m, i, k)\, {}_H p(n, k, j)$$

whatever $m, n \geqslant 0$, $i, j \notin H$, with the convention

$${}_H p(0, i, j) = \delta(i, j),\ i \notin H.$$

5.4 Consider a reversible ergodic Markov chain. Prove that if it is groupable with respect to some partition then the grouped Markov chain is also reversible.

5.5 Consider a groupable ergodic Markov chain. Are the periods of the original and the grouped Markov chains equal?

5.6 Prove that the reversed chain for an ergodic expanded Markov chain may be obtained by expanding the reversed chain for the original Markov chain.

5.7 Construct the expanded chain for the Markov chain occurring in Example 4 from 2.2. Express in terms of the expanded chain the duration τ of the epidemic. (H i n t: τ is the first passage time to the set $((0, 0), \ldots, (r, r))$.

5.8 Prove computationally the equality of the $m(i, j)$ in 4.6.1 and 5.4.4.

5.9 *Whittle's formula* [399]. Consider a Markov chain $(X(n))_{n \geqslant 0}$ and denote by $\nu_n(k, l)$ the number of values $0 \leqslant m \leqslant n-1$ such that $X(m) = k$ and $X(m+1) = l$. Clearly, $\sum_{k,l \in S} \nu_n(k, l) = n$, and

$$\sum_{l \in S} \nu_n(k, l) - \sum_{l \in S} \nu_n(l, k) = \delta(k, X(0)) - \delta(k, X(n)),\ k \in S.$$

i) Prove that the probability

$$\mathsf{P}_i(\nu_n(k, l) = n_{kl},\ k,\ l \in S),$$

where the n_{kl} are constrained by the equations $\sum_{k,l \in S} n_{kl} = n$, $\sum_{l \in S} n_{kl} - \sum_{l \in S} n_{lk} = \delta(k, i) - \delta(k, j)$, $k \in S$, for some $j \in S$ is given by

$$N_{ij}^{(n)} \frac{\prod_{k \in S} n_{k.}!}{\prod_{k,l \in S} n_{kl}!} \prod_{k,l \in S} (p(k, l))^{n_{kl}}.$$

Here $n_{k.} = \sum_{l \in S} n_{kl}$, $k \in S$, and $N_{ij}^{(n)}$ is the ijth cofactor of the matrix $(n_{ij}^*)_{i,j \in S}$ defined by

$$n_{ij}^* = \begin{cases} \delta(i, j) - n_{ij}/n_{i.}, & \text{if } n_{i.} > 0, \\ \delta(i, j), & \text{if } n_{i.} = 0. \end{cases}$$

(We use the convention $0^0 = 1$.) (H i n t [29]: induction on n.)

5.10 Show that whatever i, k, $l \in S$ and $n \geq 1$

$$\mathsf{E}_i(\nu_n(k, l)) = p(k, l) \sum_{m=0}^{n-1} p(m, i, k).$$

Relate this result to the expanded Markov chain. [H i n t: put $u_m(k, l) = 1$ if $X(m) = k$, $X(m+1) = l$, and $= 0$, otherwise, $0 \leq m \leq n-1$. Then $\sum_{m=0}^{n-1} u_m(k, l) = \nu_n(k, l)$. (For an inductive proof see [262], p. 122.)]

CHAPTER 6

Applications of Markov Chains in Psychology and Genetics

This chapter is aimed at emphasizing the fundamental part played by finite Markov chains in mathematical modelling in psychology and genetics.

These two fields were chosen on account of their present interest. On the one hand, mathematical psychology via mathematical learning theory helped in an essential manner in introducing mathematical methods in the social sciences. At the same time, the Romanian school of probability has contributed substantially to the present state of mathematical learning theory. On the other hand, from the point of view of the life sciences, our century is considered as the century of genetics.

6.1 MATHEMATICAL LEARNING THEORY

6.1.1. *Learning* is one of the most important concepts of the theory of behaviour. According to H. Piéron (*Vocabulaire de la psychologie*, Paris, 1957) learning is the adaptive modification of behaviour during repeated trials. By *mathematical learning theory* we mean the body of research methods and results concerned with the conceptual representation of learning phenomena, the mathematical formulation of assumptions or hypotheses about learning, and the derivation of testable theorems (R. R. Bush and W. K. Estes). The purpose of mathematical learning theory is to provide simple quantitative descriptions of the processes which are basic to modifications of behaviour.

Mathematical learning theory should be considered as part of what is now called *mathematical psychology*, a field that has developed rapidly during the last 30 years in the context of experimental psychology. Mathematical psychology is not identical with the utilization of mathematical methods in psychology, because this phrase usually refers to a body of statistical techniques that are often used as auxiliaries of psychology. With mathematical psychology a more advanced integration of psychology and mathematics is reached where the latter is used in a more complex manner to study the problems of the former. In this respect the term "mathematical psychology" as well as that of "experimental psychology" designates a method and not a specific domain of application.

6.1.2. Mathematical psychology expresses psychological theories in mathematical form by means of mathematical models. In a quite general context, a mathematical model is an abstract, simplified, mathematical setting related to a precise reality and created for a well-defined purpose. In particular, learning models are devices for providing simple descriptions of basic learning processes. Prior to 1953, learning models were concerned with predicting at most the mean performance curves obtained from the experiments. Although a few of the earlier learning models involved probability and random processes (the fact was unavoidable on account of the well known variability of behaviour data) no properties other than the mean curve were deduced. A main feature of the modern formulations is that they imply random processes. The start of this new development should be located in the early fifties when several papers in the American psychological literature (G. A. Miller, F. C. Frick, W. K. Estes, C. Mueller, R. R. Bush, F. Mosteller) put foward in a formalized manner the following idea: the behaviour changes observed in the course of a learning experiment can be described by mechanisms modifying the probabilities of the possible responses in terms of random events occurring from trial to trial. From this point of view, the study of learning is concerned with the trial-to-trial probability changes that characterize a random sequence (e.g., a Markov chain). The above idea suitably applied to various experimental situations has led to a large number of stochastic learning models. Since the publication of the now classical book *Stochastic models for learning* (1955) by R. R. Bush and F. Mosteller, experimental and theoretical work on stochastic learning models has been continuing *).

6.2 THE PATTERN MODEL

6.2.1. In what follows we present a stochastic model for learning based on the so-called *stimulus sampling theory* proposed by W.K. Estes in 1950. Our choice is dictated by the fact that the model leads to a finite Markov chain. For a brilliant general analysis of the impact of the idea of finite Markov chains in the psychology of learning the interested reader should consult [128].

---

*) It should be noted that the Bush-Mosteller models involve either chains with complete connections (see 5.6.3) or generalizations of them. The theory initiated by O. Onicescu and G. Mihoc (of which the American authors were unaware) thus preceded by almost two decades its first major application. In this fact one has to see a confirmation of the opinion held by the distinguished Romanian mathematician Dan Barbilian (alias the great poet Ion Barbu) as to "the curious solidarity between theoretical depth and practical efficiency so often verified in mathematics".

In the last two decades, work accomplished by Romanian probabilists was instrumental in establishing a general theory of stochastic learning models based on the concept of a random system with complete connections, which greatly extends that of a chain with complete connections (the interested reader may consult the monographs [167 and 292]).

At the level of experimental interpretation most contemporary learning theories make use of a common conceptualization of learning in terms of *stimulus*, *response*, and *reinforcement*. The first notion refers to the environment with respect to which behaviour is being observed, the second one to the class of observable behaviours whose measurable properties change during learning, and the third one to the experimental operations or events believed to be critical in producing learning. For example, in an experiment consisting in learning the English equivalents of words of a foreign language, the stimulus might amount to presentation of the printed foreign word alone, the response measure to the relative frequency with which the subject is able to supply the English equivalent from memory, and reinforcement to paired presentation of the stimulus and response words.

The starting point of Estes' theory (cf. [9]) is the representation of the totality of stimulus conditions (= stimuli) that may be effective during the course of an experiment in terms of a mathematical set (that, as a rule, is assumed to be finite). The stimulation effective on any one trial is represented by a random sample from the total population of stimuli. Here, we will restrict our attention to the case where exactly one stimulus is sampled on each trial.*) In this special case each stimulus may be interpreted as a *pattern* of stimulation, and the model is consequently referred as the *pattern model*. At the beginning of every trial each pattern is conditioned to (i.e., determines) a precisely defined response. Therefore, the response made on any one trial is precisely that to which the sampled pattern is conditioned. Next, the conditioning of the sampled pattern may change when the response made is not reinforced. To clarify this assertion let us denote by $A_1, \ldots, A_r$ the mutually exclusive and exhaustive responses available to the subject. The possible experimenter-defined outcomes of a trial (reward, punishment, etc.) are classified by their effect on response probability and are represented by mutually exclusive and exhaustive reinforcing events $E_0, E_1, \ldots, E_r$. The event $E_i (i \neq 0)$ indicates that response A_i is reinforced and this means it occurs when the outcome of a trial increases the probability of response A_i in the presence of the given stimulus provided that this probability is not already at its maximum value. The event E_0 represents any trial outcome that reinforces none of the A_i. The subject responses and the experimenter-defined outcomes are observable, but the reinforcing events are purely hypothetical.

Coming back to the change in conditioning of the sampled pattern, one postulates the existence of a conditioning parameter $0 < c \leqslant 1$ which represents the probability that the pattern, if not conditioned to A_i, becomes conditioned to A_i if the corresponding reinforcing event E_i occurs.

To summarize, the temporal order of any trial is: (state of conditioning at beginning of trial) → (sampling of a pattern) → (response) →

*) For more general sampling procedures see Exercise 6.1.

(outcome) → (reinforcing event) → (state of conditioning at beginning of next trial).

6.2.2. The axioms of the model formalize the previous ideas in a precise manner. They are divided into three groups. The first group deals with the conditioning of sampled patterns, the second one with the sampling of patterns, and the third one with responses.

Conditioning Axioms

C1. At the beginning of every trial each pattern is conditioned to exactly one response.

C2. A pattern sampled on a trial becomes conditioned with probability c to the response (if any) that is reinforced on the trial. It does not change its conditioning with probability $1-c$. (Consequently, if it is already conditioned to the reinforced response, it remains so.)

C3. If no reinforcement occurs on a trial (i.e., E_0 occurs), there is no change in conditioning on that trial.

C4. Patterns that are not sampled on a trial do not change their conditioning on that trial.

C5. The change of conditioning of the sampled pattern on a trial is (stochastically) independent of the whole previous course of the experiment.

Sampling Axioms

S1. Exactly one pattern is sampled on each trial.

S2. If there are l patterns, each has probability $1/l$ of being sampled on any trial.

Response Axiom

R1. On any trial that response is made to which the sampled pattern is conditioned.

6.2.3. Here are three examples of learning experiments to which the pattern model was more or less successfully applied.

Experiment 1 (*Probability learning*). A human subject is seated before a panel to which are affixed two lamps numbered 1 and 2. The subject's task is to predict on each trial which of the two lamps will flash. Response A_i is the prediction of light i, $i = 1,2$. After the subject has made his prediction he is permitted to observe which of the lamps flashes (the actual outcome). Under suitable conditions the flashing of a light will reinforce, in some degree, the prediction of that light, regardless of what light was predicted on that trial. The special case when the flashing of light i has some fixed probability π_i, $i = 1, 2$, which is constant over trials and does not depend on the response made is the first and most often studied version of the experiment. It was found that when the number of trials becomes large the proportion of A_i responses is very close to π_i, $i = 1, 2$. This probability matching tendency justifies the nomenclature of the experiment. The result is somewhat surprising, since the frequency of correct prediction is maximized by always predicting the light which flashes most frequently.

Experiment 2 (T-*maze learning*). On each trial an animal, say a rat, is placed at the bottom of a T-shaped allay. Eventually he enters the top of the maze and goes to the end of the left (response A_1) or right arm (response A_2). There he may receive food, an outcome which reinforces the response just made, or may simply be detained, an outcome which, in some cases, reinforces the other response. No tendency to match response to outcome probability was noticed in rats. Practically all of them develop preference for the response with the highest probability of yielding food.

Experiment 3 (*Avoidance learning*). On each trial a dog can avoid an electric shock if he jumps over a barrier (response A_1) shortly after a warning signal. Otherwise (response A_2) he must clear the barrier to escape the shock. The only possible outcomes are avoidance and shock, both appearing to reinforce jumping.

6.2.4. From now on, for the sake of simplicity, we shall limit consideration to the case $r = 2$.*) Thus, there will be two possible responses A_1 and A_2 and there reinforcing events E_0, E_1, and E_2. By *state* of the subject at the beginning of trial n, denoted by $X(n)$, we shall mean the number of patterns conditioned to response A_1 at the time. Therefore the possible subject states are the integral values $0, 1, \ldots, l$. According to Axiom S2 the conditional probability of making response A_1 on a trial, given the subject is in state i at the beginning of that trial, equals i/l, $0 \leqslant i \leqslant l$.

Let us denote by $A_s(n)$ and $E_t(n)$ the occurrence of response A_s and reinforcing event E_t on trial n, respectively, $s = 1, 2$; $t = 0, 1, 2$; $n \geqslant 0$. Clearly, $A_s(n)$ and $E_t(n)$ are random events.

Since only one pattern is sampled per trial, the subject can go from state i only to one of the three states $i-1$, i, or $i+1$ on any given trial. We now proceed to compute the conditional probabilities

$$\mathsf{P}(X(n+1) = j \mid X(n) = i), \quad j = i-1, \; i, \; i+1.$$

The tree in Figure 6.1 (6.2) indicates the transition from state i at the beginning of trial n to the new state at the beginning of trial $n+1$, given that the reinforcing event $E_1(E_2)$ occurred. Clearly,

$$\{X(n) = i\} \cap \{X(n+1) = i+1\} \subset A_2(n) \cap E_1(n), \; 0 \leqslant i < l,$$

$$\{X(n) = i\} \cap \{X(n+1) = i-1\} \subset A_1(n) \cap E_2(n), \; 0 < i \leqslant l,$$

*) For the case $r > 2$ the reader should consult [49].

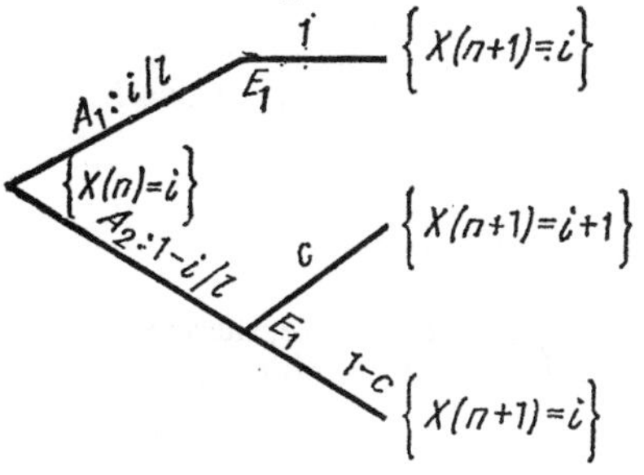

Figure 6.1

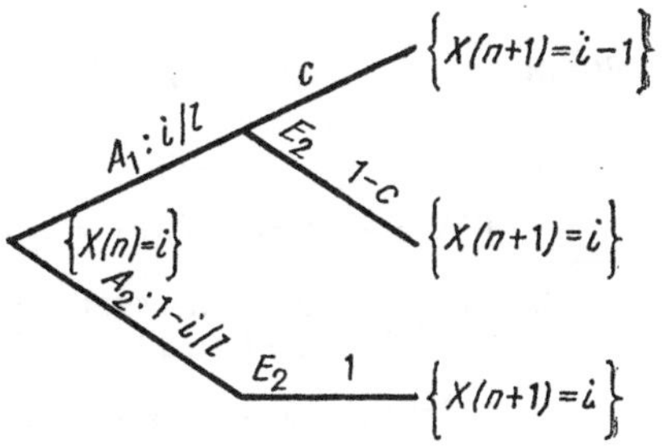

Figure 6.2

whence, on account of the conditioning axioms,

$$\mathsf{P}(X(n+1) = i+1 \mid X(n) = i)$$
$$= \mathsf{P}(\{X(n+1) = i+1\} \cap E_1(n) \cap A_2(n) \mid X(n) = i)$$
$$= \mathsf{P}(X(n+1)=i+1 \mid E_1(n) \cap A_2(n) \cap \{X(n)=i\}) \mathsf{P}(E_1(n) \mid A_2(n) \cap \{X(n)=i\}) \times$$
$$\times \mathsf{P}(A_2(n) \mid X(n) = i) = c\, \mathsf{P}(E_1(n) \mid A_2(n) \cap \{X(n) = i\}) \left(1 - \frac{i}{l}\right),$$

$$\mathsf{P}(X(n+1) = i-1 \mid X(n) = i)$$
$$= \mathsf{P}(\{X(n+1) = i-1\} \cap E_2(n) \cap A_1(n) \mid X(n) = i)$$
$$\mathsf{P}(X(n+1)=i-1 \mid E_2(n) \cap A_1(n) \cap \{X(n)=i\})\ \mathsf{P}(E_2(n) \mid A_1(n) \cap \{X(n)=i\}) \times$$
$$\times \mathsf{P}(A_1(n) \mid X(n) = i) = c\, \mathsf{P}(E_2(n) \mid A_1(n) \cap \{X(n) = i\}) \frac{i}{l}.$$

6.2.5. To ensure the Markov property for the sequence of subject states on the successive trials we should introduce a

Reinforcing Axiom

Re 1. The probability of a reinforcing event on any trial, conditional on the whole previous course of the experiment, depends at most on the response of the given trial, without depending on the trial number.

Therefore, if we set $\mathsf{P}(E_1(n) \mid A_2(n)) = \pi_{21}$, $\mathsf{P}(E_2(n) \mid A_1(n)) = \pi_{12}$, the equations of the previous subparagraph allow us to write

$$\begin{aligned} \mathsf{P}(X(n+1) = i+1 \mid X(n) = i) &= c\pi_{21}\left(1 - \frac{i}{l}\right), \quad 0 \leqslant i < l, \\ \mathsf{P}(X(n+1) = i-1 \mid X(n) = i) &= c\pi_{12}\frac{i}{l}, \quad 0 \leqslant i \leqslant l, \end{aligned} \tag{6.1}$$

whence

$$\mathsf{P}(X(n+1)=i \mid X(n)=i) = 1 - c\left[\pi_{12}\frac{i}{l} + \pi_{21}\left(1-\frac{i}{l}\right)\right], \quad 0 \leqslant i \leqslant l.$$

Clearly,

$$\mathsf{P}(X(n+1)=j \mid X(n)=i) = 0, \; j \neq i-1, \; i, \; i+1.$$

The Markov property of the sequence $(X(n))_{n \geqslant 0}$ is immediate, the axioms assumed ensuring the equality of the conditional probability

$$\mathsf{P}(X(n+1)=j \mid X(n)=i, \; X(n-1)=i', \ldots)$$

to

$$\mathsf{P}(X(n+1)=j \mid X(n)=i).$$

It is easily seen that the chain is regular if $\pi_{21}, \pi_{12} > 0$ except for the case $\pi_{12} = \pi_{21} = c = 1$ (when we obtain again the Markov chain associated with the Ehrenfest model!).

6.2.6. The fact that the Markov chain associated with the Ehrenfest model is a special case of the Markov chain associated with the pattern model is not at all fortuitous. Indeed, the latter model can be regarded as an urn scheme as follows. Assume there are l balls either black or white. (The balls are to be interpreted as a realization of the patterns.) At any time 0, 1, 2, ... a ball is chosen at random. If the drawn ball is white (response A_1), then with probability $1-b$ it is replaced and with probability $b = c\pi_{12}$ it is discarded while a black ball is replaced instead in the urn. It the drawn ball is black (response A_2), then with probability $1-a$ it is replaced and with probability $a = c\pi_{21}$ it is discarded while a white ball is replaced instead in the urn. The reader will easily check that the evolution of the number of white balls in the urn is that of a Markov chain with transition probabilities given by equations (6.1). In the special case $a = b = 1$ the urn scheme considered reduces to the Onicescu-Mihoc scheme (p. 69) whose associated Markov chain coincides with the Markov chain associated with the Ehrenfest model.

Notice that the above urn scheme interpretation of the pattern model helps to make concrete one's intuitive ideas about the learning process. At the same time it obscures certain important features of the underlying psychological theory.

6.3 THE MARKOV CHAIN ASSOCIATED WITH THE PATTERN MODEL

6.3.1. We shall prove that the transition matrix with components given by (6.1) is diagonalizable by determining its eigenvalues and the eigenvectors corresponding to them.

The left eigenvectors $\mathbf{u}' = (u(0), ..., u(l))$ satisfy the equation $\mathbf{u}'\mathbf{P} = \lambda \mathbf{u}'$, i.e., the system

$$(1-a)\, u(0) + \frac{b}{l}\, u(1) = \lambda u(0),$$

$$au(0) + \left(1 - \frac{b}{l} - \frac{l-1}{l}\, a\right) u(1) + \frac{2b}{l}\, u(2) = \lambda u(1),$$

. . . .

$$\frac{2a}{l}\, u(l-2) + \left(1 - \frac{l-1}{l}\, b - \frac{a}{l}\right) u(l-1) + bu(l) = \lambda u(l-1),$$

$$\frac{a}{l}\, u(1-l) + (1-b)\, u(l) = \lambda u(l),$$

where $a = c\pi_{21}$, $b = c\pi_{12}$.

If we consider the polynomial

$$u(z) = \sum_{r=0}^{l} u(r)\, z^r,$$

multiply the above equations by $1, z, ..., z^{l-1}, z^l$, respectively, and then sum them up we obtain

$$\frac{a}{l} \sum_{r=0}^{l} (l-r)\, u(r) z^{r+1} + \sum_{r=0}^{l} \left(1 - \frac{rb}{l} - \frac{l-r}{l}\, a\right) u(r)\, z^r + {} $$

$$+ \frac{b}{l} \sum_{r=0}^{l} r u(r)\, z^{r-1} = \lambda u(z),$$

which on account of the identities

$$\sum_{r=0}^{l} r u(r)\, z^{r+m} = z^{m+1}\, u'(z), \quad m = -1, 0, 1,$$

can be written as

$$azu(z) - \frac{a}{l}\, z^2 u'(z) + (1-a)\, u(z) - \frac{b}{l}\, zu'(z) + \frac{a}{l}\, zu'(z) + \frac{b}{l}\, u'(z) = \lambda u(z),$$

whence

$$u'(z)/u(z) = l\ (az + 1 - a - \lambda)/(z-1)\,(az+b).$$

The general solution of this differential equation is

$$u(z) = C(z-1)^{l(1-\lambda)/(a+b)}\, (az+b)^{l(a+b+\lambda-1)/(a+b)},$$

where C is an arbitrary constant. The only acceptable solutions are those yielding $u(z)$ as a polynomial of degree l in z. This happens if and only if $l\left(\dfrac{1-\lambda}{a+b}\right)=r$, yielding the eigenvalues $\lambda_r = 1-\dfrac{r}{l}(a+b)$, $0 \leqslant r \leqslant l$. Since the number of them is $l+1$ (the order of the transition matrix), we have found all the eigenvalues. It follows also that the left eigenvector $\mathbf{u}_r = (u_r(i))_{0\leqslant i\leqslant l}$ corresponding to the eigenvalues λ_r has components proportional to the coefficients of 1, $z, \dots, z^l$ in the polynomial $(z-1)^r(az+b)^{l-r}$, $0\leqslant r\leqslant l$. For example we can take

$$u_r(i) = \frac{\displaystyle\sum_{h=\max(0,\,r+i-l)}^{\min(r,\,i)} \binom{r}{h}\binom{l-r}{i-h}(-1)^{r-h}a^{i-h}b^{l+h-r-i}}{(a+b)^{l-r}}$$

$$= \frac{\pi_{21}^{i}\,\pi_{12}^{l-r-i}}{(\pi_{12}+\pi_{21})^{l-r}} \sum_{h=\max(0,\,r+i-l)}^{\min(r,\,i)} \binom{r}{h}\binom{l-r}{i-h}(-1)^{r-h}\left(\frac{\pi_{12}}{\pi_{21}}\right)^{h}, \quad 0\leqslant i,\ r\leqslant l.$$

In this way $\mathbf{u}_0 = (u_0(i))_{0\leqslant i\leqslant l}$ represents the unique stationary distribution of the chain. Its components are just the terms in the expansion of

$$\left(\frac{b}{a+b}+\frac{a}{a+b}\right)^l = \left(\frac{\pi_{12}}{\pi_{12}+\pi_{21}}+\frac{\pi_{21}}{\pi_{12}+\pi_{21}}\right)^l$$

Therefore the stationary distribution is independent of the conditioning parameter c. But the speed of convergence of the probabilities $p(n, i, j)$ to this distribution which is of the order of $\lambda_1^n = [1-cl^{-1}(\pi_{12}+\pi_{21})]^n$ does depend on c: the larger c, the more rapid the convergence.

The right eigenvectors $\mathbf{v}_r = (v_r(i))_{0\leqslant i\leqslant l}$, $0 \leqslant r \leqslant l$, can be determined by making use of the fact that the matrix whose columns they are is the inverse of the matrix whose rows are the left eigenvectors (see 1.9.6). Therefore we have

$$\sum_{r=0}^{l} v_r(i)u_r(j) = \delta(i,j),\quad 0\leqslant i, j\leqslant l.$$

Multiplying these equations by z^j and summing over j yields

$$\sum_{r=0}^{l} v_r(i)\left(\sum_{r=0}^{l} u_r(j)z^j\right) = z^i,\quad 0\leqslant i\leqslant l,$$

whence, on account of the fact that

$$\sum_{j=0}^{l} u_r(j)z^j = \frac{(z-1)^r(az+b)^{l-r}}{(a+b)^{l-r}},$$

we deduce

$$\sum_{r=0}^{l}(a+b)^r v_r(i)\left(\frac{z-1}{az+b}\right)^r = \left(\frac{a+b}{az+b}\right)^l z^i,\quad 0\leqslant i\leqslant l.$$

Putting $(z-1)/(az+b)=w$, whence $z=(1+bw)/(1-aw)$, the above equations can be written as

$$\sum_{r=0}^{l}(a+b)^r\,v_r(i)\,w^r=(1-aw)^{l-i}(1+bw)^i,\quad 0\leqslant i\leqslant l.$$

Identifying coefficients yields

$$v_r(i)=\frac{\displaystyle\sum_{h=\max(0,\,r+i-l)}^{\min(r,\,i)}\binom{l-i}{r-h}\binom{i}{h}(-1)^{r-h}a^{r-h}b^h}{(a+b)^r}$$

$$=\frac{\pi_{21}^r}{(\pi_{12}+\pi_{21})^r}\sum_{h=\max(0,\,r+i-l)}^{\min(r,\,i)}\binom{l-i}{r-h}\binom{i}{h}(-1)^{r-h}\left(\frac{\pi_{12}}{\pi_{21}}\right)^h,\quad 0\leqslant i,\ r\leqslant l.$$

The spectral representation (1.12) allows us to write down the fundamental matrix $\mathbf{Z}=\mathbf{I}+\sum_{n\geqslant 1}(\mathbf{P}^n-\mathbf{\Pi})$. Since

$$\mathbf{P}^n=\sum_{r=0}^{l}\lambda_r^n\mathbf{v}_r\mathbf{u}_r'$$

and $\mathbf{v}_0\mathbf{u}_0'=\mathbf{\Pi}$, we get

$$\mathbf{Z}=\mathbf{I}+\sum_{r=1}^{l}\frac{\lambda_r}{1-\lambda_r}\mathbf{v}_r\mathbf{u}_r'.$$

6.3.2. We proceed now to compute the probability that on trial n the subject makes response A_1, i.e., the probability of the random event $A_1(n)$. Remembering the fact that the conditional probability of making A_1 given the subject is in state i is i/l, $0\leqslant i\leqslant l$, we have

$$\mathsf{P}_{\mathbf{p}}(A_1(n))=\sum_{i=0}^{l}\mathsf{P}_{\mathbf{p}}(A_1(n)\mid X(n)=i)\,\mathsf{P}_{\mathbf{p}}(X(n)=i)=\mathbf{p}'\mathbf{P}^n\boldsymbol{\gamma},$$

where $\boldsymbol{\gamma}$ is the vector of components i/l, $0\leqslant i\leqslant l$. Thus the problem amounts to the computation of the vector $\mathbf{P}^n\boldsymbol{\gamma}$. Remark first that the components of the vector $\mathbf{P}\boldsymbol{\gamma}$ are given by

$$\sum_{j=0}^{l}p(i,j)\,\frac{j}{l}=\frac{(l-i)a}{l}\,\frac{i+1}{l}+\frac{bi}{l}\,\frac{i-1}{l}+\left[1-\frac{(l-i)a}{l}-\frac{bi}{l}\right]\frac{i}{l}$$

$$=\left(1-\frac{a+b}{l}\right)\frac{i}{l}+\frac{a}{l},\quad 0\leqslant i\leqslant l.$$

This means that

$$\mathbf{P}^n\boldsymbol{\gamma}=\mathbf{P}^{n-1}(\mathbf{P}\boldsymbol{\gamma})=\left(1-\frac{a+b}{l}\right)\mathbf{P}^{n-1}\boldsymbol{\gamma}+\frac{a}{l}\,\mathbf{e}_{l+1},$$

recursion implying

$$\mathbf{\Pi}\gamma = \lim_{n\to\infty} \mathbf{P}^n\gamma = \frac{a}{a+b}\,\mathbf{e}_{l+1} = \frac{\pi_{21}}{\pi_{21}+\pi_{12}}\,\mathbf{e}_{l+1}.$$

It is easily verified that

$$\mathbf{P}^n\gamma = \mathbf{\Pi}\gamma - \left(1 - \frac{a+b}{l}\right)^{n-1} (\mathbf{\Pi} - \mathbf{P})\,\gamma,$$

whence

$$\mathbf{p}'\mathbf{P}^n\gamma = \frac{\pi_{21}}{\pi_{21}+\pi_{12}} - \left[1 - \frac{c(\pi_{21}+\pi_{12})}{l}\right]^{n-1} \left(\frac{\pi_{21}}{\pi_{21}+\pi_{12}} - \mathbf{p}'\mathbf{P}\gamma\right).$$

Therefore the asymptotic value as $n \to \infty$ of the probability of the random event $A_1(n)$ equals $\pi_{21}/(\pi_{21}+\pi_{12})$. It is independent of both the initial probability distribution and the conditioning parameter c.

It is interesting to note that if event E_0 does not occur on any trial (i.e., on any trial one of the responses must be reinforced) then the asymptotic value as $n \to \infty$ of the probability of the random event $E_1(n)$ equals

$$\frac{\pi_{21}}{\pi_{21}+\pi_{12}}(1-\pi_{12}) + \frac{\pi_{12}}{\pi_{21}+\pi_{12}}\,\pi_{21} = \frac{\pi_{21}}{\pi_{21}+\pi_{12}},$$

This means that, asymptotically, the subject's behaviour is in complete agreement with the experimenter's action.

6.3.3. We shall now consider the case where $\pi_{21} = \pi > 0$, $\pi_{12} = 0$. (The case $\pi_{12} > 0$, $\pi_{21} = 0$ can be treated in an analogous manner.) This amounts to the impossibility of reinforcing response A_1 after response A_2 has been made. Obviously, the corresponding Markov chain is absorbing. States $0, 1, \ldots, l-1$ are transient, state l is absorbing and the matrix $\mathbf{T}$ of the transition probabilities within the set of transient states has the form

$$\mathbf{T} = \begin{array}{c} \\ 0 \\ 1 \\ 2 \\ \cdot \\ \cdot \\ \cdot \\ l-1 \end{array} \begin{array}{c} \begin{array}{ccccc} 0 & 1 & 2 & \cdots & l-1 \end{array} \\ \left(\begin{array}{ccccc} 1-c\pi & c\pi & 0 & \cdots & 0 \\ 0 & 1-\left(1-\frac{1}{l}\right)c\pi & \left(1-\frac{1}{l}\right)c\pi & \cdots & 0 \\ 0 & 0 & 1-\left(1-\frac{2}{l}\right)c\pi & \cdots & 0 \\ \cdot & \cdot & \cdot & \cdots & \cdot \\ \cdot & \cdot & \cdot & \cdots & \cdot \\ \cdot & \cdot & \cdot & \cdots & \cdot \\ 0 & 0 & 0 & \cdots & 1-\frac{c\pi}{l} \end{array}\right) \end{array}.$$

The results previously obtained allow us to compute the fundamental matrix $\mathbf{N} = (\mathbf{I}-\mathbf{T})^{-1} = \mathbf{I} + \mathbf{T} + \mathbf{T}^2 + \ldots$. Indeed, the matrix $\mathbf{T}^n$ is obtained simply by suppressing the lth row and the lth column of the matrix $\mathbf{P}^n$. It is easily seen in our special case that the eigenvalues are

$$\lambda_r = 1 - \frac{rc\pi}{l}, \quad 0 \leqslant r \leqslant l,$$

and the left and right eigenvectors corresponding to λ_r are

$$\mathbf{u}_r' = (-1)^l \left(\binom{r}{l}, \binom{r}{l-1}(-1)^1, \ldots, \binom{r}{0}(-1)^l \right),$$

$$\mathbf{v}_r' = (-1)^r \left(\binom{l}{r}, \binom{l-1}{r}, \ldots, \binom{0}{r} \right), \quad 0 \leqslant r \leqslant l,$$

if we agree that $\binom{u}{v} = 0$ for $v > u$. If we compute $\mathbf{P}^n$ by making use of the spectral representation (1.12), and implicitly $\mathbf{T}^n$, we eventually get

$$n(i, j) = \begin{cases} 0, & \text{if } i > j, \\ \dfrac{l}{l-j} \dfrac{1}{c\pi}, & \text{if } i \leqslant j. \end{cases}$$

The reader should be warned that a direct computation of the inverse matrix of $\mathbf{I} - \mathbf{T}$ is in fact much simpler. Trying this computation he will acquire the conviction that avoiding using the eigenvalues and eigenvectors may be quite advantageous in certain cases.

In the case we have considered absorption in l is certain. Consequently, in the end, the subject will certainly only make response A_1. It is thus interesting to know the mean value of the number of trials necessary to reach such a completion of learning. This mean value is nothing but the mean time to absorption in l. On account of Theorem 3.2 we have

$$\mathsf{E}_i(\nu) = \frac{l}{c\pi} \sum_{j=i}^{l-1} \frac{1}{l-j}, \quad 0 \leqslant i \leqslant l-1.$$

If we regard A_2 as an erroneous response in a given learning situation, then the probability of an error is $(l-j)/l$ conditional on subject state j. Hence, the mean number of errors during learning (i.e., until absorbtion) equals

$$\sum_{j=0}^{l-1} n(i, j) \frac{l-j}{l} = \frac{1}{c\pi} \sum_{j=i}^{l-1} \frac{l}{l-j} \frac{l-j}{l} = \frac{l-i}{c\pi}.$$

6.4 THE MENDELIAN THEORY OF INHERITANCE

6.4.1. One of the most important achievements of biology in the 19th century was the establishment of the *cellular theory*. According to this theory, all plants and animals are constituted from small fundamental units called *cells*. These are surrounded by a membrane and contain, in general, an internal formation, the *nucleus*, which is also endowed with a membrane.

The nucleus of each cell contains a fixed number of filament like structures called *chromosomes*, which can be stained at the time of cell division. Their number is characteristic of the species. In superior plants and animals any specific type of chromosome is normally present in two copies, so that pairs of homologous chromosomes exist (the so-called *diploid* case)*). In inferior plants and bacteria any specific type of chromosome is present, as a rule, in only one copy (the so-called *haploid* case). There are also cases of *polyploidy* where any specific type of chromosome is present in groups of three, four etc. copies.

6.4.2. New cells arise from old cells by a division process. Most cells are able to grow and split into two relatively equal parts giving birth to two daughter cells that, as a rule, are identical to the mother cell.

There ar two types of cellular division. The most common is *mitosis*. In this process each chromosome divides to form two chromosomes identical to the original one. When a cell splits into two new cells one chromosome of each pair so formed migrates to one of the two new nuclei. At the end, the set of chromosomes of the daughter cells is identical with that of the mother cell.

The most important exception from the general mitosis process occurs during the formation of the reproductive cells of diploid organisms. In this process called *meiosis* a diploid cell gives birth to four reproductive cells *(gametes)*, each of which contains only one member of each of the pair of chromosomes present in the original cell. The union of two gametes arising from two different organisms leads to a fertilized diploid cell — the *zygote* — in which in any pair of homologous chromosomes one of them comes from the paternal cell and the other from the maternal one. The zygote contains all the information needed for the growth and the differentiation of a vegetable or animal adult organism.

6.4.3. The most remarkable property of living cells is their ability to transmit their characters from one generation to another. The existence of heredity was long ago noticed in man, the fact that, e.g., the

*) For instance, there are 14 chromosomes in peas and 20 chromosomes in maize. Man has 46 chromosomes. As early as 1890, it was noticed that a chromosome was present, known nowadays as the X chromosome, which does not always have a morphologically identical replica. It was proved in 1905 that in women the 23rd chromosome pair consists of two X chromosomes. Instead, in men the 23rd chromosome pair consists of an X chromosome and a very different chromosome called the Y chromosome which does not exist in the female organism.

hair or eye colour is inherited by children being well known. The physical basis of this process was understood just at the turn of the century when the chromosomal theory of inheritance was established. The basic laws of heredity were stated by Gregor Mendel (1822—1884) as early as 1865 but his time was not prepared to accept them. According to Mendel, various characters are controlled by pairs of factors (that nowadays we call *genes* and which we know to be parts of chromosomes), one of them coming from the paternal organism and the other from the maternal one. The position at which a gene is located on the chromosome is called its *locus*. Any gene at a given locus may exist in several alternate forms called *alleles*. Performing crossing experiments with varieties of peas, Mendel was led to state a number of principles that we shall explain in what follows.

6.4.4. In a diploid cell the genes appear in pairs just as the chromosomes*) do, their loci on a pair of homologous chromosomes being identical. In the simplest case of a gene with two alleles, A and a, at the corresponding locus on the paired chromosomes three different pairs of genes AA, Aa, and aa can be found (one makes no distinction between Aa and aA). These are called *genotypes*. Individuals carrying the genotypes AA or aa are called *homozygotes* while those carrying the genotype Aa are called *heterozygotes* (or *hybrids*). For example, peas carry a pair of genes such that A and a control red and white blossom colour. In this case the three genotypes manifest themselves as red, pink, and white blossom colour.

By means of the corresponding chromosome the gametes receive only one of the two alleles. The homozygotes (AA or aa) produce gametes of only one kind (A or a) but the hybrids produce gametes of both kinds in equal numbers. The genotypes of the offspring obey a chance process. At any mating each parental gene has probability 1/2 of being transmitted via the corresponding gamete. For instance, the genotypes of an $Aa \times Aa$ pairing are AA, Aa, and aa with probabilities 1/4, 1/2, and 1/4. The pairing $AA \times aa$ only has hybrid descendants (Aa), etc. This case of equally likely segregation is known as *Mendel's law of segregation*.

The pairing of two individuals belonging to a given population involves a second chance process. Mendel's implicit assumption was what is nowadays known as *random mating*. This means all the pairings of parents are (stochastically) independent and equally likely. Clearly, random mating is an extreme idealization of the actual conditions prevailing in nature. It assumes a very large population in which there is no preferential selectivity.

To sum up, the genotype of a descendant is the result of four independent random choices. The genotypes of the parents can be selected in $3 \times 3 = 9$ ways and their genes in the gametes produced by the mates in $2 \times 2 = 4$ ways. Clearly, the process can be equivalently

*) We will not consider the case of the human X and Y chromosomes.

described as follows *(random union of gametes)*: the paternal and maternal genes of a descendant are selected independently and at random from the sets of all the genes carried by the paternal respectively the maternal individuals of the parental population.

6.4.5. Let us examine an important consequence of random mating. Assume the three genotypes *AA*, *Aa*, and *aa* occur in paternal and maternal individuals in the same ratios, $u : 2v : w$, where $u + 2v + w = 1$. Let $p = u + v$, $q = v + w$, hence $p + q = 1$. Then the probability of selecting an $A(a)$- gene from either the set of all the paternal genes or the set of all the maternal genes is $p(q)$. Under random mating conditions the probabilities for a descendant to have genotypes *AA*, *Aa*, and *aa* are $u_1 = p^2$, $2v_1 = 2pq$, $w_1 = q^2$ respectively. (The genotype *Aa* arises in two ways *Aa* and *aA* with probability pq each.) If we now consider a population in which the three genotypes occur in the ratios $u_1 : 2v_1 : w_1$, on account of the above we conclude that the probabilities for a descendent to have genotypes *AA*, *Aa*, and *aa* are respectively

$$\begin{aligned} u_2 &= (u_1 + v_1)^2 = p^2(p + q) = p^2 = u_1, \\ 2v_2 &= 2(u_1 + v_1)(v_1 + w_1) = 2pq(p + q)(p + q) = 2pq = 2v_1, \\ w_2 &= (v_1 + w_1)^2 = q^2(p + q) = q^2 = u_2. \end{aligned}$$

This shows that random mating yields a probability distribution of the genotypes (therefore of the genes) which remains invariant from generation to generation.*) This result, known as the Hardy-Weinberg law, was an implicit assumption in Mendel's work.

6.4.6. The reader should be warned that the Hardy-Weinberg law does not assert the constancy of the genotypic frequencies, which are random variables, but of their mean values (i.e., the corresponding genotypic probabilities). The random fluctuations of the gene frequencies will be studied later on. This aspect is even more important, since the Hardy-Weinberg law is strictly valid only in an infinite population under ideal mating conditions. Besides the mating system there are factors that affect the genetic variability of a natural population. (This variability is considered to be the very core of biological evolution.) The best known factors of this kind are *mutation*, *migration* and *selection*. Mutation consists of naturally occurring sudden and spontaneous transformations of an allelomorph into another one, migration introduces into or excludes from the population certain individuals, and selection acts through either variations in viability (the genotypes differ in their chances of survival to reproduce) and fertility (different pairs of parents, according to their genotypes, may produce differing numbers of offspring) or distortions from Mendel's law of segregation. Another

---

*) It is equally easy to prove that if the three genotypes occur in paternal and maternal individuals in different ratios then the probability distribution of the genotypes reaches equilibrium in the second generation.

less well known factor, whose significance for evolution is still controversial, is *genetic drift*, which might be defined as the effect of the finiteness of natural populations. In small populations the genetic drift may lead to the disappearance of one of the alleles A and a, so that one eventually reaches a population of homozygotes (AA or aa).

6.4.7. In the above we restricted ourselves to the case of a single locus with two possible alleles. Similar considerations may be made in the case of a single locus with r possible alleles when there are $r(r + 1)/2$ genotypes.

A new situation may appear in the case of two loci when recombination of genes may occur. For the sake of simplicity let us assume that at the first locus there are the alleles A and a and at the second locus the alleles B and b(A and B on a chromosome and a and b on the homologous one). An individual whith this genotype can produce gametes of the type AB and ab (called *parental*) and also gametes of the type Ab and aB (called *recombinant*). (This phenomenon consisting of the spliting at identical points during meiosis of a pair of homologous chromosomes and their crossed recombination is known by the name of *crossing-over*.) If all the four types of gametes are produced in equal numbers, then the loci are said to be *unlinked*. If the parental gametes are in excess over the recombinant ones, then the loci are said to be *linked*. Mendel stated the *law of independent recombination* corresponding to the case of unlinked loci. In what follows we will not consider the case of two loci. The discussion above is intended just to give an idea of the supplementary difficulties occurring in this case.

6.4.8. Quite unavoidably, our presentation of the basic ideas of the Mendelian theory of inheritance is schematical and incomplete. To fill the lacunes the reader should consult such genetics textbooks as [267 and 394].

6.5 SIB MATING

6.5.1. A classical example of violations of the conditions of random mating is that of so called sister-brother mating (sib mating).* In this mating system two individuals are mated and from their offspring two individuals of opposite sex are chosen at random. These are mated and the procedure continues indefinitely.

We shall show that the evolution of the type of mating in the successive generations is that of a Markov chain with two absorbing states and four transient states. The Markovian character being obvious, it remains to determine the transition matrix.

With the three possible genotypes AA, Aa, and aa we should distinguish six types of mating (combinations of parents) which we shall

---

* Many other examples are discussed in [187].

number as in the table below, where the probabilities of occurrence of the genotypes in the descendants are also given.

Type of mating	Genotype of descendants AA	Aa	aa
1. $AA \times AA$	1	0	0
2. $aa \times aa$	0	0	1
3. $AA \times Aa$	1/2	1/2	0
4. $Aa \times Aa$	1/4	1/2	1/4
5. $Aa \times aa$	0	1/2	1/2
6. $AA \times aa$	0	1	0

The first type of mating can only produce individuals of genotype AA and thus can only lead to a mating of the same type. The same thing is true for the second type of mating. Type 3 produces with equal probabilities individuals of genotypes AA and Aa. The possible types of mating between descendants will therefore be $AA \times AA$ (type 1) with probability 1/4, $AA \times Aa$ (type 3) with probability 1/2 and $Aa \times Aa$ (type 4) with probability 1/4. Considering in a similar manner types 4, 5, and 6 we obtain the transition matrix

$$\mathbf{P} = \begin{array}{c} \\ 1 \\ 2 \\ 3 \\ 4 \\ 5 \\ 6 \end{array} \begin{array}{c} \begin{array}{cccccc} 1 & 2 & 3 & 4 & 5 & 6 \end{array} \\ \begin{pmatrix} 1 & 0 & 0 & 0 & 0 & 0 \\ 0 & 1 & 0 & 0 & 0 & 0 \\ 1/4 & 0 & 1/2 & 1/4 & 0 & 0 \\ 1/16 & 1/16 & 1/4 & 1/4 & 1/4 & 1/8 \\ 0 & 1/4 & 0 & 1/4 & 1/2 & 0 \\ 0 & 0 & 0 & 1 & 0 & 0 \end{pmatrix} \end{array} \tag{6.2}$$

Clearly, sib mating unavoidably leads to the elimination of one of the genes A and a and, implicitly, of the heterozygous genotype Aa. Absorption in state 1(2) marks the disappearence of the gene $a(A)$.

6.5.2. The results obtained in Chapter 3 allow us to find the probability of disappearance of the gene $A(a)$, i.e., the absorption probability in state 2 (1), and the mean number of generations until the disappearance of the heterozygous genotype Aa (the mean time to absorption) starting with one of the mating types 3, 4, 5, and 6.

In our case

$$\mathbf{T} = \begin{array}{c} \\ 3 \\ 4 \\ 5 \\ 6 \end{array} \begin{array}{c} \begin{array}{cccc} 3 & 4 & 5 & 6 \end{array} \\ \begin{pmatrix} 1/2 & 1/4 & 0 & 0 \\ 1/4 & 1/4 & 1/4 & 1/8 \\ 0 & 1/4 & 1/2 & 0 \\ 0 & 1 & 0 & 0 \end{pmatrix} \end{array}$$

and the fundamental matrix will be

$$\mathbf{N} = (\mathbf{I} - \mathbf{T})^{-1} = \frac{1}{6}\begin{pmatrix} 16 & 8 & 4 & 1 \\ 8 & 16 & 8 & 2 \\ 4 & 8 & 16 & 1 \\ 8 & 16 & 8 & 8 \end{pmatrix}$$

Therefore the absorption probabilities are given by the matrix

$$\mathbf{A} = \mathbf{NR} = \frac{1}{6}\begin{pmatrix} 16 & 8 & 4 & 1 \\ 8 & 16 & 8 & 2 \\ 4 & 8 & 16 & 1 \\ 8 & 16 & 8 & 8 \end{pmatrix}\begin{pmatrix} 1/4 & 0 \\ 1/16 & 1/16 \\ 0 & 1/4 \\ 0 & 0 \end{pmatrix} = \begin{matrix} 3 \\ 4 \\ 5 \\ 6 \end{matrix}\overset{\begin{matrix} 1 & \quad\quad 2 \end{matrix}}{\begin{pmatrix} 3/4 & 1/4 \\ 1/2 & 1/2 \\ 1/4 & 3/4 \\ 1/2 & 1/2 \end{pmatrix}}$$

and the mean time to absorption by the vector

$$\mathbf{Ne} = \begin{matrix} 3 \\ 4 \\ 5 \\ 6 \end{matrix}\begin{pmatrix} 29/6 \\ 34/6 \\ 29/6 \\ 40/6 \end{pmatrix}.$$

Thus the minimum value of the probability of disappearence of the gene $a(A)$ corresponds to the initial type of mating 5(3). Also, the maximum value of the mean number of generations until disappearence of the heterozygous genotype corresponds to the initial type of mating 6. Intuitively, these findings were to be expected.

6.5.3. The spectral representation of the transition matrix $\mathbf{P}$ can easily be obtained. By the corollary to Theorem 3.3, the eigenvalue $\lambda = 1$ is of (both algebraic and geometric) multiplicity 2. The left- and right-eigenvectors corresponding to this eigenvalue are

$$\mathbf{u}_1' = (1, 0, 0, 0, 0, 0), \ \mathbf{u}_2' = (0, 1, 0, 0, 0, 0),$$

$$\mathbf{v}_1' = (1, 0, 3/4, 1/2, 1/4, 1/2), \ \mathbf{v}_2' = (0, 1, 1/4, 1/2, 3/4, 1/2).$$

The remaining eigenvalues of $\mathbf{P}$ coincide with those of the matrix $\mathbf{T}$. The characteristic equation of the latter can be written as

$$\left(\lambda - \frac{1}{2}\right)\left(\lambda^3 - \frac{3}{4}\lambda^2 - \frac{1}{8}\lambda + \frac{1}{16}\right) = 0,$$

hence the eigenvalues

$$\lambda_3 = (1 + \sqrt{5})/4, \ \lambda_4 = 1/2, \ \lambda_5 = (1 - \sqrt{5})/4, \ \lambda_6 = 1/4.$$

The left and right-eigenvectors of **P** corresponding to these eigenvalues can be chosen as follows (the reader is asked to check them):

$$\mathbf{u}_3' = \frac{1}{40}\,(-9-4\sqrt{5},\ -9-4\sqrt{5},\ 6+2\sqrt{5}, 4+4\sqrt{5}, 6+2\sqrt{5},\ 2),$$

$$\mathbf{v}_3' = (0,\ 0,\ 1, \sqrt{5}-1,\ 1, 6-2\sqrt{5}),$$

$$\mathbf{u}_4' = \frac{1}{4}\,(-1,\ 1,\ 2,\ 0,\ -2,\ 0),$$

$$\mathbf{v}_4' = (0, 0, 1, 0, -1, 0),$$

$$\mathbf{u}_5' = \frac{1}{40}\,(-9+4\sqrt{5}, -9+4\sqrt{5},\ 6-2\sqrt{5},\ 4-4\sqrt{5},\ 6-2\sqrt{5},\ 2),$$

$$\mathbf{v}_5' = (0, 0,\ 1, -\sqrt{5}-1,\ 1,\ 6+2\sqrt{5}),$$

$$\mathbf{u}_6' = \frac{1}{20}(-1,\ -1,\ 4,\ -4,\ 4,\ -2),$$

$$\mathbf{v}_6' = (0,\ 0,\ 1, -1, 1, -4).$$

Then the spectral representation (1.12) yields

$$\mathbf{P}^n = \begin{pmatrix} 1 & 0 & 0 & 0 & 0 & 0 \\ 0 & 1 & 0 & 0 & 0 & 0 \\ 3/4 & 1/4 & 0 & 0 & 0 & 0 \\ 1/2 & 1/2 & 0 & 0 & 0 & 0 \\ 1/4 & 3/4 & 0 & 0 & 0 & 0 \\ 1/2 & 1/2 & 0 & 0 & 0 & 0 \end{pmatrix} + \frac{\lambda_3^n + \lambda_5^n}{40} \begin{pmatrix} 0 & 0 & 0 & 0 & 0 & 0 \\ 0 & 0 & 0 & 0 & 0 & 0 \\ -9 & -9 & 6 & 4 & 6 & 2 \\ -11 & -11 & 4 & 16 & 4 & -2 \\ -9 & -9 & 6 & 4 & 6 & 2 \\ -14 & -14 & 16 & -16 & 16 & 12 \end{pmatrix}$$

$$+ \frac{\lambda_3^n - \lambda_5^n}{40}\sqrt{5} \begin{pmatrix} 0 & 0 & 0 & 0 & 0 & 0 \\ 0 & 0 & 0 & 0 & 0 & 0 \\ -4 & -4 & 2 & 4 & 2 & 0 \\ -5 & -5 & 4 & 0 & 4 & 2 \\ -4 & -4 & 2 & 4 & 2 & 0 \\ -6 & -6 & 0 & 16 & 0 & -4 \end{pmatrix}$$

$$+ \frac{2^{-n}}{4} \begin{pmatrix} 0 & 0 & 0 & 0 & 0 & 0 \\ 0 & 0 & 0 & 0 & 0 & 0 \\ -1 & 1 & 2 & 0 & -2 & 0 \\ 0 & 0 & 0 & 0 & 0 & 0 \\ 1 & -1 & -2 & 0 & 2 & 0 \\ 0 & 0 & 0 & 0 & 0 & 0 \end{pmatrix} + \frac{4^{-n}}{20} \begin{pmatrix} 0 & 0 & 0 & 0 & 0 & 0 \\ 0 & 0 & 0 & 0 & 0 & 0 \\ -1 & -1 & 4 & -4 & 4 & -2 \\ 1 & 1 & -4 & 4 & -4 & 2 \\ -1 & -1 & 4 & -4 & 4 & -2 \\ 4 & 4 & -16 & 16 & -16 & 8 \end{pmatrix}$$

In particular, it follows that the rate of convergence of the probabilities $p(n, i, 1)$ and $p(n, i, 2)$, $3 \leqslant i \leqslant 6$, to the absorption probabilities in states 1 and 2, respectively, is of the order of λ_3^n. In other words, in any generation the probability of mating two individuals of different genotypes or two hybrids diminishes by a factor approximately equal to λ_3:

$$\lim_{n\to\infty} \frac{\sum_{j=3}^{6} p(n, i, j)}{\sum_{j=3}^{6} p(n-1, i, j)} = \lambda_3, \quad 3 \leqslant i \leqslant 6.$$

Knowledge of eigenvalues and eigenvectors of **P** allows us to compute the quasi-stationary distributions $(u(i))_{3\leqslant i\leqslant 6}$ and $(v(i)u(i))_{3\leqslant i\leqslant 6}$ (see 3.4.2).

The vectors $\mathbf{u} = (u(i))_{3\leqslant i\leqslant 6}$ and $\mathbf{v} = (v(i))_{3\leqslant i\leqslant 6}$ are respectively the left- and right-eigenvectors of **T** corresponding to the eigenvalue λ_3, chosen so that $\sum_{i=3}^{6} u(i) = \sum_{i=3}^{6} v(i)\, u(i) = 1$. We easily find that

$$\mathbf{u}' = \frac{1}{9+4\sqrt{5}} \; (3+\sqrt{5}, 2+2\sqrt{5}, 3+\sqrt{5},\; 1),$$

$$\mathbf{v}' = \frac{9+4\sqrt{5}}{20} \, (1, \sqrt{5}-1,\; 1, 6-2\sqrt{5}),$$

Therefore the components of the vector $(v(i)u(i))_{3\leqslant i\leqslant 6}$ are

$$\frac{3+\sqrt{5}}{20}, \; \frac{8}{20}, \; \frac{3+\sqrt{5}}{20}, \; \frac{6-2\sqrt{5}}{20}.$$

Remember that $u(i)$, $3\leqslant i\leqslant 6$, represents the asymptotic probability of a mating of type i given that absorption has not yet taken place while $v(i)u(i)$, $3\leqslant i\leqslant 6$, represents the asymptotic expected proportion of matings of type i before absorption given a long time to absorption.

6.6 GENETIC DRIFT. THE WRIGHT MODEL

6.6.1. We stressed that the Hardy-Weinberg law is strictly valid only in an ideal infinite population, not subject to any outside influences, and in which random mating takes place. Any natural population, regardless of its size is, however, a finite entity. This is the main reason why in such a population only an approximate and temporary genetic equilibrium is reached. Recall that m diploid individuals in a generation develop from $2m$ gametes produced in the preceding generation. These gametes are just a sample from the set of all the parental genes and, as such, are subject to inherent sampling errors. The smaller the population (and, consequently, the sample of gametes), the larger the sampling errors. The impact of these errors on the gene frequencies in the successive generations is the core of the phenomenon of (random) genetic drift we alluded to when reviewing the factors that affect the genetic variability.

6.6.2. The basic mathematical model of random genetic drift was introduced by S. Wright [404] in 1931 and formulated in terms of finite Markov chains by G. Malécot [253] in 1944.

We shall consider the simplest case of a population consisting of diploid individuals able to produce a single type of gamete (as usually happens in plants) that reproduces by random mating but whose size is kept fixed by selecting m individuals in every generation. More precisely, we assume each individual produces infinitely many A-and/or a-gametes, the $2m$ gametes picked to form the next generation being chosen from them at random. We assume also that selection, migration and mutation are absent and that generations do not overlap.

The population is said to be in state i, $0 \leqslant i \leqslant 2m$, if and only if there are i A-genes (and, consequently, $2m-i$ a-genes). The assumptions made lead us to conclude that the population state in the next generation equals the number of successes in $2m$ independent Bernoulli trials for which the probability of success is $i/(2m)$ (see 1.3.2). Then it is clear that the evolution of the state of the population is described by a Markov chain with states $0, 1, \ldots, 2m$ and transition probabilities

$$p(i, j) = \binom{2m}{j}\left(\frac{i}{2m}\right)^j\left(1 - \frac{i}{2m}\right)^{2m-j}, \quad 0 \leqslant i,\ j \leqslant 2m,$$

(We adopt the convention that $0^0 = 1$.)

The presence of the even number $2m$ is only incidental, so that in what follows we shall consider the Markov chain $(X(n))_{n\geqslant 0}$ with states $0, 1, \ldots, m$ (m being odd or even) and transition probabilities

$$p(i, j) = \binom{m}{j}\left(\frac{i}{m}\right)^j\left(1 - \frac{i}{m}\right)^{m-j}, \quad 0 \leqslant i, j \leqslant m. \tag{6.3}$$

States 0 and m of this Markov chain are absorbing and correspond to one of the two homozygous conditions AA and aa (the attainment of which can be observed in natural populations of small size).

It is, of course, of interest to find the probabilities of absorption in 0 and m as well as the mean time to absorption, which in the present case can be called the mean time to homozygosity.

The expressions for the transition probabilities (6.3) do not allow a direct computation of the fundamental matrix of the chain. As shown in [204 and 205] it is possible, instead, to compute the n-step transition probabilities, $n \geqslant 1$.

6.6.3. To compute the n-step transition probabilities we shall first show that they can be expressed in terms of certain moments of the random variable $X(n-1)$. In what follows i will denote a nonabsorbing state, $i \neq 0, m$. By the Chapman-Kolmogorov equations we have

$$p(n, i, j) = \sum_{k=0}^{m} p(n-1, i, k)\, p(k, j)$$

$$= \sum_{k=0}^{m} p(n-1, i, k) \binom{m}{j} \left(\frac{k}{m}\right)^{j} \left(1 - \frac{k}{m}\right)^{m-j}$$

$$= \sum_{k=0}^{m} p(n-1, i, k) \binom{m}{j} \left(\frac{k}{m}\right)^{j} \sum_{l=0}^{m-j} \binom{m-j}{l} (-1)^{l} \left(\frac{k}{m}\right)^{l}$$

$$= \sum_{l=0}^{m-j} (-1)^{l} \binom{m}{j} \binom{m-j}{l} m^{-j-l} \sum_{k=0}^{m} p(n-1, i, k)\, k^{j+l},$$

whence

$$p(n, i, j) = \sum_{l=0}^{m-j} (-1)^{l} \binom{m}{j} \binom{m-j}{l} m^{-j-l} \mathsf{E}_i(X^{j+l}(n-1)), \qquad (6.4)$$

This means that to determine $p(n, i, j)$ it is sufficient to know the first m moments of $X(n-1)$. Let $\mathsf{E}_i(X^l(n)) = \mu_i^l(n)$, $1 \leqslant l \leqslant m$. We shall prove the recursion

$$\mu_i^l(n+1) = \sum_{r=1}^{l} a_{lr}((m)_r/m^r)\, \mu_i^r(n), \quad n \geqslant 0, \qquad (6.5)$$

where

$$a_{lr} = \sum_{h=0}^{r-1} \frac{(-1)^h (r-h)^{l-1}}{(r-h-1)!\, h!}, \qquad 1 \leqslant r \leqslant l,$$

are the so-called Stirling numbers of the second kind that are obtained from the identity

$$x^l = \sum_{r=1}^{l} a_{lr} (x)_r, \qquad (6.6)$$

with x a real or complex variable and $(x)_k = x(x-1) \ldots (x-k+1)$.

To prove equation (6.5) let us notice that

$$\mu_i^l(n+1) = \sum_{j=0}^{m} p(n+1, i, j)\, j^l = \sum_{j=0}^{m} \sum_{k=0}^{m} p(n, i, k)\, p(k, j)\, j^l$$

$$= \sum_{k=0}^{m} p(n, i, k) \sum_{j=0}^{m} p(k, j) j^l.$$

On account of (6.3) we have

$$\sum_{j=0}^{m} p(k, j)\, j^l = \sum_{j=0}^{m} j^l \binom{m}{j}\left(\frac{k}{m}\right)^j\left(1-\frac{k}{m}\right)^{m-j}$$
$$= \sum_{j=0}^{m}\sum_{r=1}^{l} a_{l\cdot}(j)_r \binom{m}{j}\left(\frac{k}{m}\right)^j\left(1-\frac{k}{m}\right)^{m-j}$$

[by 6.6)]

$$= \sum_{r=1}^{l} a_{lr} \sum_{j=r}^{m} (j)_r \binom{m}{j}\left(\frac{k}{m}\right)^j\left(1-\frac{k}{m}\right)^{m-j}$$
$$= \sum_{r=1}^{l} a_{lr}(m)_r \left(\frac{k}{m}\right)^r \sum_{j=r}^{m} \frac{(m-r)\,!}{(j-r)\,!\ (m-j)\,!}\left(\frac{k}{m}\right)^{j-r}\left(1-\frac{k}{m}\right)^{m-j} \tag{6.7}$$
$$= \sum_{r=1}^{l} a_{lr}(m)_r \left(\frac{k}{m}\right)^r$$

[since

$$(j)_r \binom{m}{j} = \frac{j\,!}{(j-r)\,!} \frac{m\,!}{j\,!(m-j)\,!} = \frac{(m)_r\,(m-r)\,!}{(j-r)\,!\,(m-j)\,!}$$

and

$$\sum_{j=r}^{m} \frac{(m-r)\,!}{(j-r)\,!\,(m-j)\,!}\left(\frac{k}{m}\right)^{j-r}\left(1-\frac{k}{m}\right)^{m-j}$$
$$= \sum_{h=0}^{m-r} \frac{(m-r)\,!}{h\,!(m-r-h)\,!}\left(\frac{k}{m}\right)^{h}\left(1-\frac{k}{m}\right)^{m-r-h} = \left(\frac{k}{m}+1-\frac{k}{m}\right)^{m-r} = 1\Bigg].$$

Therefore

$$\mu_i^l(n+1) = \sum_{k=0}^{m} p(n, i, k) \sum_{r=1}^{l} a_{lr}(m)_r \left(\frac{k}{m}\right)^r$$
$$= \sum_{r=1}^{l} a_{lr}((m)_r/m^r) \sum_{k=0}^{m} p(n, i, k)\, k^r = \sum_{r=1}^{l} a_{lr}((m)_r/m^r)\,\mu_i^r(n),$$

and (6.5) is proved.

Putting $\boldsymbol{\mu}_i'(n) = (\mu_i^1(n), \ldots, \mu_i^m(n))$, equation (6.5) allows us to write the matrix equation

$$\boldsymbol{\mu}_i'(n+1) = \mathbf{C}\,\boldsymbol{\mu}_i'(n), \tag{6.8}$$

where

$$\mathbf{C} = \begin{pmatrix} 1 & 0 & 0 & \cdots & 0 \\ 1 & (m)_2/m^2 & 0 & \cdots & 0 \\ 1 & a_{32}(m)_2/m^2 & (m)_3/m^3 & \cdots & 0 \\ \cdot & \cdot & \cdot & \cdots & \cdot \\ 1 & a_{m2}(m)_2/m^2 & a_{m3}(m)_3/m^3 & \cdots & (m)_m/m^m \end{pmatrix}$$

(it is easily seen that $a_{l1} = a_{ll} = 1$ whatever $l \geqslant 1$). By iteration, (6.8) leads to

$$\boldsymbol{\mu}_i(n) = \mathbf{C}^n \boldsymbol{\mu}_i(0), \tag{6.9}$$

where, obviously,

$$\boldsymbol{\mu}_i'(0) = (i, i^2, \ldots, i^m).$$

Since the matrix $\mathbf{C}$ is triangular, its eigenvalues coincide with the diagonal elements $(m)_r/m^r$, $1 \leqslant r \leqslant m$, which are all distinct. It follows that $\mathbf{C}$ is diagonalizable. We shall now show how its eigenvectors can be determined.

The left eigenvector $\mathbf{u}_r' = (u_r(1), \ldots, u_r(m))$ corresponding to the eigenvalue $\lambda_r = (m)_r/m^r$ satisfies the equation $\mathbf{u}_r'\mathbf{C} = \lambda_r \mathbf{u}_r'$, whence

$$\sum_{i=j}^{m} a_{ij}\lambda_j u_r(i) = \lambda_r u_r(j), \quad 1 \leqslant j \leqslant m.$$

It follows from the equations corresponding to the values $j > r$ that $u_r(j) = 0$ for $j > r$ so that the above equations can be written as

$$\sum_{i=j}^{r} a_{ij}\lambda_j u_r(i) = \lambda_r u_r(j), \quad 1 \leqslant j \leqslant r. \tag{6.10}$$

Taking $u_r(r) = 1$, we shall show that the $u_r(i)$, $i < r$, are given by

$$u_r(i) = \Sigma\, a_{ri_1} \frac{m^{r-i_1}(m)_{i_1}}{(m)_r - m^{r-i_1}(m)_{i_1}}\, a_{i_1 i_2} \frac{m^{r-i_2}(m)_{i_2}}{(m)_r - m^{r-i_2}(m)_{i_2}} \cdots$$

$$\cdots a_{i_n i} \frac{m^{r-i}(m)_i}{(m)_r - m^{r-i}(m)_i},$$

the sum being over all decreasing sequences $r > i_1 > i_2 > \ldots > i$. The proof amounts to checking (6.10). We have

$$\sum_{i=j}^{r} a_{ij}\lambda_j u_r i) = \lambda_j u_r(j) + \sum_{i=j+1}^{r} a_{ij}\lambda_j u_r(i)$$

$$= \lambda_j u_r(j) + \sum_{i=j+1}^{r} \sum a_{i_1} \frac{(m)_j}{m^j} a_{ri_1} \frac{m^{r-i_1}(m)_{i_1}}{(m)_r - m^{r-i_1}(m)_{i_1}} \cdots a_{i_n i} \frac{m^{r-i}(m)_i}{(m)_r - m^{r-i}(m)_i}$$

$$= \lambda_j u_r(j) + \frac{(m)_r - m^{r-j}(m)_j}{m^r} \sum_{i=j+1}^{r} \sum a_{ri_1} \frac{m^{r-i_1}(m)_{i_1}}{(m)_r - m^{r-i_1}(m)_{i_1}} \cdots$$

$$\cdots a_{ij} \frac{m^{r-j}(m)_j}{(m)_r - m^{r-j}(m)_j} = \lambda_j u_r(j) + (\lambda_r - \lambda_j)\, u_r(j) = \lambda_r u_r(j).$$

The right eigenvector $\mathbf{v}_r' = (v_r(1), \ldots, v_r(m))$ corresponding to the eigenvalue $\lambda_r = (m_r)/m^r$ satisfies the equation $\mathbf{C}\mathbf{v}_r = \lambda_r \mathbf{v}_r$, whence

$$\sum_{i=1}^{j} a_{ji}\lambda_i v_r(i) = \lambda_r v_r(j), \quad 1 \leqslant j \leqslant m.$$

It follows from the equations corresponding to the values $j < r$ that $v_r(j) = 0$ for $j < r$, so that the above equations can be written as

$$\sum_{i=r}^{i} a_{ji}\lambda_i v_r(i) = \lambda_r v_r(j), \quad r \leqslant j \leqslant m. \tag{6.11}$$

Taking $v_r(r) = 1$, we shall show that the $v_r(i)$, $i > r$, are given by

$$v_r(i) = m^{i-r}\sum a_{ii_1}\frac{(m)_{i_1}}{m^{i-r}(m)_r - (m)_i}\, a_{i_1i_2}\frac{(m)_{i_2}}{m^{i_1-r}(m)_r - (m)_{i_1}} \cdots$$
$$\cdots a_{i_nr}\frac{(m)_r}{m^{i_n-r}(m)_r - (m)_{i_n}},$$

the sum being over all decreasing sequences $i > i_1 > i_2 > \ldots > r$.

As before, the proof amounts to checking (6.11). We have

$$\sum_{i=r}^{j} a_{ji}\lambda_i v_r(i) = \lambda_j v_r(j) + \sum_{i=r}^{j-1} a_{ji}\lambda_i v_r(i)$$

$$= \lambda_j v_r(j) + \sum_{i=r}^{j-1}\sum a_{ji}\frac{(m)_i}{m^i}\, m^{i-r} a_{ii_1}\frac{(m)_{i_1}}{m^{i-r}(m)_r - (m)_i} \cdots a_{i_nr}\frac{(m)_r}{m^{i_n-r}(m)_r - (m)_{i_n}}$$

$$= \lambda_j v_r(j) + \frac{m^{j-r}(m)_r - (m)_j}{m^j}\, m^{j-r}\sum_{i=r}^{j-1}\sum a_{ji}\frac{(m)_i}{m^{j-r}(m)_r - (m)_j} \cdots$$

$$\cdots a_{i_nr}\frac{(m)_r}{m^{i_n-r}(m)_r - (m)_{i_n}} = \lambda_j v_r(j) + (\lambda_r - \lambda_j)\, v_r(j) = \lambda_r v_r(j).$$

In particular, $\mathbf{v}_1' = (1, m, \ldots, m^{m-1})$.

Let us come back to the problem of determining the probabilities $p)n, i, j)$. The spectral representation (1.12) and equation (6.9) allow us to write

$$\boldsymbol{\mu}_i(n-1) = \sum_{r=2}^{m} \lambda_r^{n-1}\mathbf{v}_r\mathbf{u}_r'\boldsymbol{\mu}_i(0)$$

or, on account of the fact that $\lambda_1 = (m)_1/m = 1$,

$$\boldsymbol{\mu}_i(n-1) = \left(\mathbf{v}_1\mathbf{u}_1' + \sum_{r=2}^{m} \lambda_r^{n-1}\mathbf{v}_r\mathbf{u}_r'\right)\boldsymbol{\mu}_i(0).$$

Therefore

$$\mathsf{E}_i(X^{j+l}(n-1)) = \mu_i^{j+l}(n-1)$$
$$= \begin{cases} 1, \text{ if } j = l = 0, \\ \sum_{k=1}^{m} v_1(j+l)\, u_1(k)\, i^k + \sum_{r=2}^{m} \lambda_r^{n-1} v_r(j+l) \sum_{k=1}^{m} u_r(k)\, i^k, \text{ if } j+l > 0. \end{cases}$$

Since $u_r(j) = 0$, $j > r$, $u_1(1) = 1$, $v_1(j+l) = m^{j+l-1}$, $j+l > 0$, we have

$$\mathsf{E}_i(X^{j+l}(n-1)) = \begin{cases} 1, \text{ if } j = l = 0, \\ im^{j+l-1} + \sum_{r=2}^{m} \sum_{k=1}^{r} \lambda_r^{n-1} v_r(j+l)\, u_r(k)\, i^k, \text{ if } j+l > 0. \end{cases} \tag{6.12}$$

Now, using equation (6.4), we get for $j > 0$

$$p(n, i, j) = \sum_{l=0}^{m-j} (-1)^l \binom{m}{j} \binom{m-j}{l} m^{-j-l} \Big(im^{j+l-1} + \sum_{r=2}^{m} \sum_{k=1}^{r} \lambda_r^{n-1} v_r(j+l)\, u_r(k)\, i^k \Big). \tag{6.13}$$

Notice that

$$\sum_{l=0}^{m-j} (-1)^l \binom{m}{j} \binom{m-j}{l} m^{-j-l} im^{j+l-1} = \frac{i}{m} \binom{m}{j} \sum_{l=0}^{m-j} (-1)^l \binom{m-j}{l}$$
$$= \frac{i}{m} \binom{m}{j} (1-1)^{m-j} = \begin{cases} 0, \text{ if } j \neq m, \\ i/m, \text{ if } j = m. \end{cases}$$

It follows that for $j > 0$

$p(n, i, j) =$

$$= \begin{cases} \binom{m}{j} \sum_{l=0}^{m-j} (-1)^l \binom{m-j}{l} m^{-j-l} \sum_{r=2}^{m} \sum_{k=1}^{r} \lambda_r^{n-1} v_r(j+l)\, u_r(k)\, i^k, \text{ if } j \neq m, \\ \frac{i}{m} + m^{-m} \sum_{r=2}^{m} \sum_{k=1}^{r} \lambda_r^{n-1} v_r(m)\, u_r(k)\, i^k, \text{ if } j = m. \end{cases}$$

For $j = 0$ equations (6.4) and (6.12) yield

$$p(n, i, 0) = 1 + \sum_{l=1}^{m} (-1)^l \binom{m}{l} m^{-l} \Big(im^{l-1} + \sum_{r=2}^{m} \sum_{k=1}^{r} \lambda_r^{n-1} v_r(l)\, u_r(k)\, i^k \Big)$$
$$= 1 - \frac{i}{m} + \sum_{l=1}^{m} (-1)^l \binom{m}{l} m^{-l} \sum_{r=2}^{m} \sum_{k=1}^{r} \lambda_r^{n-1} v_r(l)\, u_r(k) i^k.$$

6.6.4. The results we have obtained allow us to find the probabilities of absorption in 0 and m. On account of (3.5) we have

$$a(i, 0) = \lim_{n\to\infty} p(n, i, 0) = 1 - i/m,$$

$$a(i, m) = \lim_{n\to\infty} p(n, i, m) = i/m.$$

There is another very simple and elegant way of obtaining the absorption probabilities $a(i, 0)$ and $a(i, m)$ ([102], p. 399). For $l = 1$ equation (6.7) yields

$$\sum_{j=0}^{m} p(k, j) j = a_{11}(m)_1 \frac{k}{m} = k, \quad 0 \leqslant k \leqslant m. \tag{6.14}$$

(A Markov chain having this property is called a *martingale*.) From (6.14) we get by induction that

$$\sum_{j=0}^{m} p(n, k, j)\, j = k, \quad n \geqslant 1,\ 0 \leqslant k \leqslant m.$$

Since states $j = 1, \ldots, m-1$ are transient we have

$$\lim_{n\to\infty} p(n, k, j) = 0,$$

Letting $n \to \infty$ we get

$$\lim_{n\to\infty} p(n, k, m) = k/m, \ 0 \leqslant k \leqslant m,$$

i.e., the result we have already obtained.

6.6.5. Let us now turn to the time until homozygosity (absorption) ν. By equation (3.4) for all $i \neq 0, m$ we have

$$\mathsf{P}_i(\nu = n) = \sum_{j=1}^{m-1} [p(n-1, i, j) - p(n, i, j)]$$

$$= p(n, i, m) - p(n-1, i, m) + p(n, i, 0) - p(n-1, i, 0).$$

By symmetry $p(n, i, 0) = p(n, m-i, m)$. It follows that $\mathsf{P}_i\ (\nu=n) = \mathsf{P}_{m-i}\ (\nu=n)$ and that

$$\mathsf{P}_i(\nu = n) = p(n, i, m) - p(n-1, i, m) + p(n, m-i, m) - p(n-1, m-i, m).$$

Substituting the values of $p(n, i, m)$ and $p(n, m-i, m)$ determined above we get

$$\mathsf{P}_i\ (\nu= 1) = p(i, m) + p(i, 0) = m^{-m}(i^m + (m-i)^m),$$

$$\mathsf{P}_i(\nu = n) = m^{-m} \sum_{r=2}^{m} \sum_{k=2}^{r} c(i, r, k)\ \lambda_r^{n-2}, \quad n \geqslant 2,$$

where

$$c(i, r, k) = (\lambda_r - 1)\,[i^k + (m - i)^k]\, v_r(m)u_r(k).$$

In ([205], p. 44) a table is given containing the values of $\mathsf{P}_i(\nu \leqslant n)$ for $m = 12$, $1 \leqslant i \leqslant 11$, and $n = 1(5)36(10)96$.

It is now easy to derive the generating function of ν (see 1.4.6). We have

$$\mathsf{E}_i(z^\nu) = m^{-m}(i^m + (m - i)^m)z + m^{-m} \sum_{n\geqslant 2} \sum_{r=2}^{m} \sum_{k=1}^{r} c(i, r, k)\, \lambda_r^{n-2} z^n$$

$$= m^{-m} \left[(i^m + (m - i)^m)z + \sum_{r=2}^{m} \sum_{k=1}^{r} c(i, r, k) \frac{z^2}{1 - z\lambda_r} \right].$$

It follows that

$$\mathsf{E}_i(\nu) = \frac{d}{dz}\, \mathsf{E}_i(z^\nu)\,|_{z=1}$$

$$= m^{-m} \left[(i^m + (m - i)^m) + \sum_{r=2}^{m} \sum_{k=1}^{r} c(i, r, k)\, \frac{2 - \lambda_r}{(1 - \lambda_r)^2} \right].$$

Also

$$\mathsf{E}_i(\nu(\nu - 1)) = \frac{d^2}{dz^2}\, \mathsf{E}_i(z^\nu)\,|_{z=1} = m^{-m} \sum_{r=2}^{m} \sum_{k=1}^{r} c(i, r, k)\, \frac{2}{(1 - \lambda_r)^3}.$$

Hence we can obtain the variance of ν as

$$\text{var}_i(\nu) = \mathsf{E}_i(\nu(\nu - 1)) + \mathsf{E}_i(\nu) - [\mathsf{E}_i(\nu)]^2.$$

In ([205], p. 43) a table is given containing the values of $\mathsf{E}_i(\nu)$ and $\sqrt{\text{var}_i(\nu)}$ for $m = 4, 6, 8, 10, 12$, and 16, $1 \leqslant i \leqslant m/2$ (recall that for any fixed m the values corresponding to the initial states i and $m-i$ are the same).

Remark. The results concerning the time to homozygosity may be obtained using Theorem 3.2, as equation (6.13) allows us to compute the fundamental matrix of the chain.

6.6.6. The Wright model does not take into account any factor influencing the genetic variability. Factors such as mutation, migration, and selection may be incorporated by generalizing the model as follows. For mutation, it is assumed that when the next generation is formed each allelomorph may or may not mutate to the other one. Let α_1 and α_2 be the probabilities of the mutations $A \to a$ and $a \to A$, respectively. (This means A and a do not mutate with probabilities $1 - \alpha_1$ and $1 - \alpha_2$, respectively.) Next, assume that when the next generation is formed a number c_1 of A-gametes and a number c_2 of a-gametes are introduced from outside. Finally, assume that the chances of survival to reproduction of A-and a-gametes are σ_1 and σ_2, respectively. Under the above conditions the probability of selecting an A-gamete when the population state is i is given by

$$p_i = \frac{(1-\alpha_1)\,\sigma_1 i + \alpha_2\sigma_2(m-i) + c_1}{\sigma_1 i + \sigma_2(m-i) + c_1 + c_2}. \tag{6.15}$$

The corresponding probability for an a-gamete will be $1-p_i$. According to the generalized Wright model allowing for mutation, migration and selection the evolution of the number of A-genes in the population is described by a Markov chain with states $0, 1, \ldots, m$ and transition probabilities

$$p(i,j) = \binom{m}{j} p_i^j (1-p_i)^{m-j}, \qquad 0 \leqslant i,\ j \leqslant m.$$

The study of this Markov chain is quite difficult and only partial results have been obtained. Thus W. Feller [101] showed that in the case of the model only allowing for mutation ($c_1 = c_2 = 0$, $\sigma_1 = \sigma_2 = 1$) the eigenvalues of the transition matrix of the associated Markov chain are given by

$$\lambda_r = (1-\alpha_1-\alpha_2)^r \frac{(m)_r}{m^r}, \qquad 0 \leqslant r \leqslant m,$$

(with the convention that $(m)_0 = 1$). If $\alpha_1 > 0$, $\alpha_2 > 0$, $\alpha_1 + \alpha_2 \neq 1$, the eigenvalues are distinct and all the states are recurrent and form a single class (of period 2 if $\alpha_1 = \alpha_2 = 1$ and aperiodic otherwise).

S. Karlin and J. McGregor [192] showed that the Markov chain associated with the generalized Wright model is a special case of a Markov chain induced by direct product branching processes, a setting which they studied exhaustively under assumptions which allowed them to rediscover Feller's above result.

6.6.7. One of the assumptions of the Wright model is that generations do not overlap. In other words, it is assumed that all the individuals in a generation die (or at least are no longer able to reproduce) when the next generation appears, i.e., all the individuals are simultaneously replaced. P.A.P. Moran [279] modified the Wright model by assuming that at any time 0, 1, 2, ... one of the gametes chosen at random dies and is immediately replaced by a new gamete which is A or a with probabilities p_i and $1-p_i$, where p_i is given by (6.15).The number of steps corresponding to a generation in the Moran model equals m, although not all the m individuals present at, say, time 0 are replaced in the first m steps. As in the Wright model, if we define the state of the population at time n to be the number $X(n)$ of A-genes, then $(X(n))_{n\geq 0}$ is a Markov chain with states 0, 1, ..., m and transition probabilities

$$p(i, i-1) = \frac{i}{m}(1-p_i), \qquad 1 \leqslant i \leqslant m,$$

$$p(i, i) = \frac{i}{m}p_i + \left(1 - \frac{i}{m}\right)(1-p_i), \qquad 0 \leqslant i \leqslant m,$$

$$p(i, i+1) = \left(1 - \frac{i}{m}\right)p_i, \qquad 0 \leqslant i < m,$$

$$p(i, j) = 0, \text{ if } |i-j| > 1, \qquad 0 \leqslant i,\ j \leqslant m.$$

The reader should notice that this is a random walk of the type discussed in Example 2 from 2.2.

The study of the Markov chain associated with the Moran model is much simpler than that of the Markov chain associated with the Wright model. For a review of the results obtained [95, 190, 279, and 395] the reader may consult [166], II, §§ 4.1.2 and 4.1.3.

We shall just note that states 0 and m are absorbing in the Markov chain associated with the Moran model without mutation, migration and selection ($\alpha_1 = \alpha_2 = c_1 = c_2 = 0$, $\sigma_1 = \sigma_2 = 1$). As in the Wright model, this chain is a martingale (see 6.6.4), that is

$$\sum_{j=0}^{m} p(k, j)\, j = k, \qquad 0 \leqslant k \leqslant m.$$

Consequently, the probabilities of absorption in 0 and m are i/m and $1 - i/m$, respectively, if the initial state is i.

EXERCISES

6.1 *Component models* [93 and 94]. A component model is subjected to all the axioms of the pattern model (see 6.2.2) except for the sampling and response axiomes which are replaced by the following axiom: on each trial the subject draws a sample from the population of the l stimulus elements and makes a response with probability equal to the proportion of elements in the trial sample conditioned to that response. One distinguishes two general types of component models:

a) The *fixed-sample-size model* where one assumes that the sample size is a fixed number s throughout any given experiment and all possible samples of size s are sampled with equal probability;

b) The *independent sampling model* where one assumes that the stimulus elements are sampled independently on each trial, each element having some fixed probability θ of being drawn.

Let $X(n)$ denote the number of stimulus elements conditioned to response A_1 at the beginning of trial $n \geq 0$. Prove that

i) For the fixed-sample-size model the sequence $(X(n))_{n\geq 0}$ is a Markov chain with state space $(0, 1, \ldots, l)$ and transition probabilities

$$p(i, i+m) = c\,\frac{\binom{l-i}{m}\binom{i}{s-m}}{\binom{l}{s}}\left(\frac{s-m}{s}\pi_{11} + \frac{m}{s}\pi_{21}\right),$$

$$(0 \leq i < l,\ 0 < m \leq \min\ (s,\ l-i))$$

$$p(i, i-m) = c\,\frac{\binom{l-i}{s-m}\binom{i}{m}}{\binom{l}{s}}\left(\frac{m}{s}\pi_{12} + \frac{s-m}{s}\pi_{22}\right),$$

$$(0 < i \leq l,\ 0 < m \leq \min\ (i,\ s))$$

$$p(i, i) = 1 - c + c\left[\frac{\binom{i}{s}}{\binom{l}{s}}\pi_{11} + \frac{\binom{l-i}{s}}{\binom{l}{s}}\pi_{22} + \frac{i}{l}\pi_{10} + \frac{l-i}{l}\pi_{20}\right]$$

$$(0 \leq i \leq l).$$

ii) For the independent sampling model without neutral event E_0 the sequence $(X(n))_{n\geq 0}$ is a Markov chain with state space $(0, 1, \ldots, l)$ and transition probabilities

$$p(i, i+m) =$$

$$= c\binom{l-i}{m}\theta^m(1-\theta)^{l-i-m}\sum_{u=0}^{i}\left(\frac{u}{u+i}\pi_{11} + \frac{i}{u+i}\pi_{21}\right)\binom{i}{u}\theta^u(1-\theta)^{i-u}$$

$$(0 \leq i < l,\ 0 < m \leq l-i)$$

$$p(i, i-m) =$$

$$= c\binom{i}{m}\theta^m(1-\theta)^{i-m}\sum_{u=0}^{l-i}\left(\frac{i}{u+i}\pi_{12} + \frac{u}{u+i}\pi_{22}\right)\binom{l-i}{u}\theta^u(1-\theta)^{l-i-u}$$

$$(0 < i \leq l,\ 0 < m \leq i)$$

$$p(i, i) = 1 - c$$

$$+ c((1-\theta)^l + (1-\theta)^{l-i}\ \pi_{11}[1-(1-\theta)^i] + (1-\theta)^i\ \pi_{22}[1-(1-\theta)^{l-i}])$$

$$(0 \leq i \leq l).$$

6.2 Show that the Markov chain with transition matrix (6.2) is groupable with respect to the partition $S = (1, 2) \cup (3,5) \cup (4) \cup (6)$.

6.3 *Family relations.* Number the genotypes AA, Aa, aa by 1, 2, 3, respectively, and let the sequence $(X(n))_{n\geqslant 0}$ of genotypes be defined as follows: $X(0)$ being arbitrarily given, $X(n)$, $n \geqslant 1$, is the genotype of the offspring of a father of genotype $X(n-1)$ and a mother whose genotype is 1, 2, 3 with probabilities p^2, $2pq$, q^2, respectively.

i) Show that $(X(n))_{n\geqslant 0}$ is a regular Markov chain with state space (1, 2, 3) and transition matrix

$$\mathbf{P} = \begin{pmatrix} p & q & 0 \\ p/2 & 1/2 & q/2 \\ 0 & p & q \end{pmatrix}.$$

ii) Show that the stationary distribution is $\pi = (p^2, 2pq, q^2)'$ and that the chain is reversible.

iii) Interpret the n-step transition probabilities $p(n, i, j)$ also taking reversibility into account.

iv) Show that

$$\mathbf{P}^n = \mathbf{e}\pi' + \frac{1}{2^{n-1}} \begin{pmatrix} pq & q(q-p) & -q^2 \\ p(q-p)/2 & (1-4pq)/2 & q(p-q)/2 \\ -p^2 & p(p-q) & pq \end{pmatrix}$$

for all $n \geqslant 1$.

v) Compute the fundamental matrix of the chain.

(Hint: iii) $p(n, i, j)$ is the conditional probability that a descendant of the nth generation is of genotype j given that a specified ancestor was of genotype i. On account of reversibility $p(n, i, j)$ is also the conditional probability that the ancestor is of genotype j given that the nth offspring is of genotype i. For instance, $p(3, i, j)$ is the conditional probability of a great-grandson (great-grandfather) being of genotype j given that the great-grandfather (great-grandson) is of genotype i.)

6.4 The notation being that in 6.6.3 prove [204] that

$$\lim_{m\to\infty} \frac{u_r(i)}{m^{r-i}} = (-1)^{r-i} \frac{(r)_{r-i}(r-1)_{r-i}}{(2r-2)_{r-i}(r-i)\,!}, \qquad i < r.$$

$$\lim_{m\to\infty} \frac{v_r(i)}{m^{i-r}} = \frac{(i)_{i-r}(i-1)_{i-r}}{(i+r-1)_{i-r}\,(i-r)\,!}, \qquad r < i.$$

6.5 Consider the Markov chain with state space $(0, 1, \ldots, m)$ and transition probabilities

$$p(i,j) = \binom{m}{j}\left(\frac{i+r}{s}\right)^j \left(1 - \frac{i+r}{s}\right)^{m-j}, \quad 0 \leqslant i,\ j \leqslant m,$$

where s and r are nonnegative integers such that $s \geqslant m+r$. It occurs in a model for opinion change during repeated balloting [227] and coincides with the Markov chain associated with the Wright model when $r = 0$, $s = m$. Show that the eigenvalues of the transition matrix are given by $\lambda_i = (m)_i/s^i$, $0 \leq i \leq m$, where $(m)_0 = 1$.

6.6 Show [95] that the entries of the fundamental matrix $\mathbf{N} = (n(i,j))_{1 \leq i,j \leq m-1}$ of the absorbing Markov chain associated with the Moran model with selection are given by

$$n(i,j) = \frac{(1+\sigma)^{m-i} - 1}{(1+\sigma)^m - 1} \cdot \frac{(1+\sigma)^j - 1}{\sigma p(j, j-1)}, \qquad i \geq j,$$

$$n(i,j) = \frac{(1+\sigma)^m - (1+\sigma)^{m-i}}{(1+\sigma)^m - 1} \cdot \frac{(1+\sigma)^m - (1+\sigma)^j}{[(1+\sigma)^m - (1+\sigma)^{m-1}]\, p(j, j+1)},$$

$$i < j.$$

Compute the absorption probabilities and the mean time to absorption. Deduce the corresponding results for the model without selection by letting $\sigma \to 0$.

6.7 Prove [279] that the eigenvalues of the transition matrix of the Markov chain associated with the Moran model without mutation, selection and migration are given by

$$\lambda_r = 1 - r(r-1)/m^2, \quad 0 \leq r \leq m.$$

CHAPTER 7

Nonhomogeneous Markov Chains

This chapter is aimed to be an introduction to the study of nonhomogeneous finite Markov chains. Although in such chains the Markov property still holds, the transition matrix changes at each step. This increase in generality unavoidably leads to a diminution of the number of interesting results as compared to the homogeneous case.

The problems of interest in the study of nonhomogeneous Markov chains are suggested by the concepts and results from the theory of homogeneous Markov chains. A central place is therefore held by ergodicity and asymptotic behaviour.

Nowadays nonhomogeneous Markov chains are being intensively investigated. Lots of recent papers [56 — 61, 130, 131, 161, 162, and 210] confirm this assertion.

7.1 PRELIMINARIES

7.1.1. The entities from which a nonhomogeneous Markov chain is constructed (compare with 2.1.1) are a finite set S, whose elements, called *states*, will be assumed numbered in a definite manner, a probability distribution $\mathbf{p} = (p(i))_{i \in S}$ over S, whose components are called *initial probabilities*, and a sequence of stochastic matrices $\mathbf{P}_m = (p(m, i, m+1, j))_{i,j \in S}$, $m \geq 0$, whose entries are called *transition probabilities*.

A sequence of S-valued random variables $(X(n))_{n \geq 0}$ is said to be a (*nonhomogeneous finite*) *Markov chain* with *state space* S, *initial distribution* $\mathbf{p}$ and *transition matrices* $\mathbf{P}_m$, $m \geq 0$, if and only if $\mathsf{P}(X(0) = i) = p(i)$, $i \in S$, and

$$\underline{\mathsf{P}(X(m+1)=i_{m+1}|X(m)=i_m, \ldots, X(0)=i_0)=\mathsf{P}(X(m+1)=i_{m+1}|X(m)=i_m)} = p(m, i_m, m+1, i_{m+1})$$

whatever the states $i_0, \ldots, i_{m+1} \in S$ and the nonnegative integer m, whenever the left hand side is defined. (The underlined equality is the *Markov property* — see 2.1.)

7.1.2. The existence of nonhomogeneous Markov chains can be proved as has been done for homogeneous Markov chains (the case $\mathbf{P}_m = \mathbf{P}$, $m \geq 0$). The reader will be able to carry over the reasoning in 2.1.2 to the nonhomogeneous case.

7.1.3. Let us define the matrices

$$\mathbf{P}(m, m+n) = \begin{cases} \mathbf{I}, & \text{if } n = 0, \\ \mathbf{P}_m & \text{if } n = 1, \\ \mathbf{P}_m \dots \mathbf{P}_{m+n-1}, & \text{if } n>1, \ m \geqslant 0. \end{cases}$$

The entries $p(m, i, m+n, j)$, $i, j \in S$, of the matrix $\mathbf{P}(m, m+n)$ have a very simple probabilistic interpretation. If $\mathsf{P}(X(m) = i) > 0$, then

$$p(m, i, m+n, j) = \mathsf{P}(X(m+n) = j \mid X(m) = i), \ n \geqslant 0.$$

The proof by induction with respect to n is immediate. Therefore the probabilities $p(m, i, m+n, j)$ generalize the n-step transition probabilities of the homogeneous case

Notice that

$$\mathbf{P}(m, m+n) = \mathbf{P}(m, m+s)\,\mathbf{P}(m+s, m+n)$$

whatever $0 \leqslant s \leqslant n$. This identity is equivalent to the equations

$$p(m, i, m+n, j) = \sum_{k \in S} p(m, i, m+s, k)\, p(m+s, k, m+n, j), \ i, j \in S,$$

that represent the *Chapman-Kolmogorov equations* in the nonhomogeneous case.

7.1.4. A classical example of a nonhomogeneous Markov chain is associated with the so called *Pólya urn*. Consider an urn that initially contains a white and b black balls. At any time $m = 1, 2, \dots$ one ball is drawn at random, after which it is replaced while s balls of the same colour are added to the urn. Let $X(m)$ denote the number of white balls obtained in the first m drawings, $0 \leqslant X(m) \leqslant m$, and set $X(0) = 0$. If $X(m) = i$ then i white and $m - i$ black balls have been drawn in the first m drawings. Therefore the composition of the urn before the $(m+1)$ th drawing is $a + si$ white and $b + (m - i)s$ black balls. Obviously, the conditional probability distribution of $X(m+1)$ given the values of $X(0), \dots, X(m)$ depends only on the value of $X(m)$ that determines the composition of the urn after the first m drawings. We have

$$p(m, i, m+1, i) = \mathsf{P}(X(m+1) = i \mid X(m) = i) = \frac{b + (m-i)s}{a + b + ms},$$

$$p(m, i, m+1, i+1) = \mathsf{P}(X(m+1) = i+1 \mid X(m) = i) = \frac{a + is}{a + b + ms},$$

$$p(m, i, m+1, j) = 0, j \neq i, i+1,$$

for all $0 \leqslant i \leqslant m$, $m \geqslant 0$.

This example emphasizes the fact that the state space is time varying (the possible states i at time m, those for which $\mathsf{P}(X(m) = i) > 0$, are $0, 1, \ldots, m$) and, consequently, suggests an extension of the concept of a nonhomogeneous Markov chain by considering a sequence $(S_m)_{m \geqslant 0}$ of finite sets and a sequence

$$\mathbf{P}_m = (p(m, i, m+1, j))_{i \in S_m, j \in S_{m+1}}, \quad m \geqslant 0,$$

of stochastic matrices (in general rectangular) instead of a single set S and a sequence of square stochastic matrices indexed over S. In the new framework the values of $X(m)$ are the elements of S_m, $m \geqslant 0$, and the definition from 7.1.1 can be immediately reformulated.

We shall not adopt this point of view, because most of the results obtained in the case $S_m = S$, $m \geqslant 0$, can be reformulated for the case of a sequence of state spaces.

7.1.5. It can be easily verified that all the properties in 2.1.4 are still valid in the nonhomogeneous case*).

However, when we come to the strong Markov property, things become intricate and even its statement no longer has the elementary aspect met in the homogeneous case. We shall at once become convinced of that.

Let τ be a finite stopping time for a nonhomogeneous Markov chain $(X(n))_{n \geqslant 0}$, i.e., exactly as in the homogeneous case, a random variable with values 0, 1, 2, ... (the value ∞ is excluded for the sake of simplicity) such that whatever $k = 0, 1, 2, \ldots$ the random event $\{\tau = k\}$ is determined by the random variables $X(0), \ldots, X(k)$, that is, it is of the form $\{(X(0), \ldots, X(k)) \in A^{(k+1)}\}$ where $A^{(k+1)}$ is a set of $(k+1)$-tuples of states of the chain. Denote by $\mathcal{K}_\tau$ the set of all the random events A prior to τ, i.e., such that whatever $k = 0, 1, 2, \ldots$ the random event $A \cap \{\tau = k\}$ is determined by $X(0), \ldots, X(k)$. It is easily proved that $\mathcal{K}_\tau$ is a σ-algebra. If for any random event B we denote by $\mathsf{P}(B \mid \mathcal{K}_\tau)$ the random variable taking the value

$$\mathsf{P}(B \mid X(k) = i_k, \ldots, X(0) = i_0, \tau = k)$$

at all the sample points in the random event

$$\{X(0) = i_0, \ldots, X(k) = i_k, \tau = k\}, \quad i_0, \ldots, i_k \in S, \quad k = 0, 1, 2, \ldots,$$

then the *strong Markov property* for the Markov chain considered amounts to the equality

$$\mathsf{P}(X(\tau + n) = j \mid \mathcal{K}_\tau) = p(\tau, X(\tau), \tau + n, j) \tag{7.1}$$

for all $n \geqslant 0$, $j \in S$. The proof of (7.1) is beyond the scope of this book.

Notice that in the nonhomogeneous case the strong Markov property does not imply that $(X(n + \tau))_{n \geqslant 0}$ is a Markov chain (as happens in the homogeneous case). A simple example (due to U. Krengel) justifies the above assertion. Consider a nonhomogeneous Markov chain

---

*) Clearly, in the symmetric form of the Markov property we can no longer write that $\mathsf{P}(B \mid X(n) = i_n) = \mathsf{P}((X(1), X(2), \ldots) \in B^{(\infty)} \mid X(0) = i_n)$ if $B = \{(X(n+1), X(n+2), \ldots) \in B^{(\infty)}\}$. (This equality is a consequence of homogeneity.)

with two states 0 and 1, initial distribution $\mathbf{p}' = (1/2, 1/2)$ and transition probabilities at times 0, 1, and 2 given by

$$p(0, i, 1, i) = 1, p(1, i, 2, j) = 1/2, p(2, i, 3, i) = 1$$

whatever $0 \leqslant i, j \leqslant 1$ (the transition probabilities at other times are completely arbitrary). It is easily verified that $\tau = X(0)$ is a stopping time for the Markov chain considered. Setting $Y(n) = X(n + \tau), n \geqslant 0$, we have

$$\mathsf{P}(Y(0) = 0, \ Y(1) = 0) = \sum_{i=0}^{1} \mathsf{P}(X(0) = i, \ Y(0) = 0, \ Y(1) = 0)$$

$$= \mathsf{P}(X(0) = 0, X(1) = 0) + \mathsf{P}(X(0) = 1, X(1) = 0, X(2) = 0) = 1/2,$$

$$\mathsf{P}(Y(0) = 0, \ Y(1) = 0, \ Y(2) = 0) = \mathsf{P}(X(0) = 0, \ X(1) = 0, \ X(2) = 0)$$
$$+ \mathsf{P}(X(0) = 1, X(1) = 0, \ X(2) = 0, \ X(3) = 0) = 1/4,$$

$$\mathsf{P}(Y(1) = 0) = \sum_{i=0}^{1} \mathsf{P}(X(0) = i, \ Y(1) = 0)$$

$$= \mathsf{P}(X(0) = 0, X(1) = 0) + \mathsf{P}(X(0) = 1, X(2) = 0) = 1/2 + 1/4 = 3/4,$$

$$\mathsf{P}(Y(1) = 0, \ Y(2) = 0) = \mathsf{P}(X(0) = 0, \ X(1) = 0, \ X(2) = 0)$$
$$+ \mathsf{P}(X(0) = 1, \ X(2) = 0, \ X(3) = 0) = 1/4 + 1/4 = 1/2,$$

whence

$$\mathsf{P}(Y(2) = 0 \mid Y(1) = 0, \ Y(0) = 0) = 1/2,$$
$$\mathsf{P}(Y(2) = 0 \mid Y(1) = 0) = 2/3,$$

that is $(Y(n))_{n \geqslant 0}$ is not a Markov chain.

7.1.6. The problem of classifying the states of a nonhomogeneous Markov chain is not yet solved. There are some partial results [60, 61, 81, and 349] that only emphasize the difficulties to be overcome.

Instead, as is seen from the next two paragraphs, more progress has been made in studying the asymptotic behaviour of the probabilities $p(m, i, m + n, j)$ as $n \to \infty$. However, this study of ergodicity of nonhomogeneous Markov chains can be detached from the Markovian context and regarded as the study of products of stochastic matrices when the number of the factors increases unboundedly.

7.1.7. As was shown in 2.2, homogeneous Markov chains can be used for the mathematical modelling of a great variety of real situations. The homogeneity hypothesis allowed us to construct a rich theory whose conclusions can often be used as a good first approximation to reality. However, there are many real situations for which models based on homogeneous Markov chains are clearly inadequate. Introducing time-dependent transition probabilities, thus constructing models based on nonhomogeneous Markov chains, might lead to better results. Yet, the insufficient development of the theory in the nonhomogeneous case often makes it preferable to use a homogeneous Markov chain model that, although less realistic, has the advantage of being amenable to a complete analysis.

7.1.8. The concept of a nonhomogeneous Markov chain can be extended in the same manner as that of a homogeneous Markov chain (see 5.6). Thus we shall have nonhomogeneous denumerable Markov chains, nonhomogeneous multiple Markov chains, nonhomogeneous chains with complete connections, nonhomogeneous finite Markov processes.

The Pólya urn (see 7.1.4) yields an example of a nonhomogeneous chain with complete connections. If we denote by $Y(n)$, $n \geqslant 1$, the number of white balls occurring on the nth drawing (clearly, $Y(n)$ equals either 0 or 1), then the sequence of random variables $(Y(n))_{n\geqslant 0}$, $Y(0) \equiv 0$, is a chain with complete connections. (It is easily seen that $Y(0) + ... + Y(n) = X(n)$, $n \geqslant 0$.) Setting

$$p_n = \mathsf{P}(Y(n) = 1 \mid Y(n-1), ..., Y(0)), \ n \geqslant 1,$$

we have, indeed,

$$p_n = \frac{a + X(n-1)s}{a + b + (n-1)s}$$

and

$$p_{n+1} = \begin{cases} \dfrac{a + (X(n-1)+1)s}{a+b+ns}, & \text{if } Y(n) = 1, \\[2ex] \dfrac{a + X(n-1)s}{a+b+ns}, & \text{if } Y(n) = 0, \end{cases}$$

that is

$$p_{n+1} = \begin{cases} \dfrac{a+b+(n-1)s}{a+b+ns}\, p_n + \dfrac{s}{a+b+ns}, & \text{if } Y(n) = 1, \\[2ex] \dfrac{a+b+(n-1)s}{a+b+ns}\, p_n, & \text{if } Y(n) = 0. \end{cases}$$

7.2 WEAK ERGODICITY

7.2.1. A Markov chain is said to be *weakly ergodic* (in the sense of Kolmogorov [214]) if and only if

$$\lim_{n\to\infty} [p(m, i, m+n, k) - p(m, j, m+n, k)] = 0, \tag{7.2}$$

for all $m \geqslant 0$, and $i, j, k \in S$.

Equation (7.2) does not imply the existence of the limits of the probabilities $p(m, i, m+n, k)$ and $p(m, j, m+n, k)$ as $n \to \infty$ but only a tendency towards equalization of the rows of the matrices $\mathbf{P}(m, m+n)$, $m \geq 0$. By (1.15), equation (7.2) holds if and only if

$$\lim_{n\to\infty} \alpha\,(\mathbf{P}(m, m+n)) = 1$$

whatever $m \geq 0$.

Remark also that it is sufficient to require that (7.2) holds only for the values of m exceeding a given m_0. Then the validity of (7.2) for the values $l \leq m_0$ follows by multiplying it by $p(l, i', m, i)\, p(l, j', m, j)$ and summing over $i, j \in S$. This means that weak ergodicity is not altered when finitely many matrices $\mathbf{P}_m$, $m \geq 0$, are replaced by new ones.

7.2.2. Several necessary and sufficient conditions for weak ergodicity are known. The most immediate is

Theorem 7.1 [226]. *A Markov chain is weakly ergodic if and only if there exist stable matrices* $\mathbf{\Pi}(m, n)$, $m \geq 0$, $n \geq 1$ *such that*

$$\lim_{n\to\infty} (\mathbf{P}(m, m+n) - \mathbf{\Pi}(m, n)) = 0 \tag{7.3}$$

whatever $m \geq 0$.

Proof. If (7.3) holds with $\mathbf{\Pi}(m, n) = \mathbf{e}(\boldsymbol{\pi}(m, n))'$ then

$$\lim_{n\to\infty} [p(m, i, m+n, k) - \pi(m, n, k)] = 0,$$

$$\lim_{n\to\infty} [p(m, j, m+n, k) - \pi(m, n, k)] = 0$$

for all $m \geq 0$, $i, j, k \in S$, equalities that imply (7.2), i.e., weak ergodicity.

Conversely, if weak ergodicity holds set

$$\pi(m, n, k) = \frac{1}{r} \sum_{i\in S} p(m, i, m+n, k), \quad m \geq 0, \quad n \geq 1, \quad k \in S,$$

where r is the number of states of the chain. Then we have

$$\lim_{n\to\infty} [p(m, j, m+n, k) - \pi(m, n, k)]$$

$$= \lim_{n\to\infty} \left[p(m, j, m+n, k) - \frac{1}{r} \sum_{i\in S} p(m, i, m+n, k) \right]$$

$$= \lim_{n\to\infty} \frac{1}{r} \sum_{i\in S} [p(m, j, m+n, k) - p(m, i, m+n, k)] = 0$$

for all $m \geq 0$, $j, k \in S$. The proof is complete.

Corollary. *A Markov chain is weakly ergodic if and only if for all* $m \geqslant 0$ *the limit of any convergent subsequence of the sequence* $(\mathbf{P}(m, m+n))_{n\geqslant 1}$ *is a stable matrix.*

Proof. Assume the chain is weakly ergodic, and that therefore (7.3) holds. If for a fixed m and a subsequence $1 \leqslant m_1 \leqslant m_2 < \dots$ we have $\lim_{s\to\infty} \mathbf{P}(m, m+m_s) = \mathbf{A}(m)$, then the limit $\lim_{s\to\infty} \mathbf{\Pi}(m, m_s) = \mathbf{\Pi}(m)$ exists, is a stable matrix, and $\mathbf{A}(m) = \mathbf{\Pi}(m)$.

Conversely, if for a fixed m and a subsequence $1 < m_1 < m_2 < \dots$ we have $\lim_{s\to\infty} \mathbf{P}(m, m+m_s) = \mathbf{\Pi}(m)$ (stable matrix), then (7.3) holds with

$$\mathbf{\Pi}(m, n) = \mathbf{\Pi}(m)\, \mathbf{P}(m+m_s, m+n)$$

for $m_s \leqslant n < m_{s+1}$, $s \geqslant 1$, that is the chain is weakly ergodic.

7.2.3. Weak ergodicity can be characterized by means of the ergodicity coeficients α and δ as shown by the following two theorems.

Theorem 7.2 [135] *A Markov chain is weakly ergodic if and only if there exists an increasing sequence* $1 \leqslant n_1 < n_2 < \dots$ of *natural numbers such that the series* $\sum_{s\geqslant 1} \alpha(\mathbf{P}(n_s, n_{s+1}))$ *diverges.*

Proof. By a well known result in calculus the series $\sum_{s\geqslant 1} \alpha(\mathbf{P}(n_s, n_{s+1}))$ diverges if and only if

$$\prod_{s\geqslant 1} (1 - \alpha(\mathbf{P}(n_s\ n_{s+1}))) = 0. \tag{7.4}$$

Therefore, if the series diverges then by Theorem 1.11 we have $\lim_{n\to\infty} \alpha(\mathbf{P}(m, m+n)) = 1$ whatever $m \geqslant 0$, i.e., the chain is weakly ergodic.

Conversely, if the chain is weakly ergodic then $\lim_{n\to\infty} \alpha(\mathbf{P}(m, m+n)) = 1$ for all $m \geqslant 0$ and we can choose $1 \leqslant n_1 < n_2 < \dots$ such that $\alpha(\mathbf{P}(n_s, n_{s+1})) \geqslant a > 0$, $s \geqslant 1$. Therefore (7.4) holds, thus implying that the series $\sum_{s\geqslant 1} \alpha(\mathbf{P}(n_s, n_{s+1}))$ diverges. The proof is complete.

Corollary. *A Markov chain for which* $\sum_{m\geqslant 0} \alpha(\mathbf{P}_m)$ *diverges is weakly ergodic. For two state Markov chains divergence of the series* $\sum_{m\geqslant 0} \alpha(\mathbf{P}_m)$ *is not only a sufficient but also a necessary condition for weak ergodicity.*

Proof. The first assertion corresponds to the choice $n_s = s$, $s \geqslant 1$, in Theorem 7.2. The second assertion follows at once from the equality

$$1 - \alpha(\mathbf{P}(m, n)) = \prod_{l=m}^{n-1} (1 - \alpha(\mathbf{P}_l)),$$

which holds for two state Markov chains (see Theorem 1.11).

Theorem 7.3 [81]. *A Markov chain is weakly ergodic if and only if there exists an increasing sequence* $1 \leqslant n_1 < n_2 < \ldots$ *of natural numbers such that the series* $\sum_{s\geqslant 1} \delta(\mathbf{P}(n_s, n_{s+1}))$ *diverges.*

Proof. Since $\alpha \geqslant \delta$ (see (1.14)), by Theorem 7.2 the condition is sufficient. It remains to prove necessity.

If the chain is weakly ergodic then, by the corollary to Theorem 7.1 and the continuity of the coefficient δ as a function of the entries of the transition matrices, there exists for any $m \geqslant 0$ a subsequence $1 \leqslant m_1 < m_2 < \ldots$ such that $\lim_{s\to\infty} \delta(\mathbf{P}(m, m+m_s)) = 1$. The proof from now on is identical to that of the preceding theorem.

Corollary [301]. *A Markov chain for which* $\sum_{m\geqslant 0} \delta(\mathbf{P}_m)$ *diverges is weakly ergodic.*

Remark. A theorem analogous to Theorems 7.2 and 7.3 for the ergodicity coefficient μ cannot be obtained due to the fact that μ is not proper. Clearly, divergence of the series $\sum_{s\geqslant 1} \mu(\mathbf{P}(n_s, n_{s+1}))$ is a sufficient condition for weak ergodicity, which essentially goes back to A.A. Markov and S.N. Bernstein.

7.2.4. Weak ergodicity finds an interesting interpretation in the framework of Doeblin's two particle method (see 4.1.4). Namely, we have

Theorem 7.4 [81 and 135]. *A Markov chain is weakly ergodic if and only if, with probability* 1, *two independent copies of the chain which are in different states at any arbitrarily given time will eventually meet (i.e., will simultaneously reach the same state).*

Proof. If the two copies are at time n in different states $i \neq j$, then the probability that they meet at time $m+n$ equals

$$\sum_{k\in S} p(m, i, m+n, k)\, p(m, j, m+n, k)$$

$$\geqslant \sum_{k\in S} [\min(p(m, i, m+n, k), p(m, j, m+n, k))]^2$$

$$\geqslant \frac{1}{r}\left[\sum_{k\in S} [\min(p(m, i, m+n, k), p(m, j, m+n, k))\right]^2$$

$$\geqslant \frac{1}{r}\alpha^2(\mathbf{P}(m, m+n)).$$

[Here r is the number of states of the chain. We used the elementary inequality $\left(\sum_{k=1}^{r} a_k\right)^2 \leqslant r\sum_{k=1}^{r} a_k^2$, where a_k, $1 \leqslant k \leqslant r$, are real numbers.]

Consider a weakly ergodic Markov chain. Then there exists a sequence $m = m_0 < m_1 < m_2 < \dots$ such that $\alpha(\mathbf{P}(m_s, m_{s+1})) > a > 0$ for all $s > 0$. Hence

$$\prod_{s=0}^{n-1}\left(1 - \frac{1}{r}\alpha^2(\mathbf{P}(m_s, m_{s+1}))\right) \to 0$$

as $n \to \infty$. The above product is greater than the probability that the copies have not met at times $m_1, \dots, m_n$, which in turn is greater than the probability that they have never met at any time between m and m_n. Thus the latter probability tends to zero as $n \to \infty$. This means that the probability that the copies will eventually meet equals 1.

Conversely, assume that the probability is 1 that being in different states $i \neq j$ at any arbitrarily given time the two copies will eventually meet. This means that denoting by $r_{ij}(m, n)$ the probability that the copies first meet after time $m + n$, given that the two copies were in states i and j at time m, we have $\lim_{n\to\infty} r_{ij}(m, n) = 0$ whatever $m \geqslant 0$. The reasoning from 4.1.4 for the homogeneous case may be adapted without difficulty to the nonhomogeneous case and leads us to the inequality

$$|p(m, i, m + n, k) - p(m, j, m + n, k)| \leqslant r_{ij}(m, n),$$

Hence weak ergodicity holds.

7.3 UNIFORM WEAK ERGODICITY

7.3.1. A more restrictive variant of weak ergodicity is uniform weak ergodicity (see [157]). A Markov chain is said to be *uniformly weakly ergodic* if and only if whatever $i, j, k \in S$

$$\lim_{n\to\infty} [p(m, i, m + n, k) - p(m, j, m + n, k)] = 0$$

uniformly with respect to $m \geqslant 0$. Equivalently

$$\lim_{n\to\infty} \alpha(\mathbf{P}(m, m + n)) = 1 \tag{7.5}$$

uniformly with respect to $m \geqslant 0$.

Theorem 7.5 [157]. *A Markov chain is uniformly weakly ergodic if and only if there exist $n_0 \geqslant 1$ and $a > 0$ such that $\alpha(\mathbf{P}(m, m + n_0)) > a$ whatever $m \geqslant 0$.*

Proof. Necessity follows from the very definition of uniform weak ergodicity.

Conversely, if the condition holds then by Theorem 1.11 we have

$$1 - \alpha(\mathbf{P}(m, m + n)) \leqslant (1 - a)^s$$

whatever $sn_0 \leqslant n < (s+1)n_0$, $m \geqslant 0$, whence

$$\alpha(\mathbf{P}(m, m+n)) \geqslant 1 - (1-a)^{n/n_0 - 1}$$

whatever $m \geqslant 0$, $n \geqslant 1$, i.e., the Markov chain is uniformly weakly ergodic. Notice that on account of the last inequality we have

$$\sum_{k \in S} |p(m, i, m+n, k) - p(m, j, m+n, k)| \leqslant 2(1-a)^{n/n_0 - 1}$$

whatever $m \geqslant 0$, $n \geqslant 1$, $i, j \in S$. Hence by (1.16)

$$|p(m, i, m+n, k) - p(m, j, m+n, k)| \leqslant (1-a)^{n/n_0 - 1}$$

whatever $m \geqslant 0$, $n \geqslant 1$, $i, j, k \in S$, i.e., convergence to 0 is geometrically fast.

7.3.2. In practice the necessary and sufficient condition occurring in Theorem 7.5 is not easily verified. It is therefore useful to find simple sufficient conditions for uniform weak ergodicity. First, a few notions and auxiliary results. The *incidence matrix* of a stochastic matrix $\mathbf{P}$ is defined as the matrix obtained from $\mathbf{P}$ by replacing all its positive entries by ones. Clearly, whether or not a matrix is a Markov matrix or a scrambling matrix depends only on its incidence matrix. We write $\mathbf{P}_1 \sim \mathbf{P}_2$, where $\mathbf{P}_1$ and $\mathbf{P}_2$ are stochastic matrices of the same order, if and only if the two matrices have the same incidence matrix.

Proposition 7.6 [334 and 403] *Let* $\mathbf{P}_1$ *be a stochastic matrix and* $\mathbf{P}_2$ *a mixing matrix. If* $\mathbf{P}_1\mathbf{P}_2 \sim \mathbf{P}_1$ *then* $\mathbf{P}_1$ *is a Markov matrix.*

Proof. If $\mathbf{P}_1\mathbf{P}_2 \sim \mathbf{P}_1$ then $\mathbf{P}_1\mathbf{P}_2^2 \sim \mathbf{P}_1\mathbf{P}_2 \sim \mathbf{P}_1$ and, in general, $\mathbf{P}_1\mathbf{P}_2^n \sim \mathbf{P}_1$ whatever $n \geqslant 1$. By Theorem 4.3, for n sufficiently large $\mathbf{P}_2^n$ is a Markov matrix. Remark now that the product of two stochastic matrices is a Markov matrix if at least one of the factors is a Markov matrix. It follows that $\mathbf{P}_1\mathbf{P}_2^n$ is a Markov matrix and since $\mathbf{P}_1\mathbf{P}_2^n \sim \mathbf{P}_1$ the matrix $\mathbf{P}_1$ is a Markov matrix, too.

Proposition 7.7 [334 and 403]. *If* $\mathbf{P}(m, m+n)$ *is a mixing matrix for all* $m \geqslant 0$, $n \geqslant 1$ *then* $\mathbf{P}(m, m+n)$ *is a Markov matrix for all* $m \geqslant 0$, $n \geqslant t+1$, *where* t *is the number of distinct types of incidence matrices corresponding to mixing matrices.*

Proof. For all $m \geqslant 0$ there exist natural numbers n_1 and n_2, $0 < n_1 < n_2 \leqslant t+1$ such that $\mathbf{P}(m, m+n_1) \sim \mathbf{P}(m, m+n_2)$, since among the $t+1$ matrices $\mathbf{P}(m, m+1), \ldots, \mathbf{P}(m, m+t+1)$ at least two have the same incidence matrix. As $\mathbf{P}(m+n_1, m+n_2)$ is a mixing matrix, Proposition 7.6 implies that $\mathbf{P}(m, m+n_1)$ is a Markov matrix. It follows that $\mathbf{P}(m, m+n) = \mathbf{P}(m, m+n_1)\ \mathbf{P}(m+n_1, m+n)$ is a Markov matrix whatever $m \geqslant 0$, $n \geqslant t+1$.

Theorem 7.8 [339]. *If* $\mathbf{P}(m, m+n)$ *is a mixing matrix for all* $m \geqslant 0$, $n \geqslant n_0$, *and*

$$\min_{i,j \in S}{}^{+} p(m, i, m+n_0, j) \geqslant a > 0 \tag{7.6}$$

for all $m \geqslant 0$, *where* $\min^+$ *indicates that the minimum is taken only over the positive entries, then uniform weak ergodicity holds.*

Proof. By Proposition 7.7 applied to the matrices $\mathbf{P}(m, m+n_0)$, $m \geqslant 0$, the matrix $\mathbf{P}(m, m+(t+1)n_0)$ is a Markov matrix and by (7.6)

$$\mu(\mathbf{P}(m, m+(t+1)n_0)) \geqslant a^{t+1}$$

for all $m \geqslant 0$. But this means that

$$\alpha(\mathbf{P}(m, m+(t+1)n_0)) \geqslant a^{t+1} \tag{7.7}$$

for all $m \geqslant 0$ and by Theorem 7.5 uniform weak ergodicity holds.

Remark. The matrices $\mathbf{P}(m, m+n)$, $m \geqslant 0$, $n \geqslant n_0$, are mixing if e.g. the matrices $\mathbf{P}(m, m+n_0)$, $m \geqslant 0$, are scrambling. Indeed, by Theorem 1.11 this assumption implies that the matrices $\mathbf{P}(m, m+n)$, $m \geqslant 0$, $n \geqslant n_0$, are scrambling, and thus mixing. [It is easy to prove that if $\mathbf{P}(m, m+n_0)$ is a scrambling matrix then (7.6) implies that $\alpha(\mathbf{P}(m, m+n_0)) \geqslant a$.]

7.4 STRONG ERGODICITY

7.4.1 A Markov chain is said to be *strongly ergodic* if and only if for all $m \geqslant 0$, $i, j \in S$ the limits

$$\lim_{n \to \infty} p(m, i, m+n, j) = \pi(m, j) \tag{7.8}$$

exist and do not depend on $i \in S$. It is easy to prove that if (7.8) holds then the $\pi(m, j)$ are also independent of $m \geqslant 0$. Indeed, in matrix notation, (7.8) reads as

$$\lim_{n \to \infty} \mathbf{P}(m, m+n) = \mathbf{\Pi}_m (= \mathbf{e}(\pi(m, j))'_{j \in S}).$$

But $\mathbf{P}(m, m+n) = \mathbf{P}_m \mathbf{P}(m+1, m+n)$ so that $\mathbf{\Pi}_m = \mathbf{P}_m \mathbf{\Pi}_{m+1} = \mathbf{\Pi}_{m+1}$, $m \geqslant 0$, i.e., $\pi(m, j)$, $j \in S$, does not depend on m. The common value of the $\pi(m, j)$, $m \geqslant 0$, will be denoted by $\pi(j)$.

As with weak ergodicity, it is sufficient to require that (7.8) holds only for the values of m exceeding a given m_0.

It is obvious that strong ergodicity implies weak ergodicity but, in general, the converse implication is not true (see Theorem 7.11 below).

7.4.2. A Markov chain is said to be *asymptotically homogeneous* (see [21]) if and only if there exists a probability distribution $\mathbf{p}$ over S such that

$$\lim_{m \to \infty} \mathbf{p}'\mathbf{P}_m = \mathbf{p}', \tag{7.9}$$

Proposition 7.9. *For an asymptotically homogeneous Markov chain*

$$\lim_{m\to\infty} \mathbf{p}'\mathbf{P}(m, m+n) = \mathbf{p}',$$

for all $n \geqslant 1$.

Proof. For $n = 1$ the theorem is true by the very definition of asymptotic homogeneity. Assume that it holds for some $n \geqslant 1$ and let us show that it still holds for $n + 1$. We have

$$\begin{aligned}\mathbf{p}'\mathbf{P}(m, m+n+1) &= \mathbf{p}'\mathbf{P}(m, m+n)\,\mathbf{P}_{m+n} = \\ &= (\mathbf{p}'\mathbf{P}(m, m+n) - \mathbf{p}')\,\mathbf{P}_{m+n} + \mathbf{p}'\mathbf{P}_{m+n}.\end{aligned}$$

Letting $m \to \infty$, by the induction hypothesis, we get

$$\lim_{m\to\infty} \mathbf{p}'\mathbf{P}(m, m+n+1) = \lim_{m\to\infty} \mathbf{p}'\mathbf{P}_{m+n} = \mathbf{p}'$$

so that the proof is complete.

A Markov chain is said to be *asymptotically stationary* (see [226]) if and only if there exists a probability distribution $\mathbf{p}$ over S such that

$$\lim_{n\to\infty} \mathbf{p}'\mathbf{P}(m, m+n) = \mathbf{p}'$$

for all $m \geqslant 0$.

Proposition 7.10. *An asymptotically stationary Markov chain is asymptotically homogeneous.*

Proof. The result follows from the equality

$$\mathbf{p}'\mathbf{P}(m, m+n+1) = \mathbf{p}'\mathbf{P}_{m+n} + (\mathbf{p}'\mathbf{P}(m, m+n) - \mathbf{p}')\,\mathbf{P}_{m+n},$$

upon letting $n \to \infty$.

Theorem 7.11 [226]. *A Markov chain is strongly ergodic if and only if it is weakly ergodic and asymptotically stationary.*

Proof. Necessity. If strong ergodicity holds then, as we have already noted, weak ergodicity also holds. Setting $\boldsymbol{\pi} = (\pi(i))_{i\in S}$ we have

$$\lim_{n\to\infty} \boldsymbol{\pi}'\mathbf{P}(m, m+n) = \boldsymbol{\pi}'\mathbf{e}\boldsymbol{\pi}' = \boldsymbol{\pi}', \quad m \geqslant 0,$$

i.e., asymptotic stationarity (with probability distribution $\boldsymbol{\pi}$).

Sufficiency. If weak ergodicity and asymptotic stationarity (with propability distribution $\mathbf{p}$) hold, then

$$\begin{aligned}&\left| p(m, i, m+n, j) - \sum_{k\in S} p(k)\, p(m, k, m+n, j) \right| \\ &= \left| \sum_{k\in S} p(k)\,(p(m, i, m+n, j) - p(m, k, m+n, j)) \right| \\ &\leqslant \sum_{k\in S} | p(m, i, m+n, j) - p(m, k, m+n, j) | \to 0\end{aligned}$$

as $n \to \infty$. Because

$$\sum_{k \in S} p(k)\, p(m, k, m+n, j) \to p(j),\ m \geqslant 0,\ j \in S,$$

as $n \to \infty$, it follows that

$$\lim_{n\to\infty} p(m, i, m+n, j) = p(j), \quad m \geqslant 0, i, j \in S,$$

i.e., strong ergodicity.

7.4.3. In general, the converse of Proposition 7.10 is not true. A notable special case where that converse holds is given by

Proposition 7.12 [357]. *Assume that there exists $n_0 \geqslant 1$ such that $\alpha(\mathbf{P}(m, m+n_0)) \geqslant a > 0$ for all $m \geqslant 0$. Under this assumption asymptotic stationarity and asymptotic homogeneity are equivalent.*

Proof. On account of Proposition 7.10 we have to prove that asymptotic homogeneity implies asymptotic stationarity. Let us therefore assume that $\lim_{m\to\infty} \mathbf{p}'\mathbf{P}_m = \mathbf{p}'$. To begin with consider the special case $n_0 = 1$. Let $\mathbf{d}'(m) = \mathbf{p}' - \mathbf{p}'\mathbf{P}_m$, $m \geqslant 0$. Thus we have $\lim_{m\to\infty} \mathbf{d}(m) = \mathbf{0}$ and $\mathbf{d}'(m)\mathbf{e} = 0$, $m \geqslant 0$. Consider the vectors

$$\mathbf{a}'_m(n) = \mathbf{p}'\mathbf{P}_{m+n-1} - \mathbf{p}'\mathbf{P}(m, m+n),\ m \geqslant 0,\ n \geqslant 1.$$

We have $\mathbf{a}'_m(n)\mathbf{e} = 0$ and

$$\mathbf{a}'_m(n+1) = \mathbf{a}'_m(n)\mathbf{P}_{m+n} + \mathbf{d}'(m+n-1)\mathbf{P}_{m+n}.$$

Denote by $\alpha_m(n)$ and $\delta_m(n)$ the sum of the absolute values of the components of the vectors $\mathbf{a}_m(n)$ and $\mathbf{d}(m+n-1)$, respectively. On account of (1.18) we then have

$$\alpha_m(n+1) \leqslant (1-a)(\alpha_m(n) + \delta_m(n)),\ m \geqslant 0,\ n \geqslant 1,$$

where $\lim_{n\to\infty} \delta_m(n) = 0$, $m \geqslant 0$. Hence

$$\limsup_{n\to\infty} \alpha_m(n) \leqslant (1-a) \limsup_{n\to\infty} \alpha_m(n),$$

that is $\limsup_{n\to\infty} \alpha_m(n) = 0$ for all $m \geqslant 0$. This means that $\lim_{n\to\infty} \mathbf{a}_m(n) = \mathbf{0}$, $m \geqslant 0$, and since $\lim_{n\to\infty} \mathbf{p}'\mathbf{P}_{m+n-1} = \mathbf{p}'$ we get $\lim_{n\to\infty} \mathbf{p}'\mathbf{P}(m, m+n) = \mathbf{p}'$, $m \geqslant 0$, that is asymptotic stationarity holds.

Passing to the general case $n_0 \geqslant 1$, for a fixed $m \geqslant 0$ consider the stochastic matrices

$$\mathbf{Q}_l = \mathbf{P}(m + ln_0,\ m + (l+1)n_0),\ l \geqslant 0.$$

By assumption $\alpha(\mathbf{Q}_l) \geqslant a$, $l \geqslant 0$, and by Proposition 7.9 we have $\lim_{l\to\infty} \mathbf{p}'\mathbf{Q}_l = \mathbf{p}'$. On account of the special case previously considered the

Markov chain with transition matrices $\mathbf{Q}_l$, $l \geqslant 0$, is asymptotically stationary. Therefore

$$\mathbf{p}' = \lim_{n\to\infty} \mathbf{p}'\mathbf{Q}_0 \dots \mathbf{Q}_{n-1} = \lim_{n\to\infty} \mathbf{p}'\mathbf{P}(m, m + nn_0). \tag{7.10}$$

Then Proposition 7.9 yields

$$\lim_{n\to\infty} \mathbf{p}'\mathbf{P}(m + nn_0, m + nn_0 + s) = \mathbf{p}', \; 1 \leqslant s \leqslant n_0. \tag{7.11}$$

It follows from (7.10) and (7.11) that

$$\lim_{n\to\infty} \mathbf{p}'\mathbf{P}(m, m + nn_0 + s) = \mathbf{p}', \; 1 \leqslant s \leqslant n_0,$$

and the proof is complete.

7.4.4. We are now able to give sufficient conditions for strong ergodicity.

Proposition 7.13 [281]. *If* $\lim_{m\to\infty} \mathbf{P}_m = \mathbf{P}$, *where* $\mathbf{P}$ *is a mixing matrix, then strong ergodicity holds.*

Proof. $\mathbf{P}$ being mixing there exists $n_0 \geqslant 1$ such that $\alpha(\mathbf{P}^{n_0}) > 0$ Our assumption implies that

$$\lim_{m\to\infty} \mathbf{P}(m, m + n_0) = \mathbf{P}^{n_0}.$$

This means that for m sufficiently large $(m \geqslant m_0)$ we have $\alpha(\mathbf{P}(m, m + n_0)) \geqslant a > 0$ and Theorem 1.11 allows us to assert that $\alpha(\mathbf{P}(m, m + m_0 + n_0)) \geqslant a$ whatever $m \geqslant 0$ [we have $\mathbf{P}(m, m + m_0 + n_0) = \mathbf{P}(m, m + m_0)\mathbf{P}(m + m_0, m + m_0 + n_0)$]. Therefore by Theorem 7.5 uniform weak ergodicity holds. If $\boldsymbol{\pi}$ is the stationary distribution associated with $\mathbf{P}$, that is $\boldsymbol{\pi}'\mathbf{P} = \boldsymbol{\pi}'$, then $\lim_{m\to\infty} \boldsymbol{\pi}'\mathbf{P}_m = \boldsymbol{\pi}'\mathbf{P} = \boldsymbol{\pi}'$. It remains only to apply Theorem 7.11 and Proposition 7.12.

A more general result is

Theorem 7.14 [357]. *An asymptotically homogeneous Markov chain satisfying the assumptions of Theorem 7.8 is strongly ergodic.*

Proof. Use (7.7), Proposition 7.12 and Theorem 7.11.

7.5 UNIFORM STRONG ERGODICITY

7.5.1. A more restrictive variant of strong ergodicity is uniform strong ergodicity (see [157]). A Markov chain is said to be *uniformly strongly ergodic* if and only if whatever $i, j \in S$ the limits

$$\lim_{n\to\infty} p(m, i, m + n, j) = \pi(j)$$

exist uniformly with respect to $m \geqslant 0$.

Clearly, uniform strong ergodicity implies uniform weak ergodicity.

Notice that if j_0 is a state such that $\pi(j_0) > 0$, then there exists $n_0 \geqslant 1$ such that $p(m, i, m + n_0, j_0) \geqslant \pi(j_0)/2 = a > 0$ whatever $m \geqslant 0$, $i \in S$. In general, this condition is not sufficient for uniform strong ergodicity. However, in conjunction with an extra condition it ensures that type of ergodicity. Namely we have

Theorem 7.15 [157]. *For* $n \geqslant 1$ *let*

$$a_n = \sup_{i \in S,\, m',\, m'' \geqslant 0} \sum_{j \in S} |p(m'+n, i, m'+n+1, j) - p(m''+n, i, m''+n+1, j)|.$$

If the series $\sum_{n \geqslant 1} a_n$ *converges and there exist* $n_0 \geqslant 1$ *and* $j_0 \in S$ *such that* $p(m, i, m + n_0, j_0) \geqslant a > 0$ *whatever* $m \geqslant 0$, $i \in S$, *then uniform strong ergodicity holds and*

$$|p(m, i, m + n, j) - \pi(j)| \leqslant \inf_{n_0 \leqslant l < n} \left((1 - a)^{\frac{n}{l} - 1} + \frac{\sum_{s \geqslant l} a_s}{a} \right) \tag{7.12}$$

for all $m \geqslant 0$, $n \geqslant 1$, $i, j \in S$.

Proof. The first assumption implies the existence of the limit $\lim_{m \to \infty} \mathbf{P}_m = \mathbf{P}$ with a prescribed rate of convergence. The second one implies $\mathbf{P}^{n_0}$ is a Markov matrix, hence $\mathbf{P}$ is mixing. Consequently, Proposition 7.13 leads us to conclude that strong ergodicity holds. Thus it remains to prove uniformity.

By the Chapman-Kolmogorov equations, for any $h \geqslant 2$ we have

$$\begin{aligned}
&p(m' + n, i, m' + n + h, j) - p(m'' + n, i, m'' + n + h, j) \\
&= \sum_{k \in S} (p(m' + n, i, m' + n + 1, k)\, p(m' + n + 1, k, m' + n + h, j) \\
&\quad - p(m'' + n, i, m'' + n + 1, k)\, p(m'' + n + 1, k, m'' + n + h, j)) \\
&= \sum_{k \in S} (p(m' + n, i, m' + n + 1, k) - p(m'' + n, i, m'' + n + 1, k)) \times \\
&\qquad \times p(m' + n + 1, k, m' + n + h, j) \\
&+ \sum_{k \in S} (p(m'+n+1, k, m'+n+h, j) - p(m''+n+1, k, m''+n+h, j)) \times \\
&\qquad \times p(m'' + n, i, m'' + n + 1, k).
\end{aligned}$$

Setting

$$a_n(h) = \sup_{i \in S,\, m',\, m'' \geqslant 0} \sum_{j \in S} |p(m'+n, i, m'+n+h, j) - p(m''+n, i, m''+n+h, j)|$$

for $n, h \geqslant 1$, the preceding equation yields

$$a_n(h) \leqslant a_n(1) + a_{n+1}(h - 1), \quad h \geqslant 2.$$

Hence

$$a_n(h) \leqslant \sum_{s=n}^{n+h-1} a_s, \ n, h \geqslant 1. \tag{7.13}$$

since $a_n(1) = a_n$.

Again using the Chapman-Kolmogorov equations we have

$$\begin{aligned} &p(m', i, m' + n, j) - p(m'', i, m'' + n, j) \\ &= \sum_{k \in S} (p(m', i, m' + l, k) - p(m'', i, m'' + l, k))\, p(m' + l, k, m' + n, j) \\ &+ \sum_{k \in S} (p(m'+l, k, m'+n, j) - p(m''+l, k, m''+n, j))\, p(m'', i, m''+l, k) \end{aligned} \tag{7.14}$$

for any $n_0 \leqslant l < n$. Now remark that

$$p(m', i, m'+l, j_0) = \sum_{k \in S} p(m', i, m'+l-n_0, k)\, p(m'+l-n_0, k, m'+l, j_0) \geqslant a$$

for any $m' \geqslant 0$, $l \geqslant n_0$, $i \in S$. Therefore

$$\begin{aligned} &\sum_{k \in S} (p(m', i, m' + l, k) - p(m'', i, m'' + l, k))^+ \\ &= -\sum_{k \in S} (p(m', i, m' + l, k) - p(m'', i, m'' + l, k))^- \leqslant 1 - a \end{aligned} \tag{7.15}$$

for any $m', m'' \geqslant 0$, $l \geqslant n_0$, $i \in S$. Setting

$$\inf_{m \geqslant 0, i \in S} p(m, i, m + n, j) = \underline{p}(n, j), \quad \sup_{m \geqslant 0, i \in S} p(m, i, m + n, j) = \bar{p}(n, j),$$

on account of (7.13) and (7.15) equation (7.14) yields

$$\bar{p}(n, j) - \underline{p}(n, j) \leqslant 1 - a)\, (\bar{p}(n - l, j) - \underline{p}(n - l, j)) + \sum_{s \geqslant l} a_s \tag{7.16}$$

for any $n_0 \leqslant l < n$, $j \in S$. As $p(m, i, m + n, j)$ and $\pi(j)$ lie between $\underline{p}(n, j)$ and $\bar{p}(n, j)$ whatever $m \geqslant 0$, $n \geqslant 1$, $i, j, \in S$, (7.12) follows at once from (7.16).

7.5.2. It is natural to ask what the concepts of weak and strong ergodicity reduce to in the homogeneous case. First, we have

Proposition 7.16. *In the homogeneous case* $\mathbf{P}_m = \mathbf{P}$, $m \geqslant 0$, *weak and strong ergodicity are equivalent.*

Proof. We have to prove that in the homogeneous case weak ergodicity implies strong ergodicity. By Theorem 5.3 there exists at least one probability vector $\boldsymbol{\pi}$ such that $\boldsymbol{\pi}'\mathbf{P} = \boldsymbol{\pi}'$, whence $\boldsymbol{\pi}'\mathbf{P}^n = \boldsymbol{\pi}'$, $n \geqslant 1$. This shows that asymptotic stationarity holds. To reach the desired conclusion it remains only to apply Theorem 7.11.

Secondly, it is clear that a homogeneous Markov chain is strongly ergodic if and only if it is either regular or indecomposable (Theorems 4.2 and 4.3 and Exercise 5.1). Obviously, in the homogeneous case it makes no sense to speak about uniformity of ergodicity.

7.6 ASYMPTOTIC BEHAVIOUR OF NONHOMOGENEOUS MARKOV CHAINS

7.6.1. First we prove the result alluded to in 5.2.3 concerning the tail σ-algebra of a nonhomogeneous Markov chain.

Theorem 7.17 [56, 57, and 353] *The tail σ-algebra $\mathcal{T}$ of a nonhomogeneous finite Markov chain is finite, and the number of its atoms does not exceed the number of states of the chain. The atoms A_r, $r \in I$, of $\mathcal{T}$ can be represented as follows: there exists a sequence of partitions $(S(n), S_r(n), r \in I)$, $n \geqslant 0$, of the state space S such that $A_r = \lim_{n\to\infty} \{X(n) \in S_r(n)\}$, $r \in I$, and $\mathsf{P}(\lim_{n\to\infty} \{X(n) \in S(n)\}) = 0$.*

Proof. The representation of the atoms A_r, $r \in I$, of $\mathcal{T}$ (see 5.2.2) was suggested in [32] and is obtained by taking

$$S_r(n) = (i\colon \mathsf{P}(A_r \mid X(n) = i) > 1/2),\ S(n) = S - \bigcup_{r \in I} S_r(n),\quad n \geqslant 0.$$

Clearly, for any $n \geqslant 0$ the sets $S_r(n)$, $r \in I$, are pairwise disjoint. Therefore $(S(n), S_r(n), r \in I)$ is a partition of the state space for each $n \geqslant 0$. The convergence theorem stated in the proof of Theorem 5.5 allows us to assert that

$$A_r = \{\omega\colon \lim_{n\to\infty} f^n_{A_r}(\omega) > 1/2\} = \lim_{n\to\infty} \{X(n) \in S_r(n)\},\ r \in I.$$

Then it follows that

$$N = (\bigcup_{r\in I} A_r)^c = \lim_{n\to\infty} \{X(n) \in \bigcup_{r \in I} S_r(n)\}^c = \lim_{n\to\infty} \{X(n) \in S(n)\}.$$

Now we can conclude the proof of the theorem. Assume for a contradiction that there are s states and that the set I contains more than s elements. Then there would exist $r_1, \ldots, r_{s+1} \in I$ such that for some n sufficiently large we have $\mathsf{P}(X(n) \in S_{r_l}(n)) > 0$, $1 \leqslant l \leqslant s+1$. But these inequalities contradict the fact that the chain has s states. This contradiction shows that the number of the atoms of $\mathcal{T}$ does not exceed the number of states of the chain. It remains to be proved that $\mathsf{P}(N) = 0$. Assuming again for a contradiction that $\mathsf{P}(N) > 0$, there would exist $s+1$ pairwise disjoint random events $N_1, \ldots, N_{s+1} \in \mathcal{T}$ such that $\mathsf{P}(N_l) > 0$, $1 \leqslant l \leqslant s+1$. Defining the subsets $B_l(n) = (i\colon \mathsf{P}(N_l \mid X(n) = i) > 1/2)$, $1 \leqslant l \leqslant s+1$, of the state space it is easily seen that these are pairwise disjoint and, as above, $N_l = \lim_{n\to\infty} \{X(n) \in B_l(n)\}, 1 \leqslant l \leqslant \leqslant s+1$. These relations again contradict the fact that the chain has s states. Therefore one must have $\mathsf{P}(N) = 0$ and the proof is complete.

7.6.2. Theorem 7.17 allows us to prove some results by A.N. Kolmogorov [215] and D. Blackwell [32] concerning the asymptotic properties of the reverse probabilities

$$\hat{p}(n, i, m, j) = \mathsf{P}(X(m) = j | X(n) = i), \quad m \leqslant n,$$

defined for all $i \in S_n = (i : \mathsf{P}(X(n) = i) > 0)$, $j \in S$. It is easy to show that setting $\hat{\mathbf{P}}(n, m) = (\hat{p}(n, i, m, j))_{i \in S_n, j \in S_m}$, we have $\hat{\mathbf{P}}(s, n)\ \hat{\mathbf{P}}(n, m) = \hat{\mathbf{P}}(s, m)$, $m \leqslant n \leqslant s$. Let us prove that there exists a sequence of probability distributions $\mathbf{p}(n)$ over S_n, $n \geqslant 0$, such that

$$\mathbf{p}'(n)\ \hat{\mathbf{P}}(n, m) = \mathbf{p}'(m), \ 0 \leqslant m \leqslant n. \tag{7.17}$$

This follows from the fact that we can find an increasing sequence $(s_l)_{l\geqslant 1}$ of natural numbers such that the limit as $l \to \infty$ of the matrix $\hat{\mathbf{P}}(s_l, m)$ exists for all $m \geqslant 0$ (obviously, this implies that $S_{s_1} = S_{s_2} = \ldots$). If we denote the limit by $\mathbf{Q}(m)$, letting $l \to \infty$ in the matrix equation $\hat{\mathbf{P}}(s_l, n)\ \hat{\mathbf{P}}(n, m) = \mathbf{P}(s_l, m)$ we get $\mathbf{Q}(n)\ \hat{\mathbf{P}}(n, m) = \mathbf{Q}(m)$, $0 \leqslant m \leqslant n$. In other words, the rows of any given index of the matrices $\mathbf{Q}(n)$, $n \geqslant 0$, provide us with probability distributions $\mathbf{p}(n)$ over S_n, $n \geqslant 0$ satisfying (7.17). Next, if for each $n \geqslant 0$ we consider a probability distribution $\mathbf{q}(n)$ over S_n such that $\mathbf{q}'(n)\ \hat{\mathbf{P}}(n, m) = \mathbf{q}'(m)$, $0 \leqslant m \leqslant n$, then choosing a subsequence $(t_l)_{l\geqslant 1}$ of the sequence $(s_l)_{l\geqslant 1}$ such that the limit $\lim_{l\to\infty} \mathbf{q}(t_l) = \mathbf{q}$ exists and letting $l \to \infty$ in the equation $\mathbf{q}'(t_l)\ \hat{\mathbf{P}}(t_l, m) = \mathbf{q}'(m)$, we get $\mathbf{q}'\mathbf{Q}(m) = \mathbf{q}'(m)$, $m \geqslant 0$. If follows that $\mathbf{q}(m)$ is a convex linear combination of the rows of Kolmogorov's matrix $\mathbf{Q}(m)$ for all $m \geqslant 0$.

Now remark that since the Markov property is time reversible [see (2.2)′] the martingale convergence theorem alluded to in the footnote on p. 159 allows us to assert that

$$\lim_{n\to\infty} \mathsf{P}(X(m) = j | X(n)) = \lim_{n\to\infty} f^n_{\{X(m)=j\}}(\omega) = \mathsf{P}(X(m) = j) | A_r) \tag{7.18}$$

for all $\omega \in A_r$, $r \in I$, or, abusing the notation, that

$$\lim_{n\to\infty} \hat{p}(n, i, m, j) = \mathsf{P}(X(m) = j | A_r) \tag{7.19}$$

for all $i \in S_r(n)$*). Comparing (7.18) to the equation $\lim_{l\to\infty} \hat{\mathbf{P}}(s_l, m) = \mathbf{Q}(m)$, $m \geqslant 0$, we deduce that the elements of a basis of the set of solutions $\mathbf{p}(n)$, $n \geqslant 0$, of equations (7.17) are given by

$$\mathbf{p}_r(n) = (\mathsf{P}(X(n) = j \mid A_r))_{j\in S},\ r \in I.$$

Their number is thus equal to that of the atoms of $\mathcal{T}$. (The linear independence of the $(\mathbf{p}_r(m))_{m\geqslant 0}$, $r \in I$, is easily established on account of the representation of the A_r in Theorem 7.17.) In particular, the solution is unique if and only if $\mathcal{T}$ is P-trivial.

7.6.3. Theorem 7.17 can also be proved using a geometric interpretation of transition matrices, essentially dual to that in 5.1.3. More precisely, the matrix $\mathbf{P}(m, n)$ is interpreted as a linear operator $\mathbf{x} \to \mathbf{P}(m, n)\,\mathbf{x}$ over the unit cube

$$\Gamma = (\mathbf{x} = (x(i))_{i\in S}\colon\ 0 \leqslant x(i) \leqslant 1,\ i \in S).$$

The interested reader should consult [47, 210, and 353].

7.6.4. The aspect of the asymptotic behaviour of Markov chains discussed in 5.3 for the homogeneous case was taken up in the nonhomogeneous case by A.A. Markov ([260], pp. 465—509) as early as 1910. His results asserting the validity of the central limit theorem for two state Markov chains was improved several times by S.N. Bernstein (in the period 1922—1928) and made definitive by N.A. Sapogov (in 1947). The extension to the case of an arbitrary state space (implicitly to the case of an arbitrary finite number of states) was made by R.L. Dobrušin [79].

We shall state the fundamental result of the latter author. Assume whatever the natural number n that the random variables $X^{(n)}(0), \ldots, X^{(n)}(n-1)$ are the first n variables of a nonhomogeneous finite Markov chain with transition matrices $\mathbf{P}_m^{(n)}$, $m \geqslant 0$. If $f_l^{(n)}$, $0 \leqslant l \leqslant n-1$, are real valued functions defined on the state space of the chain of index n, set $T_n = \sum_{l=0}^{n-1} f_l^{(n)}(X^{(n)}(l))$, $n \geqslant 1$.

Theorem 7.18 *Assume the functions $f_l^{(n)}$ are uniformly bounded and* $\operatorname{var}\,[f_l^{(n)}(X^{(n)}(l))] \geqslant c > 0$, $0 \leqslant l \leqslant n-1$, $n \geqslant 1$. *Let*

$$\alpha_n = \min_{0\leqslant l<n-1} \alpha(\mathbf{P}_l^{(n)}).$$

If $\lim_{n\to\infty} \alpha_n n^{1/3} = \infty$ *then*

$$\lim_{n\to\infty} \mathsf{P}\left(\frac{T_n - \mathsf{E}(T_n)}{\sqrt{\operatorname{var}(T_n)}} < x\right) = \frac{1}{\sqrt{2\pi}} \int_{-\infty}^{x} e^{-u^2/2}\,du$$

for all real numbers x. In general, the result is no longer true if $\lim_{n\to\infty} \alpha_n n^{1/3} < \infty$.

*) Actually, it can be shown ([61], §2) that there exist sequences of subsets $(E_n^{(r)})_{n\geqslant 0}$, $r \in I$, of the state space such that (7.19) holds for all $i \in E_n^{(r)}$, $j \in S$, $m, n \geqslant 0$, $r \in I$.

For the proof the reader should consult [79].

V. Statulevicius [367 and 368] developed Dobrušin's results in several directions.

For the study of the applicability of the law of the iterated logarithm to nonhomogeneous Markov chains, the interested reader may consult [184, 229 and 338].

EXERCISES

7.1 *Doeblin's type A nonhomogeneous Markov chains* [81]. Assume the transition matrices $\mathbf{P}_m = (p(m, i, m+1, j))_{i,j\in S}$, $m \geqslant 0$, satisfy the following condition (A). There exists $a > 0$ such that whatever the fixed pair of states $i, j \in S$ either $p(m, i, m+1, j) \geqslant a$ whatever $m \geqslant 0$ or $p(m, i, m+1, j) = 0$ whatever $m \geqslant 0$. Show that the state classification procedure which works in the homogeneous case carries over to nonhomogeneous chains satisfying condition (A).

7.2 Construct a nonhomogeneous version of the absorbing Markov chain model for a chess match from Exercise 2.3, assuming that the probabilities are p_m, d_m, q_m, $p_m + d_m + q_m = 1$, that the mth game results in a win for $\mathfrak{A}$, a draw and a win for $\mathfrak{B}$, respectively, $m \geqslant 0$.

7.3 Prove [79] that in the case of a nonhomogeneous Markov chain with an arbitrary finite number of states divergence of the series $\sum_{m\geqslant 0} \alpha(\mathbf{P}_m)$ is a necessary and sufficient condition that any Markov chain whose sequence of transition matrices contains a subsequence identical to $(\mathbf{P}_m)_{m\geqslant 0}$ be weakly ergodic.

7.4 Prove [155] that strong ergodicity holds if and only if weak ergodicity holds and $\lim_{n\to\infty} p(m, j, m+n, j) = \pi(j)$ whatever $m \geqslant 0$ and $j \in S$.

7.5 Prove [155] that uniform strong ergodicity holds if and only if uniform weak ergodicity holds and $\lim_{n\to\infty} p(m, j, m+n, j) = \pi(j)$ uniformly with respect to $m \geqslant 0$ whatever $j \in S$.

7.6 Assume that there exist probability distributions $\mathbf{p}_m$, $m \geqslant 0$, and $\mathbf{p}$ over S such that

$$\lim_{n\to\infty} \mathbf{p}'_m \mathbf{P}(m, m+n) = \mathbf{p}'$$

whatever $m \geqslant 0$. Prove that if weak ergodicity holds then strong ergodicity holds, too. (H i n t: adapt the sufficiency part of the proof of Theorem 7.11.)

7.7 Let π_m be a probability distribution over S such that $\pi'_m \mathbf{P}_m = \pi'_m$, $m \geqslant 0$. Prove ([171], p. 160) that if weak ergodicity holds and

$$\sum_{m\geqslant 0} \sum_{i \in S} |\pi_m(i) - \pi_{m+1}(i)| < \infty,$$

then strong ergodicity holds, too. (Hint: the limit $\pi = \lim_{m\to\infty} \pi_m$ does exists. Then use the equalities

$$\pi'_m \mathbf{P}(m, m+n) - \pi' = (\pi'_m \mathbf{P}(m+r) - \pi'_{m+r})\, \mathbf{P}(m+r, m+n) + \\ + \pi'_{m+r}\, \mathbf{P}(m+r, m+n) - \pi', \quad 0 \leqslant r \leqslant n,$$

and

$$\pi'_{m+n} \mathbf{P}(m+r, m+n) = \sum_{s=m+r}^{m+n-2} (\pi'_s - \pi'_{s+1})\, \mathbf{P}(s+1, m+n) + \pi'_{m+n-1},$$

$0 \leqslant r \leqslant n-2$, $n \geqslant 2$, to verify the condition in Exercise 7.6.)

7.8 Let π_m be a probability distribution over S such that $\pi'_m \mathbf{P}_m = \pi'_m$, $m \geqslant 0$. Prove [141] that if $\lim_{m\to\infty} \pi_m = \pi > \mathbf{0}$ exists and uniform weak ergodicity holds, then uniform strong ergodicity holds, too.

7.9 Consider the nonhomogeneous two state Markov chain with transition matrices

$$\mathbf{P}_m = \begin{pmatrix} \frac{1}{2} - \frac{1}{m+2} & \frac{1}{2} + \frac{1}{m+2} \\ \frac{1}{2} - \frac{1}{m+2} & \frac{1}{2} + \frac{1}{m+2} \end{pmatrix}, \qquad m \geqslant 0.$$

Show that uniform strong ergodicity holds although the series $\sum_{n\geqslant 1} a_n$ diverges (see Theorem 7.15).

$$\left(\text{Hint: } \mathbf{P}(m, m+n) = \begin{pmatrix} \frac{1}{2} - \frac{1}{m+n+1} & \frac{1}{2} + \frac{1}{m+n+1} \\ \frac{1}{2} - \frac{1}{m+n+1} & \frac{1}{2} + \frac{1}{m+n+1} \end{pmatrix}, \quad n \geqslant 1. \right)$$

7.10 Let $(X(n))_{n\geqslant 0}$ be a homogeneous Markov chain with either countably infinitely or finitely many states on a probability space $(\Omega, \mathcal{K}, \mathsf{P})$. Let $A \in \mathcal{K}$ be a random event such that $\mathsf{P}(A) > 0$. Prove [286] that $(X(n))_{n\geqslant 0}$ is a Markov chain on $(\Omega, \mathcal{K}, \mathsf{P}_A)$ if and only if A belongs to the tail σ-algebra of the original chain. The derived chain is, in general, nonhomogeneous. (Compare with 3.2.9.)

CHAPTER 8

Markov Processes

This chapter takes up the study of Markovian dependence in the case where changes of state may occur at any time. Though there are many similarities to the case previously investigated, we are faced with a qualitatively new situation which involves the basic concepts of continuous parameter stochastic processes. Fortunately, the finite setting allows a reasonably elementary treatment which avoids the usual intricacies that lead one to ask whether the present theory of stochastic processes is not more difficult than its applications warrant. Nevertheless, we have had to give without proof a series of results (especially concerning the nonhomogeneous case).

8.1 MEASURE THEORETICAL DEFINITION OF A MARKOV PROCESS

8.1.1. The definition of a (homogeneous finite) Markov process parallels that of a (homogeneous finite) Markov chain.

Consider a stochastic process $(X(t))_{t\geqslant 0}$ (see 1.5.2) with a finite set of states S. Assume for any increasing sequence $0 = t_0 < t_1 < \dots$ $\dots < t_n < t_{n+1} < \dots$ of values of the parameter t that the sequence of random variables $(X(t_n))_{n\geqslant 0}$ has the Markov property, that is

$$\begin{aligned}\mathsf{P}(X(t_{n+1}) = i_{n+1} \mid X(t_n) = i_n, \dots, X(t_0) = i_0)\\ = \mathsf{P}(X(t_{n+1}) = i_{n+1} \mid X(t_n) = i_n)\end{aligned} \tag{8.1}$$

whatever $i_0, \dots, i_{n+1} \in S$, $n \geqslant 0$. Assume further that the last conditional probability is a function of i_n, i_{n+1} and of the difference $t_{n+1} - t_n$ only, that is

$$\mathsf{P}(X(t_{n+1}) = i_{n+1} \mid X(t_n) = i_n) = p(t_{n+1} - t_n, i_n, i_{n+1}).$$

Clearly, the matrix $\mathbf{P}(t) = (p(t, i, j))_{i,j\in S}$ is stochastic, whatever $t > 0$.

On account of the equations

$$\begin{aligned}\mathsf{P}(X(t+s) = j \mid X(0) = i) &= \sum_{k\in S} \mathsf{P}(X(t+s) = j, X(s) = k \mid X(0) = i)\\ &= \sum_{k\in S} \mathsf{P}(X(t+s) = j \mid X(s) = k)\, \mathsf{P}(X(s) = k \mid X(0) = i),\end{aligned}$$

we should have

$$p(s+t, i, j) = \sum_{k \in S} p(s, i, k)\, p(t, k, j)^{*)} \tag{8.2}$$

whatever $s, t > 0$, $i, j \in S$. In matrix notation this amounts to

$$\mathbf{P}(s+t) = \mathbf{P}(s)\,\mathbf{P}(t) (= \mathbf{P}(t)\,\mathbf{P}(s)). \tag{8.3}$$

Putting $p(0, i, j) = \delta(i, j)$, $i, j \in S^{**)}$, that is $\mathbf{P}(0) = \mathbf{I}$ (the unit matrix), equation (8.3) will have to hold whatever $s, t \geqslant 0$.

In what follows a function $\mathbf{P}(\cdot)$ associating with any $t \geqslant 0$ a stochastic matrix $\mathbf{P}(t)$ such that (8.3) holds will be called a (*homogeneous*) *transition matrix function.*

8.1.2. Clearly, one should first answer the problem of the existence of a transition matrix function $\mathbf{P}(\cdot)$. Assuming for a while that the answer is affirmative let us define

$$p_{t_1 \dots t_n}(i_1, \dots, i_n) = \sum_{i_0 \in S} p(i_0)\, p(t_1, i_0, i_1) \dots p(t_n - t_{n-1}, i_{n-1}, i_n),$$

where $(p(i))_{i \in S}$ is a given probability distribution over S. It is immediate that the family $(p_{t_1 \dots t_n})_{t_1 < \dots < t_n}$, $n \geqslant 1$, satisfies the consistency equations (1.9). Hence the existence of a stochastic process $(X(t))_{t \geqslant 0}$ for which

$$\mathsf{P}(X(t_1) = i_1, \dots, X(t_n) = i_n) = p_{t_1 \dots t_n}(i_1, \dots, i_n)$$

whatever $0 < t_1 < \dots < t_n$, $i_1, \dots, i_n \in S$ and $n \geqslant 1$ is ensured. Now, an argument similar to that in 2.1.2 shows that for such a process equation (8.1) holds, thus enabling us to call it a (*homogeneous finite*) *Markov process* with *state space* S, *initial distribution* $\mathbf{p} = (p(i))_{i \in S}$ and *transition matrix function* $\mathbf{P}(\cdot)$. As in the case of Markov chains, to emphasize its dependence on $\mathbf{p}$, the probability P will be written as $\mathsf{P}_{\mathbf{p}}$. In particular, for an initial distribution concentrated in state i, i.e., for the case where $p(j) = \delta(i, j)$, $j \in S$, the corresponding $\mathsf{P}_{\mathbf{p}}$ will be denoted by P_i and the corresponding Markov process will be said to *start* in state i (or to have i as *initial* state). Accordingly, the mean value operator under $\mathsf{P}_{\mathbf{p}}(\mathsf{P}_i)$ will be denoted by $\mathsf{E}_{\mathbf{p}}(\mathsf{E}_i)$.

8.1.3. The reader should note the major difference as compared with the discrete parameter case, namely the fact that for describing a homogeneous finite Markov process one cannot use any stochastic matrix function $\mathbf{P}(\cdot)$ but just one satisfying (8.3). Next, no particular

---

*) This equation which extends the Chapman-Kolmogorov equations was first considered by Smoluchowski.

**) Notice that this agrees with the equation

$$p(0, i, j) = \mathsf{P}(X(t) = j \mid X(t) = i).$$

time unit has special significance, in contrast to the case of Markov chains, where the time unit one played a basic part because transitions between states could only take place at integral times.

8.2 THE INTENSITY MATRIX

8.2.1. Let us return to the problem raised above concerning the existence of a transition matrix function $\mathbf{P}(\cdot)$. The way we choose to study it is to obtain the properties that such a transition matrix function should possess and on this basis to derive the general solution of the functional matrix equation (8.3).

For mathematical convenience we postulate that

$$\lim_{t\to 0} p(t, i, j) = \delta(i, j), \; i, j \in S, \tag{8.4}$$

i.e., the functions $p(\cdot, i, j)$, $i, j \in S$ are continuous at the origin.

Notice that by the Chapman-Kolmogorov equation we have $p(t, i, i) \geqslant [p(t/n, i, i)]^n$ for all $t \geqslant 0$, $i \in S$, and for all $n \geqslant 1$. Then (8.4) implies that $p(t, i, i) > 0$ whatever $t \geqslant 0$ and $i \in S$. Next, since $p(t, i, j) \geqslant p(u, i, j)\, p(t-u, j, j)$, $0 < u \leqslant t$, and $p(t-u, j, j) > 0$, we conclude that if $p(u, i, j) > 0$ then $p(t, i, j) > 0$ for all $t \geqslant u$. The last assertion can be sharpened to

Proposition 8.1. *For all $i, j \in S$ the function $p(\cdot, i, j)$ is either identically zero or always positive on* $(0, \infty)$.

Proof. On the account of the above considerations it is sufficient to prove that if $p(s, i, j) > 0$, $i \neq j$, for some $s > 0$, then for all $t > 0$ there exists $u < t$ such that $p(u, i, j) > 0$.

Since $p(m(s/m), i, j) > 0$ whatever the natural number m, there are states $i_1, \ldots, i_{m-1} \in S$ such that

$$p(s/m, i_0, i_1)\; p(s/m, i_1, i_2) \ldots p(s/m, i_{m-1}, i_m) > 0$$

with $i_0 = i$, $i_m = j$. Let us drop from the sequence $i_0, i_1, \ldots, i_{m-1}, i_m$ the states $i_a, \ldots, i_{b-1}$ if $i_a = i_b$, $a < b$. In this way a sequence of distinct i's is finally obtained. Since its length does not exceed the number of states, say r, of the process, we have $p(r(s/m), i, j) > 0$. Then if $m > rs/t$ we can take $u = rs/m < t$. The proof is complete.

8.2.2. In fact (8.3) and (8.4) imply that the $p(\cdot, i, j)$ must be uniformly continuous on $[0, \infty)$. Namely, we have

Proposition 8.2. *For all $i, j \in S$ the function $p(\cdot, i, j)$ is uniformly continuous on $[0, \infty)$. The modulus of continuity of $p(\cdot, i, j)$ does not exceed that of $p(\cdot, i, i)$ at 0.*

Proof. By (8.2) for all $t \geqslant 0$, $h > 0$ we have

$$p(t+h, i, j) - p(t, i, j) = \sum_{k \in S} p(h, i, k)\, p(t, k, j) - p(t, i, j)$$

$$= -(1 - p(h, i, i))\, p(t, i, j) + \sum_{k \neq i} p(h, i, k)\, p(t, k, j).$$

Hence, on account of the fact that $0 \leqslant p(t, i, j) \leqslant 1$ for all $t \geqslant 0$, $i, j \in S$,

$$-(1 - p(h, i, i)) \leqslant p(t+h, i, j) - p(t, i, j)$$

$$\leqslant \sum_{k \neq i} p(h, i, k) = 1 - p(h, i, i),$$

that is

$$|p(t+h, i, j) - p(t, i, j)| \leqslant 1 - p(h, i, i).$$

Next, consider an $h < 0$ such that $t + h \geqslant 0$ and set $h' = |h|$. Replacing t by $t - h' = t + h$ in the last inequality we get

$$|p(t+h, i, j) - p(t, i, j)| \leqslant 1 - p(|h|, i, i).$$

Therefore for all $t \geqslant 0$ and all u such that $t + u \geqslant 0$ we have

$$|p(t+u, i, j) - p(t, i, j)| \leqslant 1 - p(|u|, i, i)$$

and the proof is complete.

8.2.3. After establishing the uniform continuity of the $p(\cdot, i, j)$ let us consider the problem of whether or not they are differentiable. The answer is positive as asserted by

Proposition 8.3. *The transition matrix function* $\mathbf{P}(\cdot)$ *is differentiable on* $[0, \infty)$.

Proof. One can find $h > 0$ such that $\int_0^h \mathbf{P}(u)du$ is a nonsingular matrix. Indeed, by (8.4) we have

$$\lim_{h_{ij} \to 0,\, i, j \in S} (p(h_{ij}, i, j))_{i, j \in S} = \mathbf{I}.$$

Hence, on account of Proposition 8.2 there exist positive numbers h_i, $i \in S$, such that the matrix $(p(u_{ij}, i, j))_{i,j \in S}$ is nonsingular whatever $0 \leqslant u_{ij} \leqslant h_i$, $i, j \in S$. Let $h = \min_{i \in S} h_i$. Then by the mean value theorem

$$\int_0^h \mathbf{P}(u)du = h(p(u_{ij}, i, j))_{i, j \in S}$$

for certain values $u_{ij} \leqslant h \leqslant h_i$, $i, j \in S$. Hence the matrix $\int_0^h \mathbf{P}(u)du$ is nonsingular.

Next, by (8.3) we have

$$\int_0^h \mathbf{P}(t+u)\,du = \mathbf{P}(t)\int_0^h \mathbf{P}(u)\,du = \left(\int_0^h \mathbf{P}(u)\,du\right)\mathbf{P}(t),$$

whence, by substituting $t+u=v$ in the first integral,

$$\mathbf{P}(t) = \int_t^{t+h} \mathbf{P}(v)\,dv \left(\int_0^h \mathbf{P}(u)\,du\right)^{-1} = \left(\int_0^h \mathbf{P}(u)\,du\right)^{-1}\int_t^{t+h} \mathbf{P}(v\,)dv.$$

The matrix function $\mathbf{P}(\cdot)$ being continuous, the derivative of $\int_t^{t+h} \mathbf{P}(v)\,dv$ with respect to t exists and equals $\mathbf{P}(t+h)-\mathbf{P}(t)$. It follows that the derivative of $\mathbf{P}(\cdot)$ exists, too, and is given by

$$\mathbf{P}'(t) = (\mathbf{P}(t+h)-\mathbf{P}(t))\left(\int_0^h \mathbf{P}(u)\,du\right)^{-1}$$

$$= \left(\int_0^h \mathbf{P}(u)\,du\right)^{-1}(\mathbf{P}(t+h)-\mathbf{P}(t)). \tag{8.5}$$

The proof is now complete.

R e m a r k. In fact, equation (8.5) shows that $\mathbf{P}(\cdot)$ is infinitely differentiable on $[0, \infty)$.

8.2.4. In what follows the (right) derivative

$$\mathbf{P}'(0) = \mathbf{Q} = (q(i,j))_{i,\,j\in S}$$

which is called the *intensity matrix* of $\mathbf{P}(\cdot)$ is of fundamental importance. Note that for all $i \neq j$

$$q(i,j) = \lim_{h\to 0+} \frac{p(h,i,j)}{h} \geqslant 0$$

and for all $i \in S$

$$q(i,i) = \lim_{h\to 0+} \frac{p(h,i,i)-1}{h} \leqslant 0,$$

or, in matrix notation, $\mathbf{Q}_{\mathrm{dg}} \leqslant 0$, $\mathbf{Q}-\mathbf{Q}_{\mathrm{dg}} \geqslant 0$. For convenience we shall write $q(i,i) = -q(i)$ with $q(i) \geqslant 0$, $i \in S$. On account of the equalities $\sum_{j\in S} p(t,i,j) = 1$, $i \in S$, one should also have

$$\sum_{j\in S} q(i,j) = 0,\ i \in S,$$

or, in matrix notation, $\mathbf{Qe} = \mathbf{0}$.

Notice that differentiating (8.3) with respect to s yields

$$\mathbf{P}'(s+t) = \mathbf{P}'(s)\,\mathbf{P}(t) = \mathbf{P}(t)\,\mathbf{P}'(s).$$

Hence by taking $s = 0$ we get the matrix differential equations

$$\mathbf{P}'(t) = \mathbf{Q}\,\mathbf{P}(t), \quad \mathbf{P}'(t) = \mathbf{P}(t)\,\mathbf{Q},\ t \geqslant 0.$$

In fact, these are systems of linear ordinary differential equations. They are known as the "backward" and "forward" *Kolmogorov systems*, respectively, and can be solved under the initial condition $\mathbf{P}(0) = \mathbf{I}$ by standard methods to yield

$$\mathbf{P}(t) = e^{\mathbf{Q}t} = \mathbf{I} + \sum_{n \geqslant 1} \frac{\mathbf{Q}^n t^n}{n!}, \quad t \geqslant 0, \tag{8.6}$$

(see 1.7.4).

An alternative form of $\mathbf{P}(\cdot)$ can be obtained as follows. Consider the diagonal matrix function

$$\psi(t) = (\exp\,(q(i)t)\ \delta(i,j))_{i,j \in S}$$

defined on the whole real axis. Clearly,

$$(\psi(t))^{-1} = \psi(-t)$$

whatever the real number t. The backward Kolmogorov equation can be written as

$$\mathbf{P}'(t) = \mathbf{Q}_{\mathrm{dg}}\,\mathbf{P}(t) + (\mathbf{Q} - \mathbf{Q}_{\mathrm{dg}})\mathbf{P}(t)$$

whence

$$\psi(t)\,(\mathbf{P}'(t) - \mathbf{Q}_{\mathrm{dg}}\,\mathbf{P}(t)) = \psi(t)\,(\mathbf{Q} - \mathbf{Q}_{\mathrm{dg}})\mathbf{P}(t)$$

or

$$\frac{d}{dt}(\psi(t)\,\mathbf{P}(t)) = \psi(t)\,(\mathbf{Q} - \mathbf{Q}_{\mathrm{dg}})\,\mathbf{P}(t).$$

Hence the backward Kolmogorov equation assumes the integral form

$$\mathbf{P}(t) = \psi(-t)\,[\mathbf{I} + \left[\int_0^t \psi(u)(\mathbf{Q} - \mathbf{Q}_{\mathrm{dg}})\mathbf{P}(u)\,du\right].$$

It can similarly be shown that the forward Kolmogorov equation assumes the integral form

$$\mathbf{P}(t) = \left[\mathbf{I} + \int_0^t \mathbf{P}(u)\,(\mathbf{Q} - \mathbf{Q}_{\mathrm{dg}})\,\psi(u)du\right]\psi(-t).$$

The point here is that the matrices $\psi(u)$, $\mathbf{Q}-\mathbf{Q}_{dg}$ and $\psi(-t)$ are all nonnegative (unlike $\mathbf{Q}$).

Standard arguments in the theory of integral equations show that

$$\mathbf{P}(t)=\sum_{n\geqslant 0}\mathbf{P}_n(t),\ t\geqslant 0, \tag{8.7}$$

where $\mathbf{P}_0(t)=\psi(-t)$ and

$$\mathbf{P}_{n+1}(t)=\psi(-t)\left(\int_0^t \psi(u)\ (\mathbf{Q}-\mathbf{Q}_{dg})\mathbf{P}_n(u)\ du\right)$$

$$=\int_0^t \psi(-u)(\mathbf{Q}-\mathbf{Q}_{dg})\ \mathbf{P}_n(t-u)\ du,\ \ n\geqslant 0,$$

or, alternatively,

$$\mathbf{P}_{n+1}(t)=\left(\int_0^t \mathbf{P}_n(u)\ (\mathbf{Q}-\mathbf{Q}_{dg})\ \psi(u)du\right)\psi(-t),\ \ n\geqslant 0.$$

(Try to prove it by direct computation.)

The probabilistic interpretation of the $\mathbf{P}_n(\cdot)$ will be given in 8.3.3. Equation (8.7) is known as the *Feller representation* of $\mathbf{P}(\cdot)$. It has been shown in [87] how the $\mathbf{P}_n$, $n\geqslant 0$, can be expressed in closed form.

Clearly, the matrix functions $\mathbf{P}_n(\cdot)$, $n\geqslant 0$, are nonnegative so that on account of (8.7) the matrix function $\mathbf{P}(\cdot)$ is nonnegative, too. Next, on account of the fact that $\mathbf{Q}^n\mathbf{e}=\mathbf{Q}^{n-1}(\mathbf{Q}\mathbf{e})=\mathbf{0}$ for all $n\geqslant 1$, it follows from (8.6) that $\mathbf{P}(t)\mathbf{e}=\mathbf{e}$ for all $t\geqslant 0$. Finally, $\mathbf{P}(\cdot)$ satisfies (8.3) by virtue of 1.7.4. Therefore $\mathbf{P}(\cdot)$ is a transition matrix function.

8.2.5. To sum up our findings we can say that either equation (8.6) or equation (8.7) establishes a one-to-one correspondence between the set of intensity matrices $\mathbf{Q}=(q(i,j))_{i,j\in S}$ (i.e., such that $\mathbf{Q}_{dg}\leqslant 0$, $\mathbf{Q}-\mathbf{Q}_{dg}\geqslant 0$, and $\mathbf{Q}\mathbf{e}=0$) and the set of transition matrix functions satisfying (8.4). Thus the intensity matrix $\mathbf{Q}$ appears to play a part similar to that of the one step transition matrix of a Markov chain. Because of this a finite Markov process is usually described by giving the intensity matrix of its transition matrix function.

8.2.6. There are many papers (see, e.g., [50, 51, and 212]) that deal with expressing in closed form the solution (8.6) of the Kolmogorov systems. At least from a theoretical point of view this can be done using the Jordan canonical form of $\mathbf{Q}$ (see 1.9.7)

$$\mathbf{Q}=\mathbf{U}\begin{pmatrix} \mathbf{L}_{k_1}(\lambda_1) & & & \mathbf{0} \\ & \cdot & & \\ & & \cdot & \\ \mathbf{0} & & & \mathbf{L}_{k_l}(\lambda_l)\end{pmatrix}\mathbf{U}^{-1},$$

where $\lambda_1, \ldots, \lambda_l$ are the eigenvalues of $\mathbf{Q}$(0 is always an eigenvalue of $\mathbf{Q}$ because $\mathbf{Qe} = \mathbf{0}$; see also Exercise 8.1). On account of (8.6) we have

$$\mathbf{P}(t) = \mathbf{U}\begin{pmatrix} \exp(t\mathbf{L}_{k_1}(\lambda_1)) & & & \mathbf{0} \\ & \cdot & & \\ & & \cdot & \\ \mathbf{0} & & & \exp(t\mathbf{L}_{k_l}(\lambda_l)) \end{pmatrix}\mathbf{U}^{-1}.$$

Now, a matrix of the form $\exp(t\mathbf{L}_k(\lambda))$ can be expressed in a simple form as follows. Write

$$\exp(t\,\mathbf{L}_k(\lambda)) = e^{\lambda t}\exp(t[\mathbf{L}_k(\lambda) - \lambda\mathbf{I}_k])$$
$$= e^{\lambda t}\left(\mathbf{I}_k + \sum_{n\geqslant 1}\frac{t^n}{n!}[\mathbf{L}_k(\lambda) - \lambda\mathbf{I}_k]^n\right).$$

But

$$\mathbf{L}_k(\lambda) - \lambda\mathbf{I}_k = \begin{pmatrix} 0 & 1 & 0 & \cdots & 0 \\ 0 & 0 & 1 & \cdots & 0 \\ \cdot & \cdot & \cdot & \cdots & \cdot \\ \cdot & \cdot & \cdot & \cdots & \cdot \\ \cdot & \cdot & \cdot & \cdots & \cdot \\ 0 & 0 & 0 & \cdots & 1 \\ 0 & 0 & 0 & \cdots & 0 \end{pmatrix}, \qquad (\mathbf{L}_k(\lambda) - \lambda\mathbf{I}_k)^2 = \begin{pmatrix} 0 & 0 & 1 & 0 & \cdots & 0 \\ 0 & 0 & 0 & 1 & \cdots & 0 \\ \cdots & \cdots & \cdots & \cdots & \cdots & \cdots \\ \cdots & \cdots & \cdots & \cdots & \cdots & \cdots \\ 0 & 0 & 0 & 0 & \cdots & 1 \\ 0 & 0 & 0 & 0 & \cdots & 0 \\ 0 & 0 & 0 & 0 & \cdots & 0 \end{pmatrix},$$

$$\ldots, (\mathbf{L}_k(\lambda) - \lambda\mathbf{I}_k)^{k-1} = \begin{pmatrix} 0 & 0 & 0 & \cdots & 1 \\ 0 & 0 & 0 & \cdots & \cdot \\ \cdot & \cdot & \cdot & \cdots & \cdot \\ \cdot & \cdot & \cdot & \cdots & \cdot \\ \cdot & \cdot & \cdot & \cdots & \cdot \\ 0 & 0 & 0 & \cdots & 0 \\ 0 & 0 & 0 & \cdots & 0 \end{pmatrix}, \quad (\mathbf{L}_k(\lambda) - \lambda\mathbf{I}_k)^n = \mathbf{0}, \quad n \geqslant k.$$

Therefore

$$\exp(t\mathbf{L}_k(\lambda)) = e^{\lambda t}\begin{pmatrix} 1 & t/1! & t^2/2! & \cdots & t^{k-1}/(k-1)! \\ 0 & 1 & t/1! & \cdots & t^{k-2}/(k-2)! \\ 0 & 0 & 1 & \cdots & t^{k-3}/(k-3)! \\ \cdot & \cdot & \cdot & \cdots & \cdot \\ \cdot & \cdot & \cdot & \cdots & \cdot \\ \cdot & \cdot & \cdot & \cdots & \cdot \\ 0 & 0 & 0 & \cdots & 1 \end{pmatrix}.$$

In particular, if $\mathbf{Q}$ is diagonalizable then

$$\mathbf{P}(t) = \mathbf{U}(e^{\lambda_i t}\,\delta(i, j))_{i,j \in S}\mathbf{U}^{-1}.$$

8.3 CONSTRUCTIVE DEFINITION OF A MARKOV PROCESS

8.3.1. Let us discuss the probabilistic significance of the intensity matrix $\mathbf{Q}$. First we have

$$q(i) = -q(i, i) = \lim_{h \to 0+} \frac{1 - p(h, i, i)}{h}$$

$$= \lim_{h \to 0+} \frac{\mathsf{P}(X(t+h) \neq i \mid X(t) = i)}{h}, \quad t \geqslant 0,$$

whence $\mathsf{P}(X(t+h) \neq i \mid X(t) = i) = q(i)h + o(h)$, where $o(h)$ denotes a quantity such that $\lim\limits_{h \to 0+} o(h)/h = 0$. Therefore, whatever $t \geqslant 0, q(i)h + + o(h)$ is the conditional probability of no longer being in state i at time $t + h$ given that the process was in state i at time t. Hence $q(i)$ is called the intensity of passage from i.

Further, $q(i)$ admits of another interpretation in connection with the sojurn time of the process in state i, that is the time $\rho(i)$ until it first leaves i. Set $\mathsf{P}_i(\rho(i) \geqslant t) = g_i(t)$, $t \geqslant 0$. Then by the Markov property

$$g_i(t+h) = \mathsf{P}_i(\rho(i) \geqslant t+h) = \mathsf{P}_i(X(u) = i,\ 0 < u \leqslant t+h)$$

$$= \mathsf{P}_i(X(u) = i,\ 0 < u \leqslant t)\,\mathsf{P}_i(X(u) = i,\ t < u \leqslant t+h \mid X(t) = i)$$

$$= g_i(t)\,g_i(h),\ h, t \geqslant 0.$$

Since $g_i(h) = p(h, i, i) + o(h) = 1 - q(i)h + o(h)$ as $h \to 0+$ it follows that

$$\frac{g_i(t+h) - g_i(t)}{h} = -q(i)\,g_i(t) + \frac{o(h)}{h}$$

whence $g_i'(t) = -q(i)g_i(t)$*⁾. The solution of this differential equation with the obvious initial condition $g_i(0) = 1$ is $g_i(t) = \exp(-q(i)t)$, $t \geqslant 0$.

---

*⁾ This is just a heuristic "proof". The reader should first observe that we do not even know whether $\{X(u) = i,\ 0 < u \leqslant t+h\}$ is a random event (it is the intersection of uncountably many random events). Next the formal justification of the intuitive equality $g_i(h) = p(h, i, i) + o(h)$ is not simple. A rigorous derivation of the differential equation verified by g_i would involve the concept of separability of a stochastic process which is beyond the scope of this book. The constructive approach we consider later on (see 8.3.2) will dispense us with using separability.

Next, for $i \neq j$,

$$q(i, j) = \lim_{h \to 0+} \frac{p(h, i, j)}{h} = \lim_{h \to 0+} \frac{\mathsf{P}(X(t+h) = j \mid X(t) = i)}{h}, \quad t \geqslant 0,$$

whence $\mathsf{P}(X(t+h) = j \mid X(t) = i) = q(i, j)h + o(h)$. Therefore $q(i, j)$ is related to the transition from state i to state j and hence is called the intensity of transition from i to j.

Finally, for $i \neq j$ and assuming $q(i) \neq 0$ we have

$$\frac{q(i, j)}{q(i)} = \lim_{h \to 0+} \frac{p(h, i, j)}{1 - p(h, i, i)} = \lim_{h \to 0+} \frac{\mathsf{P}(X(t+h) = j \mid X(t) = i)}{\mathsf{P}(X(t+h) \neq i \mid X(t) = i)}, \quad t \geqslant 0.$$

Therefore $q(i, j)/q(i)$ might be thought of as the conditional probability of a transition at an arbitrary time from state i to state j given that a transition has in fact taken place *).

8.3.2. The heuristic considerations above suggest the following constructive definition of the Markov process associated with a given intensity matrix $\mathbf{Q} = (q(i, j))_{i, j \in S}$.

Define

$$p^*(i, j) = \begin{cases} (1 - \delta(i, j))\ q(i, j)/q(i), & \text{if } q(i) \neq 0, \\ \delta(i, j), & \text{if } q(i) = 0. \end{cases}$$

If the process starts in state i it remains there a random length of time exponentially distributed with parameter $q(i)$ (this means it remains forever in i if $q(i) = 0$) and then moves to state j with probability $p^*(i, j)$. After hitting j, the process remains there a random length of time exponentially distributed with parameter $q(j)$ and then moves to state k with probability $p^*(j, k)$, and so on. Formally, let us define positive random variables σ_n (to be interpreted as the successive sojourn times) and S-valued random variables $X^*(n)$ (to be interpreted as the states successively visited), $n \geqslant 0$, as follows: $\sigma_0 \equiv 0$, $X^*(0) = i_0$ (arbitrarily given in S), and given the preceding choices, we choose σ_{n+1} and $X^*(n+1)$, $n \geqslant 0$, so that

$$\mathsf{P}(\sigma_{n+1} < t \mid \sigma_l < t_l,\ X^*(l) = i_l,\ 0 \leqslant l \leqslant n) = 1 - \exp(-q(i_n)t),$$

$$\mathsf{P}(X^*(n+1) = j \mid \sigma_{n+1} < t_{n+1},\ \sigma_l < t_l,\ X^*(l) = i_l,\ 0 \leqslant l \leqslant n) = p^*(i, j)$$

whatever $t \geqslant 0$, $t_l \geqslant 0$, $0 \leqslant l \leqslant n$, $i_1, \ldots, i_n \in S$. The second equation implies (on taking $t_1, \ldots, t_{n+1} \to \infty$) that $(X^*(n))_{n \geqslant 0}$ is a Markov chain with state space S and transition matrix $\mathbf{P}^* = (p^*(i, j))_{i, j \in S}$. This is called the *imbedded* Markov chain and, clearly, it governs the sequence

*) The rigorous proof would also involve separability.

of successive transitions from state to state. It then follows from both equations above that *)

$$P(\sigma_n < t \mid X^*(0),\ X^*(1),\ X^*(2), \ldots) = 1 - \exp\,(-q(X^*(n-1))t)$$

whatever $n \geq 1$, $t \geq 0$, and that

$$P(\sigma_1 < t_1, \ldots, \sigma_n < t_n \mid X^*(0),\ X^*(1),\ X^*(2), \ldots)$$

$$= \prod_{l=1}^{n} P(\sigma_l < t_l \mid X^*(0),\ X^*(1),\ X^*(2), \ldots)$$

whatever $t_l \geq 0$, $1 \leq l \leq n$, $n \geq 1$. The last equation shows that the sojourn times $\sigma_1, \sigma_2, \ldots$ are conditionally independent given the Markov chain $(X^*(n))_{n \geq 0}$.

Define

$$X'(t) = X^*(n)$$

for $\sum_{m=0}^{n} \sigma_m \leq t < \sum_{m=0}^{n+1} \sigma_m$, $m \geq 0$ (see Figure 8.1). Remark that $X'(t)$ is defined for all $t \geq 0$. Indeed, the Markov chain $(X^*(n))_{n \geq 0}$ has at least one recurrent state which occurs infinitely often with probability 1. Then on account of conditional independence and exponentiality of the sojourn times $\sigma_1, \sigma_2, \ldots$ mentioned above and of the law

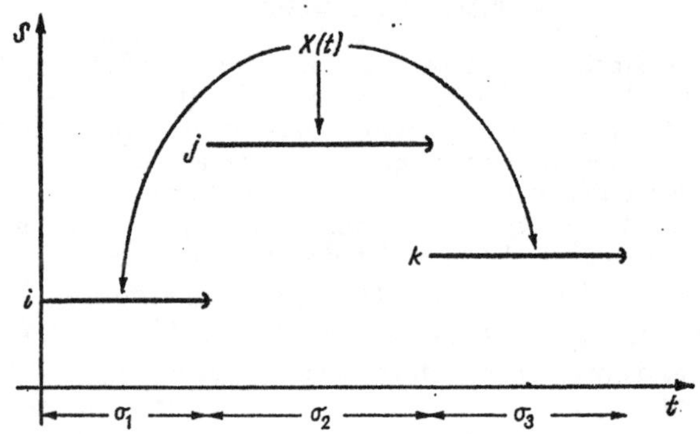

Figure 8.1

of large numbers (Theorem 1.4), we conclude that the series $\sum_{m \geq 0} \sigma_m$ diverges with probability 1. Thus whatever $t \geq 0$ there exists some integer n such that $\sum_{m=0}^{n} \sigma_m \leq t < \sum_{m=0}^{n+1} \sigma_m$, i.e., $X'(t)$ is defined for any $t \geq 0$.

---

*) This notation is a shorthand for a conditioning which prescribes values to $X^*(0), \ldots, X^*(m)$ for any fixed natural number $m \geq n - 1$.

8.3.3. Now let us prove that the process $X' = (X'(t))_{t\geqslant 0}$ we have just defined is a (homogeneous) Markov process with intensity matrix $\mathbf{Q}$ (equivalently, with transition matrix function given either by (8.6) or by (8.7)). In proving that it is Markovian, the lack of memory of the exponential density (see Proposition 1.3) is of paramount importance. On account of this property we can assert (cf. [86], p. 253) that if we choose an $s > 0$ and stop the procedure described above when the sum $\sum_{m=0}^{n} \sigma_m$ first exceeds s, so that $X'(t)$ is defined just for $t \leqslant s$, and we then restart it at $t = s$ as was done at $t = 0$, using, however, $X(s)$ as initial value with the corresponding probability distribution, then the subsequent σ's and $X^{*\prime}$ s will have the same distribution as they would have had if the procedure had not been stopped *). It then follows that, on the one hand, the conditional distribution of $X'(t_{n+1})$ given the values of $X'(t_1), \ldots, X'(t_n)$ for $t_1 < \ldots < t_n < t_{n+1}$ depends only on $X'(t_n)$, i.e., $(X'(t))_{t\geqslant 0}$ is a Markov process and, on the other hand, the conditional probability $\mathsf{P}(X'(t) = j \mid X'(s) = i)$ for $s < t$ is a function of the difference $t-s$ only, i.e., homogeneity also holds.

It is also immediate from the very definition of X' that whatever the natural number n the random process $X'' = \left(X'\left(\sum_{m=0}^{n} \sigma_m + t\right)\right)_{t\geqslant 0}$ has under the conditional probability $\mathsf{P}_{A\cap\{X^*(n)=i\}}$, where A is any random event defined in terms of $\sigma_1, \ldots, \sigma_n$, the same finite dimensional joint distributions as X' under P_i. Therefore X'' is a Markov process, too, and the transition matrix functions of X' and X'' are identical. (This is reminiscent of the strong Markov property for Markov chains (see 2.3). The study of the strong Markov property for Markov processes is beyond the scope of this book.)

To prove the fact that the intensity matrix of $(X'(t))_{t\geqslant 0}$ is $\mathbf{Q}$ we shall show that its transition matrix function is given by (8.7). Let

$$p'_n(t,i,j) = \mathsf{P}_i\left(X'(t) = j, \sum_{m=0}^{n} \sigma_m \leqslant t < \sum_{m=0}^{n+1} \sigma_m\right).$$

Therefore $p'_n(t, i, j)$ is the P_i-probability that $X'(t) = j$ and there are exactly n changes of states in the interval $[0, t]$. Then on account of the very construction of the process and the remark in the previous paragraph we have

$$p'_0(t, i, j) = \delta(i, j) \exp(-q(i)t),$$

$$p'_{n+1}(t, i, j) = \sum_{k\neq i} \int_0^t \exp(-q(i)u)q(i, k))p'_n(t-u, k, j)du, \quad n \geqslant 0.$$

The justification of the last equation is as follows. The process is in state i up to the moment u of the first change of state [probability

---

*) It is not difficult to formalize this argument. The reader may try to do it as an exercise.

density $q(i)\exp(-q(i)u)$] when it moves to state $k \neq i$ [probability $q(i, k)/q(i)$] after which it goes to state j performing exactly n changes of state [probability $p'_n(t-u, k, j)$].

Alternatively

$$p'_0(t, i, j) = \delta(i, j) \exp(-q(i)t),$$

$$p'_{n+1}(t, i, j) = \sum_{k \neq j} \int_0^t p'_n(u, i, k)\, q(k, j) \exp(-q(j)(t-u))du, \quad n \geq 0.$$

The last equation can be justified by invoking the last change of state in the interval $[0, t]$.

Clearly, $p'_n(t, i, j)$, $i, j \in S$, are precisely the entries of the matrix function $\mathbf{P}_n(\cdot)$ defined in 8.2.3 and since

$$\mathsf{P}_i(X'(t) = j) = \sum_{n \geq 0} p'_n(t, i, j)$$

the transition matrix function of $(X'(t))_{t \geq 0}$ is indeed given by (8.7) as asserted.

8.3.4. It is natural to ask what relationship exists between the Markov process $X = (X(t))_{t \geq 0}$ constructed in 8.1.2 and the Markov process $X' = (X'(t))_{t \geq 0}$ defined in 8.3.2. It can be proved that there exists a homogeneous S-valued Markov process $\widetilde{X} = (\widetilde{X}(t))_{t \geq 0}$ such that $\mathsf{P}(X(t) = \widetilde{X}(t)) = 1$ whatever $t \geq 0$ and the transitions and the sojourn times in $\widetilde{X}$ are distributed like the transitions and the sojourn times in X'.

8.4 DISCRETE SKELETONS AND CLASSIFICATION OF STATES

8.4.1. With every homogeneous finite Markov process $X = (X(t))_{t \geq 0}$ we associate a family of Markov chains $X_h = (X(nh))_{n \geq 0}$, $h > 0$. Clearly, the transition matrix of X_h is $(p_h(i, j))_{i, j \in S}$ where $p_h(i, j) = p(h, i, j)$, $i, j \in S$. Hence the entries of the n-step transition matrix of X_h are given by $p_h(n, i, j) = p(nh, i, j)$, $i, j \in S$. The Markov chain X_h is called the *discrete skeleton* (at *scale* h) of X*).

8.4.2. We shall use the concept of a discrete skeleton to study classification of states for the Markov process $X = (X(t))_{t \geq 0}$.

We say "state i *leads* to state j" (or "state j *is accessible* from state i") and write $i \to j$ if and only if there exists $t > 0$ such that $p(t, i, j) > 0$.

*) We mention in this context G. Elfving's *imbedding problem*: determine when the transition matrix of a finite Markov chain can be identified with the transition matrix of a discrete skeleton associated with a Markov process with the same state space. Raised in 1937, the problem seems to be very difficult. For partial results see [125, 177, 178, 207, 331, and 343].

Next, we say "state i *communicates* with state j" and write $i \leftrightarrow j$ if and only if $i \rightarrow j$ and $j \rightarrow i$. It follows from Proposition 8.1 that $i \rightarrow j$ or $i \leftrightarrow j$ if and only if these relations hold for any discrete skeleton X_h, $h > 0$. In particular, the grouping of states into classes of those which communicate is identical for X and all the X_h. For instance, a state i is absorbing in X if and only if $p_h(i, i) = p(h, i, i) = 1$ for some $h > 0$, hence for all $h > 0$. *)

Let us note an important simplification, namely that in any X_h all the states are return and aperiodic. This follows from (8.4) and Proposition 8.1 which implies that $p_h(i, i) = p(h, i, i) > 0$ whatever $i \in S$, $h > 0$.

8.4.3. The continuous parameter analogue of Theorem 5.1 can be easily obtained from that theorem itself by means of discrete skeletons.

Theorem 8.4 (*W. Doeblin*). *The limits*

$$\lim_{t \to \infty} p(t, i, j) = \pi(i, j), \quad i, j \in S,$$

exist, and convergence is geometrically fast. The matrix $\mathbf{\Pi} = (\pi(i, j))_{i, j \in S}$ *satisfies the equations*

$$\mathbf{\Pi P}(t) = \mathbf{P}(t)\, \mathbf{\Pi} = \mathbf{\Pi} = \mathbf{\Pi}^2, \quad t \geqslant 0.$$

Proof. By Theorem 5.1 and the aperiodicity of states in X_h the limits

$$\lim_{n \to \infty} p(nh, i, j) = \lim_{n \to \infty} p_h(n, i, j) = \pi_h(i, j) \text{ (say)}, \quad i, j \in S,$$

exist whatever $h > 0$. Therefore, given $\varepsilon > 0$ there is a natural number $n_0 = n_0(i, j)$ such that

$$|p(nh, i, j) - p(n'h, i, j)| < \varepsilon/3$$

whenever $n, n' \geqslant n_0$. By Proposition 8.2 we can choose h so that the oscillation of $p(\cdot, i, j)$ in any interval of length h is less than $\varepsilon/3$. Then given $t, t' \geqslant n_0 h$, there are natural numbers $n, n' \geqslant n_0$ such that $|t - nh| \leqslant h$, $|t' - n'h| \leqslant h$. Therefore

$$|p(t, i, j) - p(t', i, j)| = |p(t, i, j) - p(nh, i, j)| + |p(nh, i, j) - p(n'h, i, j)| + |p(n'h, i, j) - p(t', i, j)| \leqslant \varepsilon,$$

and this is equivalent to the existence of the limit $\pi(i, j)$ (in fact the proof shows that the $\pi_h(i, j)$ do not depend on h).

Next letting $s \to \infty$ in the equations

$$\mathbf{P}(s)\, \mathbf{P}(t) = \mathbf{P}(t)\, \mathbf{P}(s) = \mathbf{P}(t + s), \quad s, t \geqslant 0,$$

we get

$$\mathbf{\Pi P}(t) = \mathbf{P}(t)\, \mathbf{\Pi} = \mathbf{\Pi}, \quad t \geqslant 0.$$

---

*) Clearly, an equivalent condition is $q(i) = 0$.

The equality $\mathbf{\Pi} = \mathbf{\Pi}^2$ now follows by letting $t \to \infty$ in the equation $\mathbf{\Pi P}(t) = \mathbf{\Pi}$.

The fact that the convergence of $\mathbf{P}(\cdot)$ to $\mathbf{\Pi}$ is geometrically fast is an immediate consequence of Exercise 5.2 on account of the equality $\mathbf{P}(t) = \mathbf{P}(nh)\,\mathbf{P}(t - nh)$, $nh \leqslant t < (n+1)h$. (Actually, the exact rate of convergence of $\mathbf{P}(\cdot)$ to $\mathbf{\Pi}$ may be obtained from 8.2.6.)

Remark. The first part of Theorem 8.4 concerning the existence of the limits $\pi(i, j)$, $i, j \in S$, is a special case of the following Croftian theorem (see [208]): *Let f be a real valued continuous function defined on $[0, \infty)$ and assume that whatever $h > 0$ the limit $\lim_{n\to\infty} f(nh)$ exists and is finite. Then this limit is independent of h (and is equal to L (say)) and $\lim_{t\to\infty} f(t) = L$.*

8.4.4. As in the discrete parameter case, a probability distribution $\mathbf{p}$ over the state space S is said to be *stationary* if and only if $\mathbf{p}'\mathbf{P}(t) = \mathbf{p}'$ whatever $t \geqslant 0$. Theorem 8.4 shows that the rows of $\mathbf{\Pi}$ are stationary distributions. It is easy to show that, conversely, any stationary distribution is a convex liniar combination of the rows of $\mathbf{\Pi}$. (Compare with Theorem 5.3.)

Note also that $\mathbf{p}$ is a stationary distribution if and only if $\mathbf{p}'\mathbf{Q} = \mathbf{0}$. Indeed, if $\mathbf{p}$ is stationary we have $\mathbf{p}'\mathbf{P}(t) = \mathbf{p}'$ whatever $t \geqslant 0$. Hence differentiating with respect to t at $t = 0$ yields $\mathbf{p}'\mathbf{Q} = \mathbf{0}$. Conversely, if $\mathbf{p}'\mathbf{Q} = \mathbf{0}$ then on account of (8.6) we have $\mathbf{p}'\mathbf{P}(t) = \mathbf{p}'$ whatever $t \geqslant 0$, that is $\mathbf{p}$ is stationary. Hence $\mathbf{\Pi Q} = \mathbf{0}$. We also have $\mathbf{Q\Pi} = \mathbf{0}$. This obtains by differentiating the equation $\mathbf{P}(t)\,\mathbf{\Pi} = \mathbf{\Pi}$ with respect to t at $t = 0$.

8.4.5. Coming to the notion of recurrence we define state i to be recurrent (transient) in X if and only if it is recurrent (transient) in some X_h. Then state i is recurrent (transient) in X if and only if $\lim_{n\to\infty} p_h(n, i, i) = \lim_{n\to\infty} p(nh, i, i) > 0$ $(\lim_{n\to\infty} p_h(n, i, i) = \lim_{n\to\infty} p(nh, i, i) = 0)$. Therefore, by Theorem 8.4 state i is recurrent (transient) in S if and only if $\pi(i, i) > 0$ $(\pi(i, i) = 0)$. This shows that i is recurrent (transient) in X if and only if it is recurrent (transient) in every X_h, $h > 0$. Finally, bearing in mind the constructive definition of a Markov process we may assert that state i is recurrent (transient) if and only if the probability P_i(the set $(t : X(t) = i)$ is unbounded) equals one (zero). In terms of the imbedded Markov chain $X^* = (X^*(n))_{n\geqslant 0}$ we may also say that state i is recurrent (transient) in X if and only if it is recurrent (transient) in X^*. The reader should be warned that, unlike X_h, X^* may have periodic states. For instance if

$$\mathbf{Q} = \begin{pmatrix} -1 & 1 \\ 1 & -1 \end{pmatrix},$$

then X^* has the transition matrix

$$\mathbf{P}^* = \begin{pmatrix} 0 & 1 \\ 1 & 0 \end{pmatrix}$$

which is irreducible of period 2.

8.4.6. On account of 8.4.2 and 8.4.5 the states of a homogeneous finite Markov process can be grouped into *recurrent* and *transient* classes of states. As in 2.6 we shall distinguish *recurrent* Markov processes and *absorbing* Markov processes according as all the states are recurrent or there also exist transient states. A recurrent Markov process with only one class of states will be called *regular*. An absorbing Markov process with only one recurrent class will be called *indecomposable*.

8.5 ABSORBING MARKOV PROCESSES*)

8.5.1. As in the discrete parameter case (see Chapter 3) we shall assume the transition matrix function of the process has the canonical form

$$\mathbf{P}(\cdot) = \begin{pmatrix} \mathbf{I} & 0 \\ \mathbf{R}(\cdot) & \mathbf{T}(\cdot) \end{pmatrix}.$$

Hence

$$\mathbf{Q} = \mathbf{P}'(0) = \begin{pmatrix} 0 & 0 \\ \mathbf{R}'(0) & \mathbf{T}'(0) \end{pmatrix}.$$

Since

$$\mathbf{P}(t) = \exp(t\mathbf{Q}) = \mathbf{I} + \sum_{n\geqslant 1} \frac{t^n}{n!} \mathbf{Q}^n$$

and

$$\mathbf{Q}^n = \begin{pmatrix} 0 & 0 \\ (\mathbf{T}'(0))^{n-1}\mathbf{R}'(0) & (\mathbf{T}'(0))^n \end{pmatrix}, \quad n \geqslant 1,$$

we have

$$\mathbf{T}(t) = \mathbf{I} + \frac{t}{1!}\mathbf{T}'(0) + \frac{t^2}{2!}(\mathbf{T}'(0))^2 + \ldots = \exp(t\mathbf{T}'(0)),$$

$$\mathbf{R}(t) = \left(\mathbf{I}t + \frac{t^2}{2!}\mathbf{T}'(0) + \frac{t^3}{3!}(\mathbf{T}'(0))^2 + \ldots\right)\mathbf{R}'(0)$$

$$= \left(\int_0^t \mathbf{T}(s)\, ds\right)\mathbf{R}'(0).$$

*) The considerations in this and the next section develop ideas appearing in [196].

By Theorem 8.4 there exist a positive number a and a nonnegative matrix $\mathbf{B}$ such that

$$\mathbf{T}(t) \leqslant e^{at}\,\mathbf{B}, \quad t \geqslant 0. \tag{8.8}$$

8.5.2. The problems concerning the number and type of transitions can be solved by using the imbedded Markov chain defined in 8.3.2 whose transition matrix $\mathbf{P}^* = (p^*(i,j))_{i,j \in S}$ is also in the canonical form

$$\mathbf{P}^* = \begin{pmatrix} \mathbf{I} & 0 \\ \mathbf{R}^* & \mathbf{T}^* \end{pmatrix}.$$

Thus (see 3.1) the mean number of transitions to state $j \in T$ starting in state $i \in T$ is the ijth entry of the fundamental matrix $\mathbf{N}^* = (\mathbf{I}-\mathbf{T}^*)^{-1}$ associated with $\mathbf{P}^*$. Next, the mean number of transitions within T before absorption, starting in state $i \in T$, is the ith component of the vector $\mathbf{N}^*\mathbf{e}$. Finally, the probability $a(i, k)$ that the Markov process starting in the transient state i ends up in the absorbing state k is equal to the corresponding probability for the imbedded Markov chain. Therefore by Theorem 3.3 we have

$$\mathbf{A} = (a(i, k))_{i \in T,\, k \in S-T} = \mathbf{N}^*\mathbf{R}^*.$$

8.5.3. The problems involving time need a different approach, which we are now going to describe.

Let

$$u_t(j) = \begin{cases} 1, & \text{if } X(t) = j, \\ 0, & \text{if } X(t) \neq j, \end{cases}$$

whatever $t \geqslant 0$, $j \in T$, and set

$$\nu(j) = \int_0^\infty u_t(j)\, dt.$$

Clearly, $\nu(j)$ is the time spent in state $j \in T$ until absorption. Invoking the imbedded Markov chain shows that $\nu(j)$ is a random variable which is finite with probability 1. Notice that

$$\mathsf{E}_i(u_t(j)) = 1 \cdot \mathsf{P}_i(X(t) = j) + 0 \cdot \mathsf{P}_i(X(t) \neq j) = p(t, i, j).$$

Therefore whatever $i, j \in T$ we have

$$\mathsf{E}_i(\nu(j)) = \mathsf{E}_i\left(\int_0^\infty u_t(j)\, dt\right) = \int_0^\infty \mathsf{E}_i(u_t(j))\, dt = \int_0^\infty p(t, i, j)\, dt$$

(compare with 3.1.1) which is finite by (8.8).

Set $n(i, j) = \mathsf{E}_i(\nu(j))$, $i, j \in T$ and consider the matrix

$$\mathbf{N} = (n(i, j))_{i,j\in T} = \int_0^\infty \mathbf{T}(t)\, dt.$$

As in the discrete parameter case, $\mathbf{N}$ is called the *fundamental matrix* of the absorbing Markov process.

Proposition 8.5. *The matrix* $\mathbf{T}'(0)$ *is nonsingular and* $\mathbf{N} = -(\mathbf{T}'(0))^{-1}$.

Proof. We have

$$\mathbf{T}'(0)\int_0^t \mathbf{T}(s)\, ds = \mathbf{T}'(0)\int_0^t \exp(s\mathbf{T}'(0))\, ds$$

$$= \mathbf{T}'(0)\left(\mathbf{I}t + \sum_{n\geqslant 1} \frac{(\mathbf{T}'(0))^n}{n!} \frac{t^{n+1}}{n+1}\right) = \exp(t\mathbf{T}'(0)) - \mathbf{I} = \mathbf{T}(t) - \mathbf{I}.$$

Hence letting $t \to \infty$ we get

$$-\mathbf{T}'(0)\mathbf{N} = \mathbf{I}$$

thus completing the proof.

8.5.4. The time ν spent in the set T of transient states is called the *time to absorption*. Clearly, $\nu = \sum_{j\in T} \nu(j)$. Hence

$$\mathsf{E}_i(\nu) = \sum_{j\in T} \mathsf{E}_i(\nu(j)) = \sum_{j\in T} n(i, j), \quad i \in T.$$

Therefore

$$\mathbf{m}_1 = (\mathsf{E}_i(\nu))_{i\in T} = \mathbf{Ne}.$$

More generally, the moments of any order $s = 1, 2, \ldots$ of ν can be expressed in terms of the fundamental matrix $\mathbf{N}$. To this end notice that considering the discrete skeleton X_h at scale h with the corresponding time to absorption ν_h we have $0 \leqslant h\, \nu_h - \nu \leqslant h$. Hence by Theorem 3.2 the moments $\mathsf{E}_i(\nu^s)$ are finite for all $i \in T$, $s \geqslant 1$, and on account of the fact that

$$\lim_{h\to 0} h\, \nu_h = \nu$$

we can assert that *)

$$\lim_{h\to 0} \mathsf{E}_i((h\nu_h)^s) = \mathsf{E}_i(\nu^s), \quad i \in T.$$

Set

$$\mathbf{m}_s = (\mathsf{E}_i(\nu^s))_{i\in T}, \quad s \geqslant 1.$$

*) This is a consequence of Fatou's lemma which is to be found in any advanced course of probability theory.

Proposition 8.6. *We have*

$$\mathbf{m}_s = s\,!\,\mathbf{N}^s\mathbf{e},\ s \geqslant 1.$$

Proof. For all $i \in T$ and $s \geqslant 2$ the skeleton X_h yields the equation (see the proof of Theorem 3.2)

$$\mathsf{E}_i((h\nu_h)^s) = \sum_{l \notin T} p(h, i, l) h^s + \sum_{k \in T} p(h, i, k)\ \mathsf{E}_k((h(\nu_h + 1))^s)$$

$$= \sum_{k \in T} p(h, i, k) \left(\mathsf{E}_k((h\nu_h)^s) + \sum_{r=1}^{s-1} \binom{s}{r} \mathsf{E}_k((h\nu_h)^r) h^{s-r} \right) + h^s\mathbf{e},$$

or, in matrix notation,

$$(\mathbf{I} - \mathbf{T}(h))\ (\mathsf{E}_i((h\nu_h)^s))_{i \in T} = \sum_{r=1}^{s-1} \binom{s}{r} \mathbf{T}(h)\ \mathsf{E}_i((h\nu_h)^r) h^{s-r} + h^s\mathbf{e}.$$

Dividing both sides by h and then letting $h \to 0$ yields

$$-\,\mathbf{T}'(0)\mathbf{m}_s = s\ \mathbf{m}_{s-1},\ s \geqslant 2.$$

Hence

$$\mathbf{m}_s = s\ \mathbf{N}\mathbf{m}_{s-1},\ s \geqslant 2.$$

Thus

$$\mathbf{m}_s = s\,!\,\mathbf{N}^s\mathbf{e},\ s \geqslant 1.$$

8.5.5. The moments of any finite order of the $\nu(j)$, $j \in T$, can also be expressed in terms of the fundamental matrix $\mathbf{N}$. Since $\nu(j) \leqslant \nu$, $j \in T$, the moments $\mathsf{E}_i(\nu^s(j))$, $i, j \in T$, are finite for all $s \geqslant 1$. Set

$$\mathbf{N}_s = (\mathsf{E}_i(\nu^s(j)))_{i,j \in T}.$$

Considering again the discrete skeleton X_h at scale h with the corresponding random variables $\nu_h(j)$ (= the number of occurrences of state j until absorption in X_h) it is easily seen that $\lim_{h \to 0} h\ \nu_h(j) = \nu(j)$, $j \in T$. Hence

$$\lim_{h \to 0}\ \mathsf{E}_i((h\nu_h(j))^s) = \mathsf{E}_i(\nu^s(j))$$

for all $i, j \in T$, $s \geqslant 1$.

Proposition 8.7. *We have*

$$\mathbf{N}_s = s\,!\,\mathbf{N}\ \mathbf{N}_{\mathrm{dg}}^{s-1},\ s \geqslant 1.$$

Proof. For all i, $j \in T$ and $s \geqslant 2$ the skeleton X_h yields the equation (see the proof of Theorem 3.1)

$$\mathsf{E}_i((h\ \nu_h(j))^s) = \delta(i, j) \sum_{l \notin T} p(h, i, l) h^s + \sum_{k \in T} p(h, i, k)\ \ \mathsf{E}_k(h^s(\nu_h(j) + \delta(i, j))^s)$$

$$= \sum_{k \in T} p(h, i, k) \left(\mathsf{E}_k((h\nu_h(j))^s) + \sum_{r=1}^{s-1} \binom{s}{r} \mathsf{E}_k((h\nu_h(j))^r)\ h^{s-r}\delta(i, j) \right) + \delta(i, j)\ h^s,$$

or, in matrix notation,

$$(\mathbf{I}-\mathbf{T}(h))\,(\mathsf{E}_i((h\nu_h(j))^s))_{i,j\in T}$$
$$=\sum_{r=1}^{s-1}\binom{s}{r}h^{s-r}\left(\mathbf{T}(h)\,(\mathsf{E}_i((h\nu_h(j))^r))_{i,j\in T}\right)_{\mathrm{dg}}+h^s\mathbf{I}.$$

Dividing both sides by h and then letting $h\to 0$ yield

$$-\mathbf{T}'(0)\,\mathbf{N}_s=s(\mathbf{N}_{s-1})_{\mathrm{dg}},\quad s\geqslant 2.$$

Hence

$$\mathbf{N}_s=s\mathbf{N}(\mathbf{N}_{s-1})_{\mathrm{dg}},\quad s\geqslant 2.$$

Thus

$$\mathbf{N}_s=s\,!\,\mathbf{N}\mathbf{N}_{\mathrm{dg}}^{s-1},\quad s\geqslant 1.$$

8.5.6. We can obtain an alternative expression for the matrix $\mathbf{A}$ of absorption probabilities on account of the fact that

$$a(i,k)=\lim_{t\to\infty}p(t,i,k),\ i\in T,\ k\in S-T$$

(compare with (3.5)). Hence

$$\mathbf{A}=\lim_{t\to\infty}\mathbf{R}(t)=\lim_{t\to\infty}\left(\int_0^t\mathbf{T}(s)\,ds\right)\mathbf{R}'(0)=\mathbf{N}\mathbf{R}'(0).$$

8.6 · REGULAR MARKOV PROCESSES

8.6.1. For a regular Markov process any discrete skeleton X_h is a regular Markov chain and the imbedded Markov chain is ergodic (see 8.4.5). Consequently

$$\lim_{t\to\infty}\mathbf{P}(t)=\mathbf{\Pi}=\mathbf{e}\pi'$$

is a positive stable matrix and by Theorem 8.4 there exist a positive number a and a nonnegative matrix $\mathbf{B}$ such that

$$|\mathbf{P}(t)-\mathbf{\Pi}|\leqslant e^{-at}\,\mathbf{B},\quad t\geqslant 0, \tag{8.9}$$

where $|\mathbf{P}(t)-\mathbf{\Pi}(t)|$ is the matrix whose entries are the absolute values of the corresponding entries of the matrix $\mathbf{P}(t)-\mathbf{\Pi}$.

Notice that the limiting stable matrix $\mathbf{\Pi}^*$ corresponding to the imbedded Markov chain X^* with transition matrix $\mathbf{P}^*=-\mathbf{Q}_{\mathrm{dg}}^{-1}(\mathbf{Q}-\mathbf{Q}_{\mathrm{dg}})$ is given by $\mathbf{\Pi}^*=\mathbf{e}(\pi^*)'$, where $\pi^*=\mathbf{Q}_{\mathrm{dg}}\,\pi/(\pi'\mathbf{Q}_{\mathrm{dg}}\mathbf{e})$. The proof consists of checking the equation $(\pi^*)'\,\mathbf{P}^*=(\pi^*)'$, and is left to the reader.

It is easy to see that the above remarks also apply to an indecomposable Markov process, the only difference being that the positive components of π and π^* are those corresponding to the recurrent states of the process.

8.6.2. As for absorbing Markov processes, problems only concerning transition can be solved using the imbedded Markov chain. Thus by Theorem 4.7 the mean number of transition taken to reach j for the first time from i is the ijth entry of the matrix $(\mathbf{I} - \mathbf{Z}^* + \mathbf{E}\,\mathbf{Z}^*_{dg})(\mathbf{\Pi}^*_{dg})^{-1}$, where $\mathbf{Z}^* = (\mathbf{I} - (\mathbf{P}^* - \mathbf{\Pi}^*))^{-1}$ is the fundamental matrix of the imbedded Markov chain. Next (see 4.3.5), the (relative) frequency $\nu^*(k)/n$ of occurrence of state k in the first n transitions converges to $\pi^*(k)$ as $n \to \infty$ as stated in Theorem 1.4. Similar remarks can be made about the applicability of the central limit theorem and the law of the iterated logarithm to $\nu_n^*(k)$. Notice that the results concerning $\nu_n^*(k)$ also apply to indecomposable Markov processes.

8.6.3. Problems involving time need the kind of approach described in 8.5.3. The analogue of the fundamental matrix occurring in the discrete parameter case is the matrix

$$\mathbf{Z} = \mathbf{\Pi} + \int_0^\infty (\mathbf{P}(t) - \mathbf{\Pi})\, dt$$

whose existence is ensured by (8.9).

Proposition 8.8. *The matrix* $\mathbf{\Pi} - \mathbf{Q}$ *is nonsingular and* $\mathbf{Z} = (\mathbf{\Pi} - \mathbf{Q})^{-1}$.

Proof. Taking into account that $\mathbf{\Pi P}(s) = \mathbf{\Pi}^2 = \mathbf{\Pi}$, $\mathbf{Q\Pi} = \mathbf{0}$, and $\mathbf{QP}(s) = \mathbf{P}'(s)$ we have

$$(\mathbf{\Pi} - \mathbf{Q})\Big(\mathbf{\Pi} + \int_0^t (\mathbf{P}(s) - \mathbf{\Pi})\, ds\Big)$$

$$= \mathbf{\Pi}^2 + \int_0^t (\mathbf{\Pi P}(s) - \mathbf{\Pi}^2)\, ds - \mathbf{Q\Pi} - \int_0^t (\mathbf{QP}(s) - \mathbf{Q\Pi})\, ds$$

$$= \mathbf{\Pi} - \int_0^t \mathbf{P}'(s)\, ds = \mathbf{\Pi} - \mathbf{P}(t) + \mathbf{I}.$$

Hence letting $t \to \infty$ we get

$$(\mathbf{\Pi} - \mathbf{Q})\,\mathbf{Z} = \mathbf{I}$$

thus completing the proof.

8.6.4. As in the discrete parameter case (see 4.3.2), the mean first passage times for a regular Markov process can be expressed in terms of the fundamental matrix **Z**. Define the *first passage time* τ_j to state j in the Markov process X by

$$\tau_j = \min\ (t \geqslant 0 : X(t) = j),\ j \in S,$$

and consider also the first passage time $\tau_{h,j}$ to state j in the discrete skeleton X_h. Then we have $0 < h\,\tau_{h,j} - \tau_j \leqslant h$. Hence by Theorem 4.7 the mean values $m(i,j) = \mathsf{E}_i(\tau_j)$, $i, j \in S$, are finite and

$$\lim_{h\to 0} \mathsf{E}_i(h\,\tau_{h,j}) = \mathsf{E}_i(\tau_j),\ i, j \in S.$$

Note that $m(i,i) = \mathsf{E}_i(\tau_i) = 0$, $i \in S$, as already remarked in 4.3.3.

Proposition 8.9. *The matrix* $\mathbf{M} = (m(i,j))_{i,j\in S}$ *is given by*

$$\mathbf{M} = (\mathbf{E}\mathbf{Z}_{dg} - \mathbf{Z})\,\mathbf{\Pi}_{dg}^{-1}.$$

Proof. Let $\mathbf{Z}_h$ denote the fundamental matrix of the discrete skeleton X_h, i.e.,

$$\mathbf{Z}_h = \mathbf{I} + \sum_{n\geqslant 1} (\mathbf{P}(nh) - \mathbf{\Pi})$$

(by Theorem 8.4 we have $\mathbf{\Pi}_h = \mathbf{\Pi}$ for all $h > 0$).

Applying Theorem 4.7 to X_h yields

$$\begin{aligned}(\mathsf{E}_i(h\tau_{h,j}))_{i,j\in S} &= h(\mathbf{I} - (\mathbf{Z}_h)_{dg} + \mathbf{E}(\mathbf{Z}_h)_{dg})\,\mathbf{\Pi}_{dg}^{-1} \\ &= h\left(\mathbf{E}(\mathbf{Z}_h)_{dg} - \sum_{n\geqslant 1}(\mathbf{P}(nh) - \mathbf{\Pi})\right)\mathbf{\Pi}_{dg}^{-1}. \qquad (8.10)\end{aligned}$$

Since

$$\lim_{h\to 0} h \sum_{n\geqslant 1} (\mathbf{P}(nh) - \mathbf{\Pi}) = \int_0^\infty (\mathbf{P}(t) - \mathbf{\Pi})\,dt = \mathbf{Z} - \mathbf{\Pi}$$

and

$$\lim_{h\to 0} h\,\mathbf{E}(\mathbf{Z}_h)_{dg} = \mathbf{E}\left(\int_0^\infty (\mathbf{P}(t) - \mathbf{\Pi})\,dt\right)_{dg} = \mathbf{E}\mathbf{Z}_{dg} - \mathbf{\Pi},$$

letting $h \to 0$ in (8.10) we get

$$\mathbf{M} = (\mathbf{E}\mathbf{Z}_{dg} - \mathbf{Z})\,\mathbf{\Pi}_{dg}^{-1},$$

as asserted.

8.6.5. Now we can compute the mean $r(i)$ of the time elapsed between an entry into state i and the next return to i. This time interval consists of an exponentially distributed sojourn time in i (with mean $1/q(i)$) ending in a jump to some $j \in S$, $j \neq i$, followed by a first passage from j to i.

Therefore

$$r(i) = 1/q(i) + \sum_{j \neq i} (q(i, j)/q(i))\ m(j, i).$$

Hence by Proposition 8.9

$$r(i) = 1/q(i) + 1/(\pi(i)\, q(i)) \sum_{j \neq i} q(i, j)\ (z(i, i) - z(j, i))$$

$$= \frac{1}{q(i)} \left[1 + \frac{1}{\pi(i)} \left(z(i, i)\, q(i) - \pi(i) + 1 - z(i, i) q(i) \right) \right] = \frac{1}{\pi(i)\, q(i)} .$$

(We have used the fact that $\mathbf{QZ} = \mathbf{\Pi} - \mathbf{I}$ —see Exercise 8.4.)

Clearly, the mean value of the *Smoluchowskian recurrence time* θ_i of state i (i.e., the time interval between an exit from i and the next return to i — see 4.3.3) is given by

$$\mathsf{E}(\theta_i) = r(i) - \frac{1}{q(i)} = \frac{1 - \pi(i)}{\pi(i)\, q(i)}$$

(compare with 4.3.3).

8.6.6. As in 4.3.5 we may inquire about the existence of the limit

$$\sigma^2(f) = \lim_{t \to \infty} \frac{1}{t} \operatorname{var}_{\mathbf{p}} \int_0^t f(X(s))\, ds, \tag{8.11}$$

where f is a real valued function defined on the state space S of either a regular or an indecomposable Markov process. Writing $t = nh$ and

$$\int_0^t = \int_0^h + \int_h^{2h} + \dots + \int_{(n-1)h}^{nh}$$

the problem can be reduced to the one investigated in 4.3.5. The reader will be able to show the limit (8.11) exists and equals $\sum_{i, j \in S} f(i)\ c(i, j) f(j)$ whatever the initial distribution $\mathbf{p}$, where

$$c(i, j) = \pi(i)\, z(i, j) + \pi(j)\, z(j, i) - 2\pi(i)\ \pi(j), \quad i, j \in S.$$

Using the above discretization procedure and the theory alluded to in the footnote on p. 140 one can show that

$$\lim_{t \to \infty} \frac{1}{t} \int_0^t f(X(s))\, ds = \sum_{i \in S} \pi(i)\, f(i)$$

under any $\mathsf{P}_{\mathbf{p}}$, as stated in Theorem 1.4. Then, if $\sigma(f) > 0$ we have

$$\lim_{t \to \infty} \mathsf{P}_{\mathbf{p}} \left(\frac{\int_0^t f(X(s))\, ds - t \sum_{i \in S} \pi(i)\, f(i)}{\sigma(f)\, \sqrt{t}} < x \right) = \frac{1}{\sqrt{2\pi}} \int_{-\infty}^{x} e^{-u^2/2}\, du$$

for all real numbers x and initial distributions $\mathbf{p}$. Finally, the limits superior or inferior of the ratio

$$\frac{\int_0^t f(X(s))\,ds - t\sum_{i\in S}\pi(i)\,f(i)}{\sigma(f)\sqrt{2\ln\ln t}}$$

are equal to 1 and -1, respectively, under any $\mathsf{P}_{\mathbf{p}}$ as stated in Theorem 1.6.

The special case

$$f(i) = \begin{cases} 1, & \text{if } i = k, \\ 0, & \text{if } i \neq k, \end{cases}$$

where k is an arbitrarily given recurrent state, is very important. In this case $\int_0^t f(X(s))\,ds/t$ is the proportion of time the process spends in state k during the time interval $[0, t]$*). Therefore $\pi(k)$ appears as the limit value of this proportion as $t \to \infty$.

8.7 BIRTH AND DEATH PROCESSES

8.7.1. The continuous parameter analogue of the random walks considered in Examples 1 and 2 from 2.2 is the so called (Markovian) *birth and death process*. For such a process only unit transitions are allowed. More precisely, if at time t the process is in state i, $0 < i < l$, after an exponentially distributed length of time it may move to either state $i+1$ (a "birth" has occurred) or state $i-1$ (a "death" has occurred). States 0 and l play the part of barriers and as such any one of them may be either absorbing or reflecting**). Setting $q(i, i+1) = b_i$, $0 \leqslant i < l$, and $q(i, i-1) = d_i$, $0 < i \leqslant l$, the intensity matrix $\mathbf{Q}$ of the process can be written as

$$\mathbf{Q} = \begin{array}{c} \\ 0 \\ 1 \\ \cdot \\ \cdot \\ \cdot \\ l-1 \\ l \end{array}\begin{array}{c} \begin{array}{ccccccc} 0 & 1 & 2 & \cdots & l-2 & l-1 & l \end{array} \\ \left(\begin{array}{ccccccc} -b_0 & b_0 & 0 & \cdots & 0 & 0 & 0 \\ d_1 & -(b_1+d_1) & b_1 & \cdots & 0 & 0 & 0 \\ \cdot & \cdot & \cdot & \cdots & \cdot & \cdot & \cdot \\ \cdot & \cdot & \cdot & \cdots & \cdot & \cdot & \cdot \\ \cdot & \cdot & \cdot & \cdots & \cdot & \cdot & \cdot \\ 0 & 0 & 0 & \cdots & d_{l-1} & -(d_{l-1}+b_{l-1}) & b_{l-1} \\ 0 & 0 & 0 & \cdots & 0 & d_l & -d_l \end{array}\right) \end{array}.$$

The b_i are called the *birth rates* while the d_i are called the *death rates* ***). The barrier $0(l)$ is *absorbing* or *reflecting* according as $b_0 = 0$ $(d_l = 0)$ or $b_0 \neq 0$ $(d_l \neq 0)$.

---

*) The *occupation time* $\int_0^t f(X(s))\,ds$ (i.e., the time spent in state k during the interval $[0, t]$) has been investigated by many authors. See e.g. [304, 309, and 363].

**) Of course, the state space of a birth and death process may be any finite set of consecutive integers. Our choice is matter of convenience.

***) In particular, if all the b_i are null the process is said to be a *pure death process* while if all the d_i are null the process is said to be a *pure birth process*.

If $b_i d_i > 0$, $0 < i < l$, the process is regular or absorbing according as $b_0 d_l \neq 0$ or $b_0 d_l = 0$.

Assuming that $b_i + d_i > 0$, $0 < i < l$, the transition matrix of the imbedded Markov chain can be written as

$$\mathbf{P}^* = \begin{array}{c} 0 \\ 1 \\ \cdot \\ \cdot \\ \cdot \\ l-1 \\ l \end{array} \overset{\begin{array}{ccccccc} 0 & 1 & 2 & \cdots\ l-2 & l-1 & l \end{array}}{\begin{pmatrix} r_0 & p_0 & 0 & \cdots & 0 & 0 & 0 \\ q_1 & 0 & p_1 & \cdots & 0 & 0 & 0 \\ \cdot & \cdot & \cdot & \cdots & \cdot & \cdot & \cdot \\ \cdot & \cdot & \cdot & \cdots & \cdot & \cdot & \cdot \\ \cdot & \cdot & \cdot & \cdots & \cdot & \cdot & \cdot \\ 0 & 0 & 0 & \cdots & q_{l-1} & 0 & p_{l-1} \\ 0 & 0 & 0 & \cdots & 0 & q_l & r_l \end{pmatrix}},$$

where

$$p_i = \frac{b_i}{b_i + d_i}, \quad q_i = \frac{d_i}{b_i + d_i}, \quad 0 < i < l,$$

and r_0 and r_l are either 0 or 1, $p_0 = 1 - r$, $q_l = 1 - r_l$ (compare with Example 2 from 2.2).

8.7.2. The particular form of the intensity matrix makes it possible to indicate an effective procedure for expressing the transition matrix function $\mathbf{P}(\cdot) = (p(\cdot, i, j))_{0 \leqslant i, j \leqslant l}$ of a birth and death process with positive parameters in terms of the exponentials $e^{\lambda_r t}$, $0 \leqslant r \leqslant l$, where the λ_r are the eigenvalues of $\mathbf{Q}$ (cf. 8.2.6). The results below are a special case of much more general results (see [188, 197]).

Consider the polynomials Q_i, $0 \leqslant i \leqslant l$, defined recursively as follows

$$\begin{aligned} Q_0(x) &= 1, \\ -xQ_0(x) &= -b_0 Q_0(x) + b_0 Q_1(x), \\ -xQ_i(x) &= d_i Q_{i-1}(x) - (d_i + b_i) Q_i(x) + b_i Q_{i+1}(x), \ 1 \leqslant i < l. \end{aligned}$$

It can be shown that the zeros of the Q_i are all real and positive and the zeros of Q_i and Q_{i+1} interlace. Next, the zeros $\lambda_1, \ldots, \lambda_l$ of the polynomial $(d_l + \lambda)\, Q_l(-\lambda) - d_l\, Q_{l-1}(-\lambda)$ are all real and negative except for $\lambda_1 = 0$. They are precisely the eigenvalues of $\mathbf{Q}$. There are real numbers $c_1 (=1), \ldots, c_l$ such that

$$\sum_{r=1}^{l} c_r Q_i(-\lambda_r)\, Q_j(-\lambda_r) = \delta(i, j)/\pi(j)$$

and

$$p(t, i, j) = \pi(j) \sum_{r=1}^{l} c_r e^{\lambda_r t} Q_i(-\lambda_r)\, Q_j(-\lambda_r)$$

for all $0 \leqslant i, j \leqslant l$, $t \geqslant 0$, where

$$\pi(0) = \frac{1}{1 + \sum_{r=1}^{l} (b_0 \dots b_{r-1})/(d_1 \dots d_r)},$$

$$\pi(j) = \frac{(b_0 \dots b_{j-1})/(d_1 \dots d_j)}{1 + \sum_{r=1}^{l} (b_0 \dots b_{r-1})/(d_1 \dots d_r)}, \quad 1 \leqslant j \leqslant l.$$

(Note that the above formula can be proved simply by verifying the backward Kolmogorov system.)

It is easy to verify that $\boldsymbol{\pi} = (\pi(i))_{0 \leqslant i \leqslant l}$ is the stationary distribution of the process.

8.7.3. The birth and death process was introduced in 1939 by W. Feller [98]. It has since been used to construct stochastic models for population growth, queues, inventories, epidemics, and several other situations.

We shall give a few examples. Many others are to be found in [25, 27, 102, 166, 232, 278, and 371].

Example 1 (*Continuous time Ehrenfest model*). There are m balls distributed between two urns. We say the system is in state i if and only if there are i balls in the first urn, $0 \leqslant i \leqslant m$. At random times separated by intervals which are independently exponentially distributed with parameter μm, a ball is chosen at random (the probability of choosing any one of the m balls is $1/m$), removed from its urn, and then placed in the first urn with probability p, or in the second urn with probability $q = 1 - p$ $(0 < p < 1)$. Therefore, given that a drawing is made when the system is in state i the probabilities are

$$\frac{i}{m} q, \quad \frac{i}{m} p + \frac{m-i}{m} q, \quad \frac{m-i}{m} p$$

that the subsequent state will be $i-1$, i, $i+1$, respectively. Let us define $X(t) =$ state of system at time $t \geqslant 0$. Since the probability of performing a drawing in any interval $(t, t+h)$ equals $\mu m h + o(h)$, we conclude that $(X(t))_{t \geqslant 0}$ is a birth and death process with state space $(0, 1, \dots, m)$, birth rates $b_i = \mu p(m-i)$, $0 \leqslant i < m$, and death rates $d_i = \mu q i$, $0 < i \leqslant m$.

Example 2 (*Gregarious behaviour of elephants* [150]). Assume a given area contains l animals grouped into herds. The chance that two herds amalgamate (= meet and join) increases with the number of herds, and we may and shall consider it to be proportional to the excess above unity of the number of herds. Therefore it will be assumed that the probability is $d(m-1)h + o(h)$ that an amalgamation occurs in the interval $(t, t+h)$ given that there are m herds at time t. Next, the chance that a herd of size l_i, $1 \leqslant i \leqslant m$, separates into two herds increases with l_i und we also may and shall consider it to be proportional to the excess above unity of the size of the herd. Therefore it will be assumed the probability that one of the herds existing at time t separates during the interval $(t, t+h)$, given that there are m herds at time t, is $b\sum_{i=1}^{m}(l_i - 1)\, h + o(h) = b(l-m)\, h + o(h)$.

Finally, assuming that the life time of any grouping into herds is exponentially distributed, one can assert that, setting $X(t)$ = number of herds at time t, the process $(X(t))_{t\geqslant 0}$ is a birth and death process with state space $(1, \ldots, l)$, birth rates $b_i = b(l-i)$, $1 \leqslant i < l$, and death rates $d_i = d(i-1)$, $1 < i \leqslant l$.

Example 3 (*The simple epidemic*). The simple epidemic describes the infection of a closed group of l susceptibles into which a infectives are introduced. This seems to be a reasonable approximation to the early stages of some (mild) upper respiratory infections in which there is no isolation of infectives. The basic assumption is that the chance of any susceptible's becoming infected in a short time interval is jointly proportional to the number of infectives in circulation and the length of the interval. Hence the chance of one new infection in the whole group in a short time interval is proportional to the product of the number of infectives and the number of susceptibles present as well as to the length of the interval. Then, setting $X(t)$ = number of susceptibles at time t and assuming that the time intervals between successive infections are exponentially distributed, the process $(X(t))_{t\geqslant 0}$ is a pure death process with state space $(0, 1, \ldots, l)$ and death rates $d_i = \beta(l-i+a)i$, $0 < i \leqslant l$.

8.8 EXTENDING THE CONCEPT OF A HOMOGENEOUS FINITE MARKOV PROCESS

8.8.1. As in the discrete parameter case (see 5.6.1), the concept of a homogeneous finite Markov process can be extended by dropping out various of the assumptions underlying it and replacing them by more general ones. We shall discuss here the following situations: 1) The state space S is no longer a finite set but a countably infinite one; 2) At the expense of the Markov property itself, the distributions of the sojourn times are no longer assumed to be exponential (see the constructive definition in 8.3.2).

8.8.2. Clearly, the considerations in 8.1.2 are also valid when S is a countably infinite set. Thus we may speak of a denumerable Markov process with given transition matrix function $\mathbf{P}(\cdot) = (p(\cdot, i, j))_{i,j\in S}$. Since Proposition 8.1 still holds for denumerable Markov processes (in this setting it is known as Lévy-Austin-Ornstein theorem), the classification of states via discrete skeletons for these processes can be set up and the discussion in 5.6.2 carried over.

It can be shown that when S is countably infinite the derivative $\mathbf{P}'(0) = \mathbf{Q} = (q(i,j))_{i,j \in S}$ also exists, but in this case, although the $q(i,j)$, $i \neq j$, are finite, some or all the $q(i,i)$ may be infinite and

$$\sum_{j \neq i} q(i,j) \leqslant -q(i,i) \leqslant \infty, \quad i \in S. \tag{8.12}$$

Next, the Kolmogorov systems are no longer necessarily valid (even when all the $q(i,i)$ are finite). In general, they should be replaced by the matrix inequalities $\mathbf{P}'(t) \geqslant \mathbf{Q}\mathbf{P}(t)$ and $\mathbf{P}'(t) \geqslant \mathbf{P}(t)\mathbf{Q}$.

The most unpleasant feature is that $\mathbf{Q}$ no longer determines $\mathbf{P}(\cdot)$ uniquely. It can be proved that if all the $q(i,i)$ are finite then there exists a transition matrix function $\mathbf{P}_{\min}(\cdot)$ (that may be substochastic, i.e., $\mathbf{P}_{\min}(\cdot)\,\mathbf{e} \leqslant \mathbf{e}$) such that $\mathbf{P}'_{\min}(0) = \mathbf{Q}$ and for which both Kolmogorov systems hold. In fact $\mathbf{P}_{\min}(\cdot)$ is given by the Feller representation (8.7) which, of course, makes sense for a countably infinite S. Whatever the transition matrix function $\mathbf{P}(\cdot)$ such that $\mathbf{P}'(0) = \mathbf{Q}$, one then has $\mathbf{P}(\cdot) \geqslant \mathbf{P}_{\min}(\cdot)$. For instance, if $\sum_{j \in S} q(i,j) = 0$, $i \in S$, and $\mathbf{P}_{\min}(\cdot)$ is substochastic then there exists infinitely many such $\mathbf{P}(\cdot)$.

The Markov process $X_{\min}$ corresponding to $\mathbf{P}_{\min}(\cdot)$ can be constructed as in 8.3.2. If $\mathbf{P}_{\min}(\cdot)$ is substochastic the series $\sum_{m \geqslant 0} \sigma_m$ of the sojourn times diverges with probability smaller than 1 and for $t \geqslant \sum_{m \geqslant 0} \sigma_m \neq \infty$ one defines $X_{\min}(t) = c$, where c is a fictitious state adjoined to S. All the processes with the same $\mathbf{Q}$ agree with $X_{\min}$ up to time $\sum_{m \geqslant 0} \sigma_m$.

Feller's *complete construction problem* consists of describing all the (perhaps substochastic) transition matrix functions $\mathbf{P}(\cdot)$ such that $\mathbf{P}'(0) = \mathbf{Q}$, where the entries of $\mathbf{Q}$ satisfy (8.12). It has been completely solved only for certain classes of matrices $\mathbf{Q}$. The most recent progress concerns the case where all the $q(i,i)$ are infinite. See [398].

The standard reference for denumerable Markov processes is the monograph [52]. See also [110] and [166], I, pp. 251—258.

8.8.3. In many applied problems, the hypothesis of exponential sojourn times implicit in Markov processes (see 8.3.1 and 8.3.2) is not satisfied. This suggests the introduction of a new class of stochastic processes, the semi-Markov processes, that generalizes both finite Markov chains and processes. Roughly speaking, a semi-Markov process moves among a finite or countably infinite set S of states with the successive states visited forming a Markov chain, and with the length of stay in a given state being random, with a distribution which may depend on the present state as well as on the one to be visited next. Formally, generalizing the procedure described in 8.3.2, let us define positive random variables σ_n (to be interpreted as the successive sojourn times) and S-valued random variables $X^*(n)$ (to be interpreted as the states successively visited), $n \geq 0$, as follows: $\sigma_0 \equiv 0$, $X^*(0) = i_0$ (arbitrarily given in S), and given the preceding choices, we chose σ_{n+1} and $X^*(n+1)$, $n \geq 0$, so that

$$\mathsf{P}(X^*(n+1) = j,\ \sigma_{n+1} < t \mid \sigma_l < t_l,\ X^*(l) = i_l,\ 0 \leq l \leq n)$$
$$= \mathsf{P}(X^*(n+1) = j,\ \sigma_{n+1} < t \mid X^*(n) = i_n) = F(i_n,\ j,\ t)$$

whatever $t \geq 0$, $t_l \geq 0$, $0 \leq l \leq n$, $i_1, \ldots, i_n \in S$. Here $\mathbf{F}(\cdot) = (F(i, j, \cdot))$ is a matrix function whose entries are nonnegative, non-decreasing functions defined on $[0, \infty)$ and satisfying $\mathbf{F}(0) = \mathbf{0}$, $\mathbf{F}(\infty)\,\mathbf{e} = \mathbf{e}$. Clearly, the Markov processes correspond to the choice

$$F(i,\ j,\ t) = p^*(i, j)(1 - \exp(-q(i)t)),\ i, j \in S,$$

(see 8.3.2). As in the Markovian case it is easy to show that $(X^*(n))_{n \geq 0}$ is a Markov chain with state space S and transition matrix $\mathbf{F}(\infty)$. Also, the sojourn times $\sigma_1, \sigma_2, \ldots$ are conditionally independent given the Markov chain $(X^*(n))_{n \geq 0}$.

Define

$$X(t) = X^*(n)$$

for $\sum_{m=0}^{n} \sigma_m \leq t < \sum_{m=0}^{n+1} \sigma_m$, $m \geq 0$. It can be shown as in the Markovian case that, when S is finite, $X(t)$ is defined for all $t \geq 0$. The process $(X(t))_{t \geq 0}$ so defined is called a *semi-Markov process* with *state space* S and *transition kernel* $\mathbf{F}(\cdot)$.

It is now clear that the novel feature of a semi-Markov process is the freedom allowed in the choice of the sojourn time distributions. Nevertheless, this destroys the Markov property which holds now only at the transition times σ_1, $\sigma_1 + \sigma_2$, $\sigma_1 + \sigma_2 + \sigma_3$, ...

The reader interested in the theory of semi-Markov processes and their very numerous applications may consult [55, 223, 224, 370, and 381].

8.8.4. It is quite natural to ask whether there is, in the continuous parameter case, any concept corresponding to multiple Markovian dependence (see 5.6.3). This is still an open problem. It is discussed in [144, 163, and 165].

8.9 NONHOMOGENEOUS MARKOV PROCESSES

8.9.1. A *nonhomogeneous* S-valued Markov process is a process $(X(t))_{t \geq 0}$ satisfying (8.1). The conditional probability $\mathsf{P}(X(t_{n+1}) = i_{n+1} \mid X(t_n) = i_n)$ is, in general, a function of t_n, i_n, t_{n+1}, and i_{n+1}, that is

$$\mathsf{P}(X(t_{n+1}) = i_{n+1} \mid X(t_n) = i_n) = p(t_n, i_n, t_{n+1}, i_{n+1}).$$

For any fixed $0 \leq s < t$ the matrix $\mathbf{P}(s, t) = (p(s, i, t, j))_{i, j \in S}$ is stochastic.

On account of the equations

$$\mathsf{P}(X(t) = j \mid X(s) = i) = \sum_{k \in S} \mathsf{P}(X(t) = j,\ X(u) = k \mid X(s) = i)$$

$$= \sum_{k \in S} \mathsf{P}(X(t) = j \mid X(u) = k)\, \mathsf{P}(X(u) = k \mid X(s) = i)$$

we must have

$$p(s, i, t, j) = \sum_{k \in S} p(s, i, u, k)\, p(u, k, t, j)$$

whatever $0 \leqslant s < u < t,\ i, j \in S$. In matrix notation this amounts to

$$\mathbf{P}(s, t) = \mathbf{P}(s, u)\, \mathbf{P}(u, t). \tag{8.13}$$

Putting $p(s, i, s, j) = \delta(i, j),\ s \geqslant 0,\ i, j \in S$, that is $\mathbf{P}(s, s) = \mathbf{I}$, equation (8.13) has to hold for $0 \leqslant s \leqslant u \leqslant t$. In what follows a function $\mathbf{P}(\cdot, \cdot)$ associating with any $0 \leqslant s \leqslant t$ a stochastic matrix $\mathbf{P}(s, t)$ such that (8.13) holds will be called a (*nonhomogeneous*) *transition matrix function*. Assuming the existence of a transition matrix function $\mathbf{P}(\cdot, \cdot)$ let us define analogously to 8.1.2

$$p_{t_1 \ldots t_n}(i_1, \ldots, i_n) = \sum_{i_0 \in S} p(i_0)\, p(0, i_0, t_1, i_1) \ldots p(t_{n-1}, i_{n-1}, t_n, i_n)$$

where $(p(i))_{i \in S}$ is a given probability distribution over S. From now on, what we have said in 8.1.2 about the existence of a homogeneous Markov process applies also to the nonhomogeneous case (of course, in the considerations there $\mathbf{P}(\cdot)$ should be replaced by $\mathbf{P}(\cdot, \cdot)$). The remarks in 8.1.3 also apply in the nonhomogeneous setting.

8.9.2. The problem of the existence of a transition matrix function $\mathbf{P}(\cdot, \cdot)$ is solved as follows. Consider a measurable *intensity matrix function* $\mathbf{Q}(\cdot) = (q(\cdot, i, j))_{i, j \in S}$. This means that for almost all $t \geqslant 0$*) the matrix $\mathbf{Q}(t) = (q(t, i, j))_{i\, j \in S}$ is an intensity matrix, i.e., $\mathbf{Q}(t) - (\mathbf{Q}(t))_{\text{dg}} \geqslant \mathbf{0}$, $(\mathbf{Q}(t))_{\text{dg}} \leqslant \mathbf{0}$, and $\mathbf{Q}(t)\, \mathbf{e} = \mathbf{0}$. Set $q(\cdot, i, i) = -q(\cdot, i)$ and assume that $\int_0^t q(u, i)\, du < \infty$ for all $t \geqslant 0,\ i \in S$. Under these assumptions there exists a unique transition matrix function $\mathbf{P}(\cdot, \cdot)$ such that

$$\lim_{\substack{h+h' \to 0+ \\ (h \equiv 0 \text{ or } h' \equiv 0)}} \frac{\mathbf{P}(t-h,\ t+h') - \mathbf{I}}{h + h'} = \mathbf{Q}\,(t) \tag{8.14}$$

for all $t \notin E$, where $E \subset [0, \infty)$ is a set of Lebesgue measure zero. **) Moreover, $\mathbf{P}(\cdot, \cdot)$ satisfies the integral matrix equations

$$\mathbf{P}(s, t) = \mathbf{I} + \int_s^t \mathbf{Q}(u)\, \mathbf{P}(u, t)\, du \tag{8.15}$$

---

*) The meaning of "almost all $t \geqslant 0$" is "all $t \geqslant 0$ except perhaps a set of Lebesgue measure 0".

**) If $\mathbf{Q}(\cdot)$ is continuous then E is the empty set.

and

$$\mathbf{P}(s, t) = \mathbf{I} + \int_s^t \mathbf{P}(s, u)\, \mathbf{Q}(u)\, du \tag{8.16}$$

for all $0 \leqslant s \leqslant t$. Hence $\mathbf{P}(s, \cdot)$ is absolutely continuous on $[s, \infty)$ for all $s \geqslant 0$ and $\mathbf{P}(\cdot, t)$ is absolutely continuous on $[0, t]$ for all $t \geqslant 0$. Then it follows from (8.14) that $\mathbf{P}(\cdot, \cdot)$ satisfies the matrix differential equations

$$\frac{\partial \mathbf{P}(s, t)}{\partial s} = -\mathbf{Q}(s)\, \mathbf{P}(s, t), \quad 0 \leqslant s \leqslant t, \ s \notin E,$$

$$\frac{\partial \mathbf{P}(s, t)}{\partial t} = \mathbf{P}(s, t)\, \mathbf{Q}(t), \qquad 0 \leqslant s \leqslant t, \ t \notin E,$$

known as the "backward" and "forward" Kolmogorov equations, respectively.

The analogue (due to B. Hostinsky [153]) of the representation formula (8.6) is obtained from (8.15) and (8.16), and reads as follows

$$\mathbf{P}(s, t) = \mathbf{I} + \int_s^t \mathbf{Q}(u)\, du + \sum_{n \geqslant 2} \int_s^t ds_1 \int_{s_1}^t ds_2 \ldots \int_{s_{n-1}}^t \mathbf{Q}(s_1) \ldots \mathbf{Q}(s_n)\, ds_n$$

$$= \mathbf{I} + \int_s^t \mathbf{Q}(u)\, du + \sum_{n \geqslant 2} \int_s^t dt_1 \int_s^{t_1} dt_2 \ldots \int_s^{t_{n-1}} \mathbf{Q}(t_n) \ldots \mathbf{Q}(t_1)\, dt_n$$

for all $0 \leqslant s \leqslant t$.

The corresponding *Feller representation* of $\mathbf{P}(\cdot, \cdot)$ is as follows. Consider the diagonal matrix

$$\psi(s, t) = \left(\exp\left(-\int_0^t q(u, i)\, du \right) \delta(i, j) \right)_{i, j \in S}, \quad 0 \leqslant s \leqslant t.$$

Then $\mathbf{P}(s, t) = \sum_{n \geqslant 0} \mathbf{P}_n(s, t)$, $0 \leqslant s \leqslant t$, where $\mathbf{P}_0 = \psi$ and

$$\mathbf{P}_{n+1}(s, t) = \int_s^t \psi(s, u) (\mathbf{Q}(u) - (\mathbf{Q}(u))_{dg})\, \mathbf{P}_n(u, t)\, du, \ n \geqslant 0,$$

or, alternatively,

$$\mathbf{P}_{n+1}(s, t) = \int_s^t \mathbf{P}_n(s, u) (\mathbf{Q}(u) - (\mathbf{Q}(u))_{dg})\, \psi(u, t)\, du, \ n \geqslant 0.$$

For the proof of the above assertions the reader is referred to [166], I, pp. 206—218 and 231—233.

Remark. The problem of determining $\mathbf{P}(\cdot, \cdot)$ from $\mathbf{Q}(\cdot)$ is still open when the condition $\int_0^t q(u, i)\, du < \infty$, $t \geqslant 0$, $i \in S$, does not hold.

8.9.3. Now let us assume, conversely, that $\mathbf{P}(\cdot,\cdot)$ is a transition matrix function such that $\mathbf{P}(s,\cdot)$ is absolutely continuous on $[s, \infty)$ for all $s \geq 0$. Under these assumptions R. L. Dobrušin [77] proved that for all $0 \leq s < t$ the limit

$$\lim_{\max(t_l - t_{l-1}) \to 0} \sum_{l=1}^{n} \mathbf{P}(t_{l-1}, t_l) - \mathbf{I}) = \mathbf{A}(s, t),$$

where $s = t_0 < t_1 < \ldots < t_n = t$, exists and can be written as

$$\mathbf{A}(s, t) = \int_s^t \mathbf{Q}(u)\, du,$$

where $\mathbf{Q}(\cdot)$ is an intensity matrix function. Moreover, $\mathbf{Q}(\cdot)$ satisfies equation (8.14). Hence on account of 8.9.2 the transition matrix function $\mathbf{P}(\cdot,\cdot)$ must satisfy equations (8.15) and (8.16) as well as the Kolmogorov equations.

8.9.4. There are transition matrix functions $\mathbf{P}(\cdot,\cdot)$ which need not be absolutely continuous in the first or the second variable. See, e.g., [108], pp. 203 and 242. M. Fréchet ([108], pp. 219—248) determined the general solution of (8.13) which is continuous in both variables. For a more recent treatment see J. Aczél [1].

G. Goodman [125] raised the problem of whether, given a transition matrix function $\mathbf{P}(\cdot,\cdot)$ which is continuous in both variables, it is possible to introduce a new time scale [i.e., to make a substitution $s = \varphi(\tilde{s})$, $t = \varphi(\tilde{t})$, where φ is a continuous strictly increasing function on $[0, \infty)$ such that $\varphi(0) = 0$] with the effect of rendering

$$\tilde{\mathbf{P}}(\tilde{s}, \tilde{t}) = \mathbf{P}(\varphi(\tilde{s}), \varphi(\tilde{t}))$$

absolutely continuous in each of the new variables $\tilde{s}$ and $\tilde{t}$.

Assuming that $\mathbf{P}(s, t) = \mathbf{I}$ if and only if $s = t$ the answer is affirmative and an admissible function φ is given by

$$\varphi(\cdot) = -\log \det \mathbf{P}(0,\cdot).$$

(In fact, in this case the entries of $\mathbf{P}(\cdot,\cdot)$ are rendered Lipschitzian in each variable.) Then the intensity matrix function $\tilde{\mathbf{Q}}(\cdot)$ of $\tilde{\mathbf{P}}(\cdot,\cdot)$ satisfies $\mathbf{e}'(\tilde{\mathbf{Q}}(t))_{\mathrm{dg}}\mathbf{e} = -1$ for almost all $\tilde{t} \in [0, \varphi^{-1}(\infty))$. The new time scale has an interesting probabilistic interpretation. It is to be found in [125], pp. 177—179.

8.9.5. For further considerations concerning the possibility of determining nonhomogeneous finite Markov processes by means of suitable infinitesimal characteristics we refer the reader to [78, 351, and 352].

8.9.6. Let $\mathbf{Q}(\cdot)$ be an intensity matrix function fulfilling the conditions assumed in 8.9.2. A constructive definition of the nonhomogeneous Markov process associated with $\mathbf{Q}(\cdot)$ paralleling that given in 8.3.2 for the homogeneous case is available. Define

$$p_t^*(i,j) = \begin{cases} (1-\delta(i,j))\, q(t,i,j)/q(t,i), & \text{if } q(t,i) \neq 0, \\ \delta(i,j), & \text{if } q(t,i) = 0. \end{cases}$$

If the process starts in state i it remain there for a (possibly infinite) random length of time σ_1 with distribution function $1 - \exp\left(-\int_0^t q(u,i)\,du\right)$, $t \geqslant 0$ $\left[\text{this means that it remains forever in } i \text{ with probability } 1 - \exp\left(-\int_0^\infty q(u,i)\,du\right)\right]$ and then moves to state j with probability $p_{\sigma_1}^*(i,j)$. After hitting j, the process remains there for a (possibly infinite) random length of time σ_2 with distribution function $1 - \exp\left(-\int_0^t q(\sigma_1 + u, j)\,du\right)$, $t \geqslant 0$, and then moves to state k with probability $p_{\sigma_1+\sigma_2}^*(j,k)$. After hitting k, the process remains there for a (possibly infinite) random length of time σ_3 with distribution function $1 - \exp\left(-\int_0^t q(\sigma_1 + \sigma_2 + u, k)\,du\right)$, $t \geqslant 0$, and then moves to state l with probability $p_{\sigma_1+\sigma_2+\sigma_3}^*(k,l)$, and so on. (Of course, this can be put in formal terms.) It can be shown that the series $\sum_{m\geqslant 1} \sigma_m$ is divergent with probability 1 so that the process is defined for all $t \geqslant 0$. In the above context whatever $n \geqslant 0$ the conditional probability that the process performs exactly n changes of states in the interval $[s,t]$ and is in state j at time t, given that it was in state i at time s, is precisely the ijth entry of the matrix $\mathbf{P}_n(s,t)$ defined in 8.9.2. It should be noted that the sojourn times $\sigma_1, \sigma_1, \ldots$ are no longer conditionally independent given the successive states occupied by the process, and also that these no longer form a Markov chain (as was the case in 8.3.2).

For proofs the reader is referred to [166], I, p. 221—233.

Remark. In particular, the above construction applies to the nonhomogeneous Markov process $(\tilde{X}(\tilde{t}))_{0 \leqslant \tilde{t} \leqslant \varphi^{-1}(\infty)}$ with transition matrix function $\tilde{\mathbf{P}}(\cdot,\cdot)$ (see 8.9.4). Hence the construction of the original process $(X(t))_{t\geqslant 0}$ (where $X(t) = \tilde{X}(\varphi^{-1}(t))$, $t \geqslant 0$) whose transition matrix function $\mathbf{P}(\cdot,\cdot)$ was only assumed to be continuous in both variables, is obtained at once.

8.9.7. Extensions similar to those discussed in 8.8 have been considered for the nonhomogeneous case, too. See [126 and 169].

EXERCISES

8.1 Prove that the nonnull eigenvalues of an intensity matrix have negative real parts.

(H i n t: set $a = \max_{i \in S} q(i)$ and consider the stochastic matrix $\mathbf{P} = \mathbf{I} + a^{-1}\mathbf{Q}$. Then the eigenvalues λ of $\mathbf{Q}$ and λ' of $\mathbf{P}$ are connected by the equation $\lambda = a(\lambda' - 1)$. Now use the fact that $|\lambda'| \leqslant 1$ — see 1.11.2.)

8.2 *Quasi-stationary distributions* [69]. Consider an absorbing Markov process. Assume that the matrix $\mathbf{T}'(0) = (q(i, j))_{i, j \in T}$ is irreducible. Show that $\mathbf{T}'(0)$ has an eigenvalue λ_1 with maximal real part which is real, negative, and of (algebraic) multiplicity 1. Corresponding to it there are positive left and right eigenvectors $\mathbf{u} = (u(i))_{i \in T}$ and $\mathbf{v} = (v(i))_{i \in T}$ (choose them such that $\mathbf{u}'\mathbf{e} = 1$ and $\mathbf{u}'\mathbf{v} = 1$). Prove that

$$\mathbf{T}(t) = \exp(\mathbf{T}'(0)t) = e^{t\lambda_1}\, \mathbf{v}\mathbf{u}' + o(e^{ct}),$$

where $c < \lambda_1$. Deduce that whatever the probability distribution $\mathbf{p}$ over T

$$\lim_{t \to \infty} \mathsf{P}_{\mathbf{p}}(X(t) = j \mid \nu > t) = u(j), \qquad j \in T,$$

$$\lim_{s \to \infty} \lim_{t \to \infty} \mathsf{P}_{\mathbf{p}}(X(s) = j \mid \nu > t) = v(j)\, u(j), \quad j \in T.$$

Interpret the probability distribution $(v(i)u(i))_{i \in T}$ in a manner analogous to 3.4.4.

(H i n t: set $a = \max_{i \in T} q(i)$ and consider the nonnegative matrix $\mathbf{T} = \mathbf{I} + a^{-1}\, \mathbf{T}'(0)$ for which $\mathbf{T}\mathbf{e} \leqslant \mathbf{e}$. Now the existence of λ_1 and of the eigenvectors $\mathbf{u}$ and $\mathbf{v}$ follows from 3.4.2.)

8.3 Show that the transition matrix function of a two state Markov process with intensity matrix

$$\mathbf{Q} = \begin{pmatrix} -a & a \\ b & -b \end{pmatrix},$$

where $a,\ b \geqslant 0,\ a + b > 0$, is given by

$$\mathbf{P}(t) = \frac{1}{a + b} \begin{pmatrix} b + a \exp(-a-b)t & a - a \exp(-a-b)t \\ b - b \exp(-a-b)t & a + b \exp(-a-b)t \end{pmatrix}.$$

Compute the fundamental matrix of a regular two state Markov process.

8.4 Consider a two state Markov process with intensity matrix

$$\mathbf{Q} = \begin{pmatrix} -u & u \\ v & -v \end{pmatrix}.$$

Prove that the distribution function of the occupation time of state 1 during the time interval [0, 1] (see 8.6.6) coincides with the latter limit distribution function occurring in 5.3.2. (Hint: [304].)

8.5 Show that the fundamental matrix $\mathbf{Z}$ of a regular or indecomposable Markov process satisfies the equations $\mathbf{QZ} = \mathbf{ZQ}$, $\boldsymbol{\pi}'\mathbf{Z} = \boldsymbol{\pi}'$ $\mathbf{Ze} = \mathbf{e}$, $\mathbf{\Pi} - \mathbf{QZ} = \mathbf{I}$.

8.6 Consider a regular or indecomposable Markov process. The notation being that in 8.6.1 prove [3] that

$$\pi(i) = \left(\pi^*(i) \prod_{k \neq i} q(k)\right) \Big/ \left(\sum_{l \in S} \pi^*(l) \prod_{k \neq l} q(k)\right), \; i \in S.$$

Show that $\pi = \pi^*$ if and only if $q(i) = \sum_{j \in S} \pi(j) q(j)$ for all $i \in S$ such that $\pi(i) \neq 0$.

8.7 Consider a regular Markov process. Set

$$\mathbf{W} = (\mathsf{E}_i(\tau_j^2))_{i,j \in S} \text{ (clearly, } \mathsf{E}_i(\tau_i^2) = 0, \; i \in S).$$

Prove [196] that

$$\mathbf{W} = \mathbf{M}(2\mathbf{Z}_{dg}\mathbf{\Pi}_{dg}^{-1} - \mathbf{I}) + 2(\mathbf{ZM} - \mathbf{E}(\mathbf{ZM})_{dg}).$$

(Hint: apply the method used in 8.6.3 taking into account the corresponding result (see 4.3.2) for regular Markov chains.)

8.8 *Continuation.* Prove [196] that the second moment of the time taken between an entry into state i and the next return to i is given by

$$r_2(i) = \frac{2}{\pi^2(i)\, q^2(i)} [(z(i, i) - \pi(i))\, q(i) + \pi(i)].$$

(Hint: 8.6.5.)

8.9 *First passage time distributions.* Consider a Markov process with intensity matrix $\mathbf{Q}$. Let $F_j(.; i_1, \ldots, i_n)$, $i_1 \neq i_2, \ldots, i_{n-1} \neq i_n$, $i_r \neq j$, $1 \leqslant r \leqslant n$, denote the distribution function of the sum of n independent exponential random variables with parameters $q(i_1), \ldots, q(i_n)$. Show that for all $i \neq j$

$$\mathsf{P}_i(\tau_j < t) = p^*(i, j)\, F_j(t; i)$$
$$+ \sum_{n \geqslant 2} \sum_{i_1, \ldots, i_{n-1}} p^*(i, i_1) \ldots p^*(i_{n-1}, j)\, F_j(t; i, i_1, \ldots, i_{n-1}).$$

Prove that the derivative

$$f(t, i, j) = \frac{d}{dt} \mathsf{P}_i(\tau_j < t)$$

exists, is continuous and satisfies the *first entrance formula*

$$p(t, i, j) = \int_0^t f(s, i, j)\, p(t-s, j, j)\, ds.$$

Deduce that the Laplace transforms $L_f^{ij}(\lambda) = \int_0^\infty e^{-\lambda u} f(u, i, j) du$, $i \neq j$ and $L_p^{ij}(\lambda) = \int_0^\infty e^{-\lambda u} p(u, i, j)\, du$, $i, j \in S$, are connected by the equation

$$L_p^{ij}(\lambda) = L_f^{ij}(\lambda)\, L_p^{jj}(\lambda), \quad i \neq j$$

(compare with Exercise 2.7). (Hint: 8.3.2 and 8.3.3.)

8.10 *Continuation.* Prove ([189], p. 1131) that for a birth and death process (see 8.7)

$$L_f^{ij}(\lambda) = \begin{cases} Q_i(-\lambda)/Q_j(-\lambda), & \text{if } j > i, \\ Q^*_{l-i}(-\lambda)/Q^*_{l-j}(-\lambda), & \text{if } j < i, \end{cases}$$

where the Q_r^* are the polynomials corresponding to the birth and death process with parameters $b_k^* = d_{l-k}$, $d_k^* = b_{l-k}$. Deduce that

$$L_f^{0j}(\lambda) = \frac{(-1)^j\, b_0 \ldots b_{j-1}}{\prod_{k=1}^{j} (\lambda + x_k)},$$

where $x_1, \ldots, x_j$ are the roots of Q_j. Hence

$$f(t, 0, j) = (-1)^j\, b_0 \ldots b_{j-1} \sum_{k=1}^{j} \frac{\exp(-x_k t)}{\prod_{l \neq k} (-x_k + x_l)}.$$

8.11 *Reversible Markov processes.* Consider a regular Markov process with stationary distribution π. Show that whatever $s \leqslant t$

$$\mathsf{P}_\pi(X(s) = j \mid X(t) = i) = \frac{\pi(j)}{\pi(i)}\, p(t - s, j, i), \quad i, j \in S.$$

A regular Markov process is said to be *reversible* (compare with 4.5.1) if and only if

$$p(t, i, j) = \frac{\pi(j)}{\pi(i)}\, p(t, j, i), \quad t \geqslant 0,\ i, j \in S.$$

Prove that reversibility holds if and only if $(\mathbf{\Pi}_{\mathrm{dg}}\mathbf{Q})' = \mathbf{\Pi}_{\mathrm{dg}}\mathbf{Q}$. Deduce that reversibility holds if and only if the imbedded Markov chain is reversible. (Hint: show that the equality $(\mathbf{\Pi}_{\mathrm{dg}}\mathbf{Q})' = \mathbf{\Pi}_{\mathrm{dg}}\mathbf{Q}$ implies that $(\mathbf{\Pi}_{\mathrm{dg}}\mathbf{Q}^n)' = \mathbf{\Pi}_{\mathrm{dg}}\mathbf{Q}^n$ for all $n \geqslant 1$.)

8.12 *Continuation.* Prove that the eigenvalues of the intensity matrix of a reversible Markov process are real and the matrix is diagonalizable. (Hint: see Exercise 4.17.)

8.13 Show that the birth and death process considered in 8.7.1 is reversible.

8.14 Compute the transition matrix functions of the pure birth process and the pure death process. (Hint: use the first entrance formula — see Exercise 8.8 above.)

8.15 *Continuous time Ehrenfest model.* Prove [191] that the transition probabilities $p(t, i, j)$, $0 \leqslant i, j \leqslant m$, of the birth and death process associated with the model (see 8.7.3) are given by

$$p(t, i, j) = \binom{m}{j} p^j q^{m-j} \sum_{r=0}^{m} e^{-\mu r t} K_i(r)\, K_j(r) \left(\frac{p}{q}\right)^{m-r}$$

where the $K_i(r)$ are the Krawtchouk polynomials with parameters p and m defined by

$$K_i(r) = \frac{1}{\binom{m}{i}} \sum_{u=0}^{i} (-1)^u \binom{r}{u} \binom{m-r}{i-u} \left(\frac{q}{p}\right)^i, \quad 0 \leqslant r \leqslant m.$$

Deduce that

$$p(t, i, j) = \binom{m}{j} p^j q^{m-j} + O(e^{-\mu t}), \quad 0 \leqslant i, j \leqslant m,$$

as $t \to \infty$.

8.16 *Continuation.* Show that the birth and death process occurring in Example 2 from 8.7.3 is the one associated with the continuous time Ehrenfest model with parameters $m = l - 1$, $p = b/(b + d)$, $\mu = b + d$.

8.17 *The repairman problem* [10]. There are n identical machines which operate independently. Assume that they are backed up by m spare machines of the same type and that there is a repair facility capable of repairing s machines simultaneously. When one of the operating machines fails it is sent to the repair facility and replaced by one of the spare machines. The operating machines are assumed to break down according to some failure distribution F. If fewer than s machines are being repaired, the failed machine enters the repair facility. Repair times on different machines are assumed to be independent, identically distributed random variables with distribution G. On the other hand, if s machines are being repaired, the failed machine joins a queue, and whenever a repairman becomes free, one of the machines in the queue immediately takes him over. Let us define $X(t)$=number of machines undergoing or waiting for repair at time $t \geqslant 0$.

i) Prove that if F and G are exponential with parameters λ and μ, respectively, then $(X(t))_{t \geqslant 0}$ is a birth and death process with state space $(0, 1, \ldots, m + n)$, birth rates $b_i = n\lambda$ or $(m + n - i)\lambda$ according as

$0 \leqslant i \leqslant m$ or $m < i < m + n$, and death rates $d_i = i\mu$ or $s\mu$ according as $0 < i \leqslant \min(s, m + n)$ or $\min(s, m + n) < i \leqslant m + n$.

ii) Find the stationary distribution $(\pi(i))_{0 \leqslant i \leqslant m+n}$ of the process. Derive the asymptotic probability of total congestion of the repair facility $\left(= \sum_{i=s}^{m+n} \pi(i) \right)$ and the asymptotic probability of total failure $(= \pi(m + n))$.

iii) Show that in the case $m = 0$, $n = s$, the model is identical to the continuous time Ehrenfest model.

8.18 *A queueing system with finite waiting room* [371]. Consider a queueing system consisting of s servers and a waiting room of size r. Customers requiring service arrive in the system. Whenever there is at least one free server an incoming customer receives service immediately, while if all the servers are engaged he cannot be served and is either rejected or put in a queue in the waiting room according as this is or is not full. Whenever a server becomes free, one of the customers from the queue immediately takes him over. Assume that: a) Customers arrive at random times separated by intervals which are independent exponential random variables with parameter λ; b) Service times are exponential random variables with parameter μ. They are independent of each other and also of the arrivals of customers; c) Customers in the queue may leave the system only after being served. Let us define $X(t)$ = number of customers in the system (including those served and waiting) at time $t \geqslant 0$.

i) Prove that $(X(t))_{t \geqslant 0}$ is a birth and death process with state space $(0, 1, \ldots, r + s)$, birth rates $b_i = \lambda$, $0 \leqslant i < r + s$, and death rates $d_i = i\mu$ or $s\mu$ according as $0 < i \leqslant s - 1$ or $s \leqslant i \leqslant r + s$.

ii) Prove Erlang's formula for the stationary distribution

$$\pi(j) = \begin{cases} \pi(0)\, \rho^j \dfrac{s^j}{j\,!}, & 1 \leqslant j < s, \\[2ex] \pi(0)\, \rho^j \dfrac{s^s}{s\,!}, & s \leqslant j \leqslant r + s, \end{cases}$$

where $\rho = \lambda/(s\mu)$ is the so-called traffic intensity, and

$$\pi(0) = \begin{cases} \left(\displaystyle\sum_{i=0}^{s-1} \frac{(\rho s)^i}{i\,!} + \rho^s \frac{s^s}{s\,!} \frac{1 - \rho^{r+1}}{1 - \rho} \right)^{-1}, & \text{if } \rho \neq 1, \\[3ex] \left(\displaystyle\sum_{i=0}^{s} \frac{s^i}{i\,!} + r \frac{s^s}{s\,!} \right)^{-1}, & \text{if } \rho = 1. \end{cases}$$

Remark. In the case where $r = 0$ the system can be interpreted as an inventory model with time lag (see [342]).

8.19 *Groupable Markov processes.* Consider a Markov process with state space S. Let $S = S_1 \cup \dots \cup S_q$ be a partition of S into pairwise disjoint sets. Define a stochastic process $(\hat{Y}(t))_{t\geqslant 0}$ by

$$Y(t) = \hat{k} \text{ if and only if } X(t) \in S_k,$$

$t \geqslant 0$, $1 \leqslant k \leqslant q$. Prove that $(Y(t))_{t\geqslant 0}$ is a homogeneous Markov process if and only if $\mathbf{CBQC} = \mathbf{QC}$, where the matrices $\mathbf{B}$ and $\mathbf{C}$ are defined in 5.4.2. The transition matrix function of the new process is then given by $\hat{\mathbf{P}}(\cdot) = \mathbf{BP}(\cdot)\mathbf{C}$. What can be said about the relationship between groupability of the original process and groupability of the imbedded Markov chain? (Hint: show that the equality $\mathbf{CBQC} = \mathbf{QC}$ implies that $\mathbf{CBQ}^n\mathbf{C} = \mathbf{Q}^n\mathbf{C}$ for all $n \geqslant 1$.)

8.20 Consider a finite Markov process (either homogeneous or nonhomogeneous). Let $\mathcal{X}^t$ denote the smallest σ-algebra containing all the random events of the form

$$\{X(t_0) = i_0, \dots, X(t_n) = i_n\},$$

where $t_0 = t < t_1 < \dots < t_n$, $i_0, \dots, i_n \in S$, $n \geqslant 1$. Define the tail σ-algebra $\mathcal{T}$ of the process as the intersection of the σ-algebras $\mathcal{X}^t$, $t \geqslant 0$. Prove [57] that $\mathcal{T}$ coincides with the σ-algebra of any Markov chain $(X(s_n))_{n\geqslant 0}$, where $0 = s_0 < s_1 < \dots$. Deduce the results corresponding to Theorems 5.5, 5.7, and 7.17.

Historical Notes

The concept of Markov dependence first*) appears in an explicit form in a paper [258] of the Russian mathematician A. A. Markov (1856 – 1922). In a series of papers (see [260]) starting with the one quoted above he studied various properties of sequences of dependent random variables which in his honour are nowadays called finite Markov chains. Markov's intention was to generalize classical properties of sequences of independent random variables to sequences not fulfilling the independence assumption. Clearly, the extension of properties holding for independent random variables would not be possible or would be quite limited in the case of too general a concept of dependence. On the other hand, the concept had to be "natural", that is effectively detectable in applications as varied as possible. The Markovian concept of dependence fulfils all these requirements. As to applications, Markov himself studied the succession of consonants and vowels in the literary works "Evgeni Onegin" by Puškin and "Detskie gody Bagrova-vnuka" by Aksakov (see [259], pp. 566 – 581) and concluded that it may be considered as a homogeneous two state simple Markov chain**).

Almost at the same time, studying the problem of shuffling the card deck, the French mathematician Henri Poincaré came across sequences of random variables which are in fact Markov chains with doubly stochastic transition matrices. Nevertheless, Poincaré did not undertake a systematic study of those sequences so that the recognition accorded to Markov is entirely justified.

At present there is a huge literature on finite Markov chains (the bibliography of this book represents a very small sample from what has been written on the subject). We limit ourselves here to mentioning only books or monographs associated with significant developments in the history of finite Markov chains. These are the works by A. A. Markov [259], B. Hostinsky [152], R. von Mises [277], M. Fréchet [108], O. Onicescu and G. Mihoc [295 and 297], S. N. Bernstein [21], V. I. Romanovski [324], W. Feller [102], J. G. Kemeny and J. L. Snell [195], and K. L. Chung [52].

The first to use the Perron-Frobenius theorems in the study of Markov chains was R. von Mises. He was independently followed by V. I. Romanovski.

The method used in this book (for the homogeneous case) which does not appeal to the theory of nonnegative matrices, was devised by J. Hadamard and developed by A. N. Kolmogorov and W. Doeblin.

The first results for the nonhomogeneous case were obtained by Markov himself. Important contributions have been subsequently made by S. N. Bernstein, W. Doeblin, Ju. V. Linnik, and R. L. Dobrušin.

The important early contributors to the continuous parameter case were A. N. Kolmogorov (who was the first to define the concept of continuous parameter Markovian dependence), B. Pospišil, W. Feller, and W. Doeblin.

---

*) Sequences of random variables which are Markov chains in the presently accepted sense were considered long before this. An example might be the Markov chain associated with the Bernoulli model (Example 2 from 2.2). Other examples are discussed in [146] (with reference to [28]) and [249].

**) It seems that a more adequate model is a double Markov chain (see [324], p. 419). For recent applications in linguistics the reader should consult [117].

In this historical review we want to stress the importance of the contributions to the theory of finite Markov chains and their generalizations made by the founders of the Romanian probability school, Octav Onicescu and Gheorghe Mihoc. Their papers outlined the definitive aspect of the asymptotic theory of partial sums associated with homogeneous finite Markov chains. At the same time, they were the initiators of research concerning the extension of the concept of Markovian dependence.

Bibliography

1. Aczel, J. *Lectures on functional equations and their applications.* New York, Academic Press, 1966.
2. Adke, S. R. *Some finite time results for finite Markov chains.* J. Indian Statist. Assoc. **2** (1964), 75–79.
3. Andrieu, Colette; Theodorescu, R. *Sur les probabilités absolues stationnaires pour des processus de Markov à temps continu.* Atti Accad. Naz. Lincei Rend. Cl. Sci. Fis. Mat. Natur. (8) **54** (1973), 892–897.
4. Anisimov, V. V. *Multidimensional limit theorems for finite state Markov chains.* Dokl. Akad. Nauk SSSR **204** (1972), 519–521 (Russian).
5. Arrow, K. J.; Karlin, S.; Scarf, H. *Studies in the mathematical theory of inventory and production.* Stanford, Calif., Stanford Univ. Press, 1958.
6. Arrow, K. J.; Karlin, S.; Scarf, H. (Eds.) *Studies in applied probability and management science.* Stanford, Calif., Stanford Univ. Press, 1962.
7. Arrow, K. J.; Karlin, S.; Suppes, P. (Eds.) *Mathematical methods in the social science, 1959.* Stanford, Calif., Stanford Univ. Press, 1960.
8. Ash, R. B. *Basic probability theory.* New York, Wiley, 1970.
9. Atkinson, R. C.; Estes, W. K. *Stimulus sampling theory.* In: Luce, R. D.; Bush, R. R.; Galanter, E. (Eds.). *Handbook of mathematical psychology, Vol. II.* New York, Wiley, 1963, pp. 121–268.
10. Barlow, R. E. *Repairman problems.* In [6], pp. 18–33.
11. Barlow, R. E. *Applications of semi-Markov processes to counter problems.* In [6], pp. 34–62.
12. Barlow, R. E.; Proschan, F. *Mathematical theory of reliability.* New York, Wiley, 1965.
13. Barnett, V. D. *The joint distribution of occupation totals for a simple random walk.* J. Austral. Math. Soc. **4** (1964), 518–528.
14. Bartholomew, D. J. *Stochastic models for social processes, 2nd Ed.* New York, Wiley, 1973.
15. Bartlett, M. S. *On theoretical models for competitive and predatory biological systems.* Biometrika **44** (1957), 27–42.
16. Bartoszyński, R. *Model of circulation and exchange of banknotes.* Zastos. Mat. **13** (1972/73), 1–22.
17. Beljaev, Ju. K.; Rykova, L. V. *Kolmogorov's nonparametric test for samples from finite populations.* Dokl. Akad. Nauk SSSR **210** (1973), 1261–1264 (Russian).
18. Bellman, R. *Introduction to matrix analysis.* New York, Mc. Graw-Hill, 1960.
19. Bellman, R.; Harris, T. *Recurrence times for the Ehrenfest model.* Pacific J. Math. **1** (1951), 179–193.
20. Berman, Elizabeth *Regular stochastic matrices and digraphs.* J. Appl. Probability **10** (1973), 241–243.
21. Bernštein, S. N. *Probability theory, 4th Ed.* Moscow–Leningrad, Gostehizdat., 1946 (Russian).
22. Bernštein, S. N. *Complete works, Vol. IV.* Moscow, Nauka, 1964 (Russian).
23. Beyer, W. A.; Waterman, M. S. *Symmetries for conditioned ruin problems.* Math. Mag. **50** (1977), 42–45.
24. Bežaeva, Z. I. *Ergodic properties of conditional Markov chains.* Teor. Verojatnost. i. Primenen. **19** (1974), 547–557 (Russian).
25. Bharucha-Reid, A. T. *Elements of the theory of Markov processes and their applications.* New York, Mc. Graw-Hill, 1960.

26. Bhat, B. B. *Some properties of regular Markov chains.* Ann. Math. Statist. **32** (1961), 59–71.
27. Bhat, U. N. *Elements of applied stochastic processes.* New York, Wiley, 1972.
28. Bienaymé, I. J. *De la loi de multiplication et de la durée des familles.* Soc. Philomath. Paris Extraits, Sér. 5(1845), 37–39. (Reprinted in English as an Appendix to [201].)
29. Billingsley, P. *Statistical methods in Markov chains.* Ann. Math. Statist. **32** (1961), 12–40.
30. Bily, J. *Eine Markoffsche Kette, die zu einer Faltung zweier binomischen Verteilungen und zu kumulierten binomischen Verteilung führt.* Časopis Pěst. Mat. **84** (1959), 327–334.
31. Bithell, J. F. *Some generalized Markov chain occupancy processes and their application to hospital admission systems.* Rev. Inst. Internat. Statist. **39** (1971), 170–184.
32. Blackwell, D. *Finite nonhomogeneous Markov chains.* Ann. of Math. **46** (1945), 594–599.
33. Blackwell, D.; Freedman, D. *The tail σ-field of a Markov chain and a theorem of Orey.* Ann. Math. Statist. **35** (1964), 1921–1925.
34. Blagoveščenski, Ju. N. *On ergodicity for a series scheme of Markov chains with a finite number of states and discrete time.* Izv. Akad. Nauk UzSSR. Ser. Fiz–Mat. Nauk **4** (1960) no. 3, 7–15 (Russian).
35. Blasi, A. *On a random walk between a reflecting and an absorbing barrier.* Ann. Probability **4** (1976), 695–696.
36. Blumen, Isadore; Kogan, M.; McCarthy, P. J. *The industrial mobility of labor as a probability process.* Ithaca, N. Y., Cornell Univ. Press, 1955.
37. Bosch, K. *Notwendige und hinreichende Bedingungen dafür, dass eine Funktion einer homogenen Markoffschen Kette Markoffsch ist.* Z. Wahrscheinlichkeitstheorie verw. Gebiete **31** (1975), 199–202.
38. Bosso, J. A.; Sorrarain, O. M.; Favret, E. E. A. *Application of finite absorbing Markov chains to sib mating populations with selection.* Biometrics **25** (1969), 17–26.
39. Bowerman, B.; David, H. T.; Isaacson, D. *The convergence of Cesaro averages for certain nonstationary Markov chains.* Stochastic Processes Appl. **5** (1977), 221–230.
40. Breny, H. *Cheminements conditionnels de chaînes de Markov absorbantes.* Ann. Soc. Sci. Bruxelles Sér. I **76** (1962), 81–87.
41. Brookner, E. *Recurrent events in a Markov chain.* Information and Control **9** (1966), 215–229.
42. Bui, Trong Lieu *Estimations pour des processus de Markov.* Publ. Inst. Statist. Univ. Paris **11** (1962), 71–188.
43. Buiculescu, Mioara *Quasi-stationary distributions for continuous time Markov processes with a denumerable set of states.* Rev. Roumaine Math. Pures Appl. **17** (1972), 1013–1023.
44. Buiculescu, Mioara, *Limiting conditional probabilities for denumerable Markov chains.* In *Proc. Fourth Conf. Probability Theory (Braşov, 1971).* Bucharest, Editura Acad. R.S.R., 1973, pp. 121–128.
45. Burke, C. J.; Rosenblatt, M. *A Markovian function of a Markov chain.* Ann. Math. Statist. **29** (1958), 1112–1122.
46. Cannings, C. *The latent roots of certain Markov chains arising in genetics: a new approach. I–II.* Advances in Appl. Probability **6** (1974), 260–290; **7** (1975), 264–282.
47. Caubet, J. P. *Semi-groups généralisés de matrices positives.* Ann. Inst. H. Poincaré Sect. B **1** (1964/65), 239–310.
48. Cavender, J. A. *Quasi-stationary distributions of birth-and-death processes.* Advances in Appl. Probability **10** (1978), 570–586.
49. Chia, A. B. *Spectral representations of multi-element pattern models.* J. Mathematical Psychology **7** (1970), 150–162.
50. Chiang, C. L. *A solution of Kolmogorov differential equations — a preliminary report.* Bull. Inst. Internat. Statist., **45** (1973), book 1, 264–270.
51. Chiang, C. L.; Raman, S. *On a solution of Kolmogorov differential equations.* In *Proc. Fourth Conf. Probability Theory (Braşov, 1971).* Bucharest, Editura Acad. R.S.R., 1973, pp. 129–136.

52. Chung, K. L. *Markov chains with stationary transition probabilities, 2nd Ed.* Berlin, Springer, 1967.
53. Chung, K.L. *Elementary probability theory with stochastic processes.* Berlin, Springer, 1974.
54. Chung, K.L. *Approximation of a continuous parameter Markov chain by discrete skeletons.* Bull. London Math. Soc. **7** (1975), 71–76.
55. Çinlar, E. *Markov renewal theory: a survey.* Management Sci. **21** (1975), 727–752.
56. Cohn, H. *On the tail σ-algebra of the finite inhomogeneous Markov chains.* Ann. Math. Statist. **41** (1970), 2175–2176.
57. Cohn, H. *On the tail events of a Markov chain.* Z. Wahrscheinlichkeitstheorie verw. Gebiete **29** (1974), 65–72.
58. Cohn, H. *A ratio limit theorem for the finite nonhomogeneous Markov chains.* Israel J. Math. **19** (1974), 329–334.
59. Cohn, H. *Finite nonhomogeneous Markov chains: asymptotic behaviour.* Advances in Appl. Probability **8** (1976), 502–516.
60. Cohn, H. *Countable nonhomogeneous Markov chains: asymptotic behaviour.* Advances in Appl. Probability **9** (1977), 542–552.
61. Cohn, H. *On a paper by Doeblin on nonhomogeneous Markov chains.* Report No 7, 1978, Univ. of Melbourne.
62. Collins, L.; Drewett, R.; Ferguson, R. *Markov models in geography.* Statistician **23** (1974), 179–209.
63. Conlisk, J. *Interactive Markov chains.* J. Mathematical Sociology **4** (1976), 157–185.
64. Cuculescu, I. *A simple proof of the Perron formula.* An. Univ. C.I. Parhon Ser. Sti.-Nat. Mat.-Fiz. No. 25 (1960), 7–8 (Romanian).
65. Cullmann, G. *Initiation aux chaînes de Markov. Méthodes et applications.* Paris, Masson, 1975.
66. Cyert, R.; Davidson, H.J.; Thompson, G.L. *Estimation of the allowance for doubtful accounts by Markov chains.* Management Sci. **8** (1962), 287–303.
67. Darroch, J.N.; Morris, K.W. *Some passage-time generating functions for discrete-time and continuous-time finite Markov chains.* J. Appl. Probability **4** (1967), 496–507.
68. Darroch, J.N.; Seneta, E. *On quasi-stationary distributions in absorbing discrete-time finite Markov chains.* J. Appl. Probability **2** (1965), 88–100.
69. Darroch, J.N.; Seneta, E. *On quasi-stationary distribution in absorbing continuous-time finite Markov chains.* J. Appl. Probability **4** (1967), 192–196.
70. Darroch, J.N.; Whitford, H.J. *Exact fluctuation results for Markov-dependent coin-tossing.* J. Appl. Probability **9** (1972), 158–168.
71. Das, S.C. *A note on the structure of a certain stochastic model.* Sankhyā Ser. A **22** (1960), 345–350.
72. Davis, A.S. *Markov chains as random-input automata.* Amer. Math. Monthly **68** (1961), 264–267.
73. Decell, H. P.; Odell, P. L. *On the fixed point probability vector of regular or ergodic transition matrices.* J. Amer. Statict. Assoc. **62** (1967), 600–602.
74. Derman, C. *Finite state Markovian decision processes.* New York, Academic Press, 1970.
75. Dharmadhikari, S.W. *A characterization of a class of functions of finite Markov chains.* Ann. Math. Statist. **36** (1965), 524–528.
76. Dobrušin, R.L. *Limit theorems for two state Markov chains.* Izv. Akad. Nauk SSSR Ser. Mat. **17** (1953), 291–329 (Russian).
77. Dobrušin, R.L. *Generalization of Kolmogorov's equations for Markov processes with a finite number of possible states.* Mat. Sb. (N.S.) **33 (75)** (1953), 567–596 (Russian).
78. Dobrušin, R.L. *Conditions of regularity of Markov processes with a finite number of possible states.* Mat. Sb. (N.S.) **34 (76)** (1954), 541–556 (Russian).
79. Dobrušin, R.L. *The central limit theorem for nonhomogeneous Markov chains. I–II.* Teor. Verojatnost. i Primenen. **1** (1956), 72–88, 365–425 (Russian).
80. Doeblin, W. *Sur les propriétés asymptotiques de mouvements régis par certains types de chaînes simples.* Bull. Math. Soc. Roumaine Sci. **39** (1937) no. 1, 57–115; no. 2, 3–61.

81. Doeblin, W. *Le cas discontinu des probabilités en chaîne.* Publ. Fac. Sci. Univ. Masaryk **236** (1937), 3–13.
82. Doeblin, W. *Exposé de la theorie des chaînes simples constantes de Markoff à un nombre fini d'états.* Rev. Math. de l'Union Interbalkanique **2** (1938), 77–105.
83. Doeblin, W. *Sur l'équation matricielle $A^{(t+s)} = A^{(t)} A^{(s)}$ et ses applications aux probabilités en chaîne.* Bull. Sci. Math. (2) **62** (1938), 21–32; **64** (1940), 35–37.
84. Doeblin, W. *Sur certains mouvements aléatoires discontinus.* Skand. Aktuarietidskr. 22 (1939), 211–222.
85. Doeblin, W. *Eléments d'une théorie générale des chaînes simples constantes de Markoff.* Ann. École Norm. Sup. (3) **37** (1940), 61–111.
86. Doob, J.L. *Stochastic processes.* New York, Wiley, 1953.
87. Döring, U. *Ausführung der Feller'schen Konstruktion von Markoff-Prozessen mit abzähbarem Zustandsraum und Anwendungen auf Geburts und Todesprozesse.* Dissertation, Universität Karlsruhe, 1973.
88. Draper, J.C.; Nolin, J.H. *A Markov chain analysis of brand preferences.* J. Advertising Res. **4** (1964) no. 3, 33–38.
89. Dwass, M. *Simple random walk and rank order statistics.* Ann. Math. Statist. **38** (1967), 1042–1054.
90. Dym, H. *A note on limit theorems for the entropy of Markov chains.* Ann. Math. Statist. **37** (1966), 522–524.
91. Enns, E.G. *The distributions of a number of random variables associated with the absorbing Markov chain.* Austral. J. Statist. **14** (1972), 79–83.
92. Erikcson, R.V. *Functions of Markov chains.* Ann. Math. Statist. **41** (1970), 843–850.
93. Estes, W.K. *Component and pattern models with Markovian interpretations.* In Bush, R.R.; Estes, W.K. (Eds.). *Studies in mathematical learning theory.* Stanford, Calif., Stanford Univ. Press, 1959, pp. 9–52.
94. Estes, W.K.; Burke, C.J. *A theory of stimulus variability in learning.* Psychol. Rev. **60** (1953), 276–286.
95. Ewens, W.J. *The mean time for absorption in a process of genetic type.* J. Austral. Math. Soc. 3 (1963), 375–383.
96. Feichtinger, G. *Stochastische Modelle demographischer Prozesse.* Lecture Notes in Operations Res. and Math. Systems, Vol. 44. Berlin, Springer, 1971.
97. Feichtinger, G. *Markovian models for some demographic processes.* Statist. Hefte **14** (1973), 310–334.
98. Feller, W. *Die Grundlagen der Volterraschen Theorie des Kampfes ums Dasein in wahrscheinlichkeitstheoretischer Behandlung.* Acta Biotheoretica **5** (1939), 11–40.
99. Feller, W. *On the integro-differential equations of purely discontinuous Markoff processes.* Trans. Amer. Math. Soc. **48** (1940), 488–515. Errata Ibid **58** (1945), 474.
100. Feller, W. *The problem of n liars and Markov chains.* Amer. Math. Monthly **58** (1951), 606–608.
101. Feller, W. *Diffusion processes in genetics.* In *Proc. Second Berkely Sympos. Math. Statist. and Probability.* Berkeley, Univ. of California Press, 1951, pp. 227–246.
102. Feller, W. *An introduction to probability theory and its applications, Vol. I, 3rd Ed.* New York, Wiley, 1968.
103. Ferschl, F. *Markovketten.* Lecture Notes in Operations Res. and Math. Systems, Vol. 35. Berlin, Springer 1970.
104. Finucan, H.M. *A rule of thumb for Markov chains.* Math. Gaz. **57** (1973), 203–205.
105. Fortet, R.M. *Problèmes de statistique concernant des processus de Markov.* In *Trans. Second Prague Conf. Information Theory, Statist. Decision Functions, Random Processes.* Prague, Publ. House Czechoslovak Acad. Sci., 1960, pp. 159–175.
106. Foster, F.G. *A Markov chain derivation of discrete distributions.* Ann. Math. Statist. **23** (1952), 624–627.
107. Franckx, E. *Chaînes de Markoff et échelles numériques.* Trabajos Estadist. **1** (1950), 147–156.
108. Fréchet, M. *Recherches théoriques modernes sur le calcul des probabilités. Second livre. Méthodes des fonctions arbitraires. Théorie des événements en chaîne dans le cas d'un nombre fini d'états possible.* Paris, Hermann, 1938. (Reprinted in 1952 with a new supplement and a note of Paul Lévy.)

109. Freedman, D.A. *Bernard Friedman's urn.* Ann. Math. Statist. **36** (1965), 956–970.
110. Freedman, D.A. *Markov chains.* San Francisco, Holden-Day, 1971.
111. Friedman, B. *A simple urn model.* Comm. Pure Appl. Math. **2** (1949), 59–70.
112. Funke, Ursula H. *Mathematical models in marketing: a collection of abstracts.* Lecture Notes in Econ. and Math. Systems, Vol. 132., Berlin, Springer, 1976.
113. Gabriel, K.R. *The distribution of the number of successes in a sequence of dependent trials.* Biometrika **46** (1959), 454–460.
114. Gani, J. *The condition of regularity in simple Markov chains.* Austral. J. Phys. **9** (1956), 387–393.
115. Gani, J. *Recent advances in storage and flooding theory.* Advances in Appl. Probability **1** (1969), 90–110.
116. Gani, J. *Formulae for projecting enrollments and degrees awarded in universities.* J. Roy. Statist. Soc. Ser. A **126** (1963), 400–409.
117. Gani, J. *Stochastic models for type counts in a literary text.* In Gani, J. (Ed.) *Perspectives in probability and statistics: papers in honour of M. S. Bartlett.* London, Academic Press, 1975, pp. 313–323.
118. Gani, J.; Jerwood, D. *Markov chain methods in chain binomial epidemic models.* Biometrics **27** (1971), 591–603.
119. Gantmacher, F.R. *Applications of the theory of matrices.* New York, Interscience, 1959.
120. Gheorghe, F.C. *Some considerations on absorbing finite state Markov chains.* Bull. Math. Soc. Sci. Math. R.S. Roumanie **17 (65)** (1973), 387–393.
121. Gheorghe, F.C.; Obreja, G.; Şerbu, Mihaela *Using Markov chains to evaluate the OC and ASN functions for a sequential control plan.* Stud. Cerc. Mat. **27** (1975), 303–312 (Romanian).
122. Ghermănescu, M. *On Markov's chains.* Bul. Şti. Secţ. Şti. Mat. Fiz. Acad. R.P.R. **8** (1956), 101–114 (Romanian).
123. Gilbert, E.J. *On the identifiability problem for functions of finite Markov chains.* Ann. Math. Statist. **30** (1959), 688–697.
124. Good, I.J. *The frequency count of a Markov chain and the transition to continuous time.* Ann. Math. Statist. **32** (1961), 41–48.
125. Goodman, G.S. *An intrinsic time forn onstationary finite Markov chains.* Z. Wahrscheinlichkeitstheorie verw. Gebiete **16** (1970), 165–180.
126. Goodman, G.S.; Johansen, S. *Kolmogorov's differential equations for non-stationary, countable state Markov process with uniformly continuous transition probabilities.* Proc. Cambridge Philos. Soc. **73** (1973), 119–138.
127. Gordon, P. *Théorie des chaînes de Markov finies et ses applications.* Paris, Dunod, 1965.
128. Greeno, J.G. *Representation of learning as discrete transition in a finite state space.* In Kranz, D.H.; Atkinson, R.C.; Luce, R.D.; Suppes, P. (Eds.) *Contemporary developments in mathematical psychology, Vol. I.* Freeman, San Francisco, 1974, pp. 1–43.
129. Griffeath, D. *A maximal coupling for Markov chains.* Z. Wahrscheinlichkeitstheorie verw. Gebiete **31** (1975), 95–106.
130. Griffeath, D. *Uniform coupling of nonhomogeneous Markov chains.* J. Appl. Probability **12** (1975), 753–762.
131. Griffeath, D. *Coupling methods for Markov processes.* Ph. D. Thesis, Cornell University, 1975.
132. Gussel', M.; Popov, Ju. V.; Sragovič, V.G. *Adaptive control of partially observable Markov chains with rewards.* Dokl. Akad. Nauk SSSR **237** (1977), 767–769 (Russian).
133. Hadamard, J. *Sur le battage des cartes et ses relations avec la Mécanique Statistique.* In *Atti Congr. Internaz. Mat., Bologna* 1928, *Vol. V.* Bologna, Zanichelli, 1932, pp. 133–139.
134. Hajnal, J. *The ergodic properties of nonhomogeneous finite Markov chains.* Proc. Cambridge Philos. Soc. **52** (1956), 67–77.
135. Hajnal, J. *Weak ergodicity in nonhomogeneous Markov chains.* Proc. Cambridge Philos. Soc. **54** (1958), 233–246.

136. Hajnal, J. *On products of nonnegative matrices.* Math. Proc. Cambridge Philos. Soc. **79** (1976), 521—530.
137. Hanen, A. *Théorèmes limites pour une suite de chaînes de Markov.* Ann. Inst. H. Poincaré **18** (1963), 197—301.
138. Harary, F.; Lipstein, F.; Styan, G.P.H. *A matrix approach to nonstationary chains.* Operations Res. **18** (1970), 1168—1181.
139. Hardin, J.C.; Sweet, A.L. *A note on absorption probabilities for a random walk between a reflecting and an absorbing barrier.* J. Appl. Probability **6** (1969), 224—226.
140. Hardin, J.C.; Sweet, A.L. *Moments of the time to absorption in the random walk between a reflecting and an absorbing barrier.* SIAM Rev. **12** (1970), 140—142.
141. Hartfiel, D.J. *A result concerning strongly ergodic nonhomogeneous Markov chains.* Linear Algebra and Appl. **9** (1974), 169—174.
142. Hartfiel, D.J. *Two theorems generalizing the mean transition probability results in the theory of Markov chains.* Linear Algebra and Appl. **11** (1975), 181—187.
143. Hennequin, P.-L.; Tortrat, A. *Théorie des probabilités et quelques applications.* Paris, Masson, 1965.
144. Henze, E.; Massé, J.-C.; Theodorescu, R. *On multiple Markov chains.* J. Multivariate Anal. **7** (1977), 589—593.
145. Herniter, J.D.; Howard, R.A. *Stochastic marketing models.* In Hertz, D.B.; Eddison, R.T. (Eds) *Progress in operations research, Vol.II.* New York, Wiley, 1964, pp. 33—96.
146. Heyde, C.C.; Seneta, E. *I.J. Bienaymé: statistical theory anticipated.* Berlin, Springer, 1977.
147. Hoem, J.M. *Partial and purged Markov chains.* Skand. Aktuarietidskr. **52** (1969), 147—155.
148. Hoem, J.M. *A Markov chain model of working life tables.* Scand. Actuar. J. **1977**, no. 1, 1—20.
149. Höglund, T. *Central limit theorems and statistical inference for finite Markov chains.* Z. Wahrscheinlichkeitstheorie verw. Gebiete **29** (1974), 123—151.
150. Holgate, P. *The size of elephant herds.* Math. Gaz **51** (1967), 302—304.
151. Holland, P.W.; Leinhardt, S. *A dynamic model for social networks.* J. Mathematical Sociology **5** (1977), 5—20.
152. Hostinsky, B. *Méthodes générales du calcul des probabilités.* Mém. Sci. Math., 52. Paris, Gauthier-Villars, 1931.
153. Hostinsky, B. *Sur une classe d'équations fonctionelles.* J. Math. Pures Appl. **16** (1937), 267—284.
154. Hostinsky, B. *Revue des travaux publiés en* 1935—1948 *sur les chaînes de Markoff et problèmes voisins.* Časopis Pěst. Mat. Fys. **74** (1949), 48—62.
155. Huang, C.; Isaacson, D. *Ergodicity using mean visit times.* J. London Math. Soc. (2) **14** (1976), 570—576.
156. Huang, C.; Isaacson, D.; Vinograde, B. *The rate of convergence of certain nonhomogeneous Markov chains.* Z. Wahrscheinlichkeitstheorie verw. Gebiete **35** (1976) 141—146.
157. Iosifescu, M. *Conditions nécessaires et suffisantes pour l'ergodicité uniforme des chaînes de Markoff variables et multiples.* Rev. Roumaine Math. Pures Appl. **11** (1966), 325—330.
158. Iosifescu, M. *Mathematical theory of learning.* Gaz. Mat. Ser. A **74** (1969) 65—72 (Romanian).
159. Iosifescu, M. *Sur les chaînes de Markov multiples.* Bull. Inst. Internat. Statist. **43** (1969), book 2, 333—335.
160. Iosifescu, M. *An extension of the renewal equation.* Z. Wahrscheinlichkeitstheorie verw. Gebiete **23** (1972), 148—152.
161. Iosifescu, M. *On finite tail σ-algebras.* Z. Wahrscheinlichkeitstheorie verw. Gebiete **24** (1972), 159—166.
162. Iosifescu, M. *On two recent papers on ergodicity in nonhomogeneous Markov chains.* Ann. Math. Statist. **43** (1972), 1732—1736.
163. Iosifescu, M. *On multiple Markovian dependence.* In *Proc. Fourth Conf. Probability Theory* (Braşov, 1971). Bucharest, Editura Acad. RSR, 1973, pp. 65—71.
164. Iosifescu, M. *The tail structure of nonhomogeneous finite state Markov chains: a survey.* In *Banach Center Publications, Vol. 5 (Probability Theory).* Warszawa, PWN, 1979, pp. 125—132.

165. Iosifescu, M. *Review* # 1661 *to* [144]. Math. Rev. **57** (1979), 223.
166. Iosifescu, M.; Tăutu, P. *Stochastic processes and applications in biology and medicine. I. Theory; II. Models.* Bucharest & Berlin, Editura Acad. RSR & Springer, 1973.
167. Iosifescu, M.; Theodorescu, R. *Random processes and learning.* Berlin Springer, 1969.
168. Iosifescu, M.; Tăutu, P.; Theodorescu, R. *Some considerations on mathematical theory of learning.* Revista de psihologie **13** (1967), 163—177 (Romanian).
169. Iosifescu-Manu, Adela. *Nonhomogeneous semi-Markov processes.* Stud. Cerc. Mat. **24** (1972), 529—533 (Romanian).
170. Isaacson, D.; Madsen, R. *Positive columns for stochastic matrices.* J. Appl. Probability **11** (1974), 829—835.
171. Isaacson, D.L.; Madsen, R.W. *Markov chains: theory and applications.* New York Wiley, 1976
172. Jacobs, K. *Fastperiodische diskrete Markoffsche Prozesse von endlicher Dimension.* Abh. Math. Sem. Univ. Hamburg **21** (1957), 194—246.
173. Jacobs, K. *Markoffsche Prozesse mit monomialer Selbsteuerung.* Arch. Math. **8** (1957), 298—308.
174. Jacobs, K. *Zur Theorie der Markoffschen Prozesse.* Math. Ann. **133** (1957), 375—399.
175. Jacobsen, M. *A characterization of minimal Markov jump processes.* Z. Wahrscheinlichkeitstheorie verw. Gebiete **23** (1972), 32—46.
176. Jensen, A.; Kendall, D. *Denumerable Markov processes with bounded generators: a routine for calculating* $p_{ij}(\infty)$. J. Appl. Probability **8** (1971), 423—427.
177. Johansen, S. *A central limit theorem for finite semigroups and its application to the imbedding problem for finite state Markov chains.* Z. Wahrscheinlichkeitstheorie verw. Gebiete **26** (1973), 171—190.
178. Johansen, S.; Ramsey, F.L. *A bang-bang representation for* 3×3 *embeddable stochastic matrices.* Z. Wahrscheinlichkeitstheorie verw. Gebiete **47** (1979), 107—118.
179. Johnson, N.L.; Kotz, S. *Urn models and their application: an approach to modern discrete probability theory.* New York, Wiley, 1977.
180. Jury, E.I. *Theory and application of the Z-transform method.* New York, Wiley, 1964.
181. Kac, M. *Random walk and the theory of Brownian motion.* Amer. Math. Monthly **54** (1947), 369—391.
182. Kac, M. *On the notion of recurrence in discrete stochastic processes.* Bull. Amer. Math. Soc. **53** (1947), 1002—1010.
183. Kac, M. *Probability.* Amer. Sci. **211** (1964), 102—118.
184. Kadyrov, U. *On the theory of nonhomogeneous Markov chains.* Izv. Akad. Nauk UzSSR Ser. Fiz.-Mat. Nauk **5** (1961) no. 1, 22—38 (Russian).
185. Kamat, A.R. *A stochastic model for progress in a course of education.* Sankhyā Ser. B **30** (1968), 25—32.
186. Kane, J. *Dynamics of the Peter principle.* Management Sci. **16** (1970), B-800—B-811.
187. Karlin, S. *Equilibrium behaviour of population genetic models with nonrandom mating. I—II.* J. Appl. Probability **5** (1968), 231—313; 487—566.
188. Karlin, S.; Mc Gregor, J.L. *The differential equations of birth and death processes and the Stieltjes moment problem.* Trans. Amer. Math. Soc. **85** (1957), 489—546.
189. Karlin, S.; McGregor, J.L. *Coincidence properties of birth and death processes.* Pacific J. Math. **9** (1959), 1109—1140.
190. Karlin, S.; Mc Gregor, J.L. *On some stochastic models in genetics.* In Gurland, J. (Ed.) *Stochastic models in medicine and biology.* Madison, Univ. Wisconsin Press, 1964, pp. 245—271.
191. Karlin, S.; Mc Gregor, J.L. *Ehrenfest urn models.* J. Appl. Probability **2** (1965), 352—376.
192. Karlin, S.; Mc Gregor, J.L. *Direct product branching processes and related induced Markoff chains. I. Calculations of rates of approach to homozygosity.* In *Bernoulli, Bayes, Laplace (Berkeley Seminar).* Berlin, Springer, 1965, pp. 111—145.
193. Karpelevič, F.I. *On the eigenvalues of nonnegative matrices.* Izv. Akad. Nauk SSSR Ser. Mat. **15** (1951), 361—383 (Russian).
194. Keilson, J.; Syski, R. *Compensation measures in the theory of Markov chains.* Stochastic Processes Appl. **2** (1974), 59—72.
195. Kemeny, J.G.; Snell, J.L. *Finite Markov chains.* Princeton, Van Nostrand, 1960.

196. Kemeny, J.G.; Snell, J.L. *Finite continuous time Markov chains.* Teor. Verojatnost. i Primenen, **6** (1961), 110–115.
197. Kendall, D.G. *Geometric ergodicity and the theory of queues.* In [7], pp. 176–195.
198. Kendall, D.G. *Information theory and the limit theorem for Markov chains and processes with a countable infinity of states.* Ann. Inst. Statist. Math. **15** (1963), 137–143.
199. Kendall, D.G. *Discussion to* [354]. J. Roy. Statist, Soc. Ser. B **28** (1966), 266–269.
200. Kendall, D.G. *Renewal sequences and their arithmetic.* In *Symposium on probabilithy methods in analysis* (*Loutraki*, 1966). Lecture Notes in Math., Vol. 31. Berlin, Springer, 1967, pp. 147–175.
201. Kendall, D.G. *The genealogy of genealogy: branching processes before and (after)* 1873. Bull. London Math. Soc. **7** (1975), 225–253.
202. Kendall, D.G. *Some problems in mathematical genealogy.* In Gani, J. (Ed.) *Perspectives in probability and statistics: papers in honour of M.S. Bartlett.* London, Academic Press, 1975, pp. 325–345.
203. Kesten, H. *Occupation times for Markov and semi-Markov chains.* Trans. Amer. Math. Soc. **103** (1962), 82–112.
204. Khazanie, R.G. *An indication of the asymptotic nature of the Mendelian Markov process.* J. Appl. Probability **5** (1968), 350–356.
205. Khazanie, R.G.; Mc Kean, H.E. *A Mendelian Markov process with binomial transition probabilities.* Biometrika **53** (1966), 37–48.
206. Khazanie, R.G.; Mc Kean, H.E. *A Mendelian Markov process with multinomial transition probabilities.* J. Appl. Probability **3** (1966), 353–364.
207. Kingman, J.F.C. *The imbedding problem for finite Markov chains.* Z. Wahrscheinlichkeitstheorie verw. Gebiete **1** (1962), 14–24.
208. Kingman, J.F.C. *Ergodic properties of continuous time Markov processes and their discrete skeletons.* Proc. London Math. Soc. **13** (1963), 593–604.
209. Kingman, J.F. C. *Regenerative phenomena.* London, Wiley, 1972.
210. Kingman, J.F.C. *Geometrical aspects of the theory of nonhomogeneous Markov chains.* Math. Proc. Cambridge Philos. Soc. **77** (1975), 171–183.
211. Kingman, J.F.C.; Williams, D. *The combinatorial structure of nonhomogeneous Markov chains.* Z. Wahrscheinlichkeitstheorie verw. Gebiete **26** (1973), 77–86.
212. Kistauri, E.I. *Solution of Kolmogorov's system of differential equations for the finite homogeneous Markov process by means of eigenvalues and eigenmatrices.* Sakharth. SSR Mecn. Akad. Moambe **82** (1976), 337–340 (Georgian).
213. Klotz, J. *Markov Bernoulli trials.* Ann. Statist. **1** (1973), 373–379.
214. Kolmogorov, A.N. *Uber die analytischen Methoden in der Wahrscheinlichkeitsrechnung.* Math. Ann. **104** (1931), 415–458.
215. Kolmogorov, A.N. *Zur Theorie der Markoffschen Ketten.* Math. Ann. **112** (1936), 155–160.
216. Kolmogorov, A.N. *Zur Umkehrbarkeit der statistischen Naturgesetze.* Math. Ann. **113** (1937), 716–722.
217. Kolmogorov, A.N. *Markov chains with denumerably many states.* Bjull. MGU **1** (1937) no. 3, 1–16 (Russian).
218. Kolmogorov, A.N. *A local limit theorem for classical Markov chains.* Izv. Akad. Nauk SSSR Ser. Mat. **13** (1949), 281–300 (Russian).
219. Koopman, B.O. *A generalization of Poisson distribution for Markoff chains.* Proc. Nat. Acad. Sci. USA **36** (1950), 202–207.
220. Koopman, B.O. *A law of small numbers in Markoff chains.* Trans. Amer. Math. Soc. **70** (1951), 277–290.
221. Kordonski, H.B. *An application of Markov chains theory to statistical quality control.* Vestnik Leningrad. Univ. **10** (1955) no. 11, 75–78 (Russian).
222. Korneev, C.A. *On mean recurrence time and mean sojourn time of an irreducible finite Markov chain in a given set of states.* Avtomat. i Vyčisl. Tehn. 1970, no. 5, 14–17 (Russian).
223. Koroljuk, V.S. *Semimarkovian models of compound stochastic systems.* In *Second Internat. Summer School on Probability Theory and Math. Statist.* (*Varna,* 1976). Sofia, Izd. B'lgar. Akad. Nauk, 1977, pp. 115–185 (Russian).
224. Koroljuk, V.S.; Turbin, A.F. *Semi-Markov processes and their applications.* Kiev, Naukova dumka, 1976 (Russian).

225. Koutský, Z. *Einige Eigenschaften der modulo k addierten Markowschen Ketten.* In *Trans. Second Prague Conf. Information Theory, Statist. Decision Functions, Random Processes.* Prague, Publ. House Czechoslovak Acad. Sci., 1960, pp. 263–278.

226. Koźniewska, I. *Ergodicité et stationnarité des chaînes de Markov variables à un nombre fini d'états possibles.* Colloq. Math. **9** (1962), 333–345.

227. Kreweras, G. *Un modèle d'évolution de l'opinion exprimée par des votes successifs.* Publ. Inst. Statist. Univ. Paris **12** (1963), 3–44. (Reprinted in English in Lazarsfeld, P.F.; Henry, N.W. (Eds.) *Readings in mathematical social science.* Cambridge, Mass., The M.I.T. Press, 1966, pp. 174–190.)

228. Krisanov, A.I.; Kuzin, L.T.; Letunov, Ju. P. *Some asymptotical estimates for convergence of discrete Markov chains.* Izv. Akad. Nauk SSSR Tehn. Kibernet. **9** (1971) no. 6, 143–150 (Russian).

229. Kučkarov, Ja.H. *A law of the iterated logarithm for nonhomogeneous Markov chains.* Litovsk. Mat. Sb. **5** (1965), 575–583; **6** (1966), 132 (Russian).

230. Kullback, S. *An extension of an information-theoretic derivation of certain limit relations for a Markov chain.* SIAM J. Control **5** (1967), 51–53.

231. Kůrka, P. *Time hierarchy in finite filtered Markov chains.* In *Trans. Eighth Prague Conf. Information Theory, Statist. Decision Functions, Random Processes, Vol. B.* Prague, Academia, 1978, pp. 9–22.

232. Lahres, H. *Einführung in die diskreten Markoff-Prozesse und ihre Anwendungen.* Braunschweig, Fr. Vieweg & Sohn, 1963.

233. Larisse, J. *Sur une convergence presque sûre d'un produit infini de matrices stochastiques ayant le même nombre de classes ergodiques.* C.R. Acad. Sci. Paris **262** (1966), 913–915.

234. Larisse, J.; Schützenberger, M.P. *Sur certaines chaînes de Markov non homogènes.* Publ. Inst. Statist. Univ. Paris **13** (1964), 57–66.

235. Lee, T.C.; Judge, G.G.; Zellner, A. *Estimating the parameters of the Markov probability model from aggregate series data.* Amsterdam, North Holland, 1970.

236. Lehman, R.S.; Weiss, G.H. *A study of the restricted random walk.* J. Soc. Indust. Appl. Math. **6** (1958), 257–278.

237. LeMaire, B. *État actuel et perspectives de l'utilisation des modèles markoviens pour l'étude du comportement des consommateurs.* Math. Sci. Humaines **50** (1975), 51–79.

238. Letac, N. *Barycentres de matrices idempotentes stochastiques.* Ann. Sci. Univ. Clermont No 51 (1974), Math., 9e fasc., 51–61.

239. Letac, G. *Chaînes de Markov sur les permutations.* Montréal, Les Presses de l'Univ. de Montréal, 1978.

240. Lévy, P. *W. Doeblin (V. Doblin)* (1915–1940). Rev. Hist. Sci. Appl. **8** (1955), 107–115.

241. Lindqvist, B. *How fast does a Markov chain forget the initial state? A decision theoretical approach.* Scand. J. Statist. **4** (1977), 145–152.

242. Lindqvist, B. *On the loss of information incurred by lumping states of a Markov chain.* Scand. J. Statist. **5** (1978), 92–98.

243. Lindqvist, B. *A decision theoretical characterization of weak ergodicity.* Z. Wahrscheinlichkeitstheorie verw. Gebiete **44** (1978), 155–158.

244. Linnik, Ju. V. *On the theory of nonhomogeneous Markov chains.* Izv. Akad. Nauk SSSR Ser. Mat. **13** (1949), 65–94 (Russian).

245. Linnik, Ju. V. *An application of Markov chain theory to the arithmetic of quaternions.* Uspehi Mat. Nauk **9** (1954) no. 4 (62), 203–210 (Russian).

246. Linnik, Ju.V. *Markov chains in the analytic arithmetic of quaternions and matrices.* Vestnik Leningrad. Univ. **11** (1956) no. 13, 63–68 (Russian).

247. Linnik, Ju. V.; Sapogov, N.A. *Multidimensional integral and local laws for nonhomogeneous Markov chains.* Izv. Akad. Nauk SSSR Ser. Mat. **13** (1949), 533–566 (Russian).

248. Ljubič, Ju.I. *Linear Bernšteĭn populations.* Teor. Funkciĭ Funkcional. Anal. i Priložen. Vyp. 22 (1975), 107–111 (Russian).

249. Loeffel, H. *Glücksspiel und Markovketten.* Elem. Math. **29** (1974), 142–149.

250. Louchard, G. *Recurrence times and capacities for finite ergodic chains.* Duke Math. J. **33** (1966), 13–21.

251. Madsen, R.W. *Decidability of* $\alpha(\mathbf{P}^k) > 0$ *for some k.* J. Appl. Probability **12** (1975), 333–340.

252. Maibaum. G.; Mühlmann, P. *Uber mehrdimensionale stochastische Ketten und stochastische Automaten*. Math. Nachr. **65** (1975), 247—257.
253. Malécot, G. *Sur un problème de probabilité en chaîne que pose la génétique*. C. R. Acad. Sci. Paris **219** (1944), 279—281.
254. Maliţa, M. ; Zidăroiu, C. *Mathematical models of the educational system*. Bucharest, Editura didactică şi pedagogică, 1972 (Romanian).
255. Malyšev, V.A. *On the poles of rational generating functions. Probability of occurrence of a combination*. Litovsk. Mat. Sb. **5** (1965), 585—591 (Russian).
256. Mandl, P. *On the asymptotic behaviour of probabilities within groups of states of a homogeneous Markov chain*. Časopis Pěst. Mat. **84** (1959), 140—149 (Russian).
257. Mandl, P. *On the adaptive control of finite state Markov processes*. Z. Wahrscheinlichkeitstheorie verw. Gebiete **27** (1973), 263—276.
258. Markov, A.A. *Extension of the law of large numbers to dependent variables*. Izv. Fiz-Mat. Obšč. pri Kazansk. Univ. (2 Ser.) **15** (1906) no. 4, 135—156 (Russian).
259. Markov, A.A. *Probability calculus, 4th Ed*. Moscow, Gosizdat, 1924 (Russian).
260. Markov, A.A. *Selected works*. Moscow, Izd. Akad. Nauk SSSR, 1951 (Russian).
261. Marlin, P.E. *On the ergodic theory of Markov chains*. Operations Res. **21** (1973), 617—622.
262. Martin, J.J. *Bayesian decision problems and Markov chains*. New York, Wiley, 1967.
263. Matthews, J.P. *A central limit theorem for absorbing Markov chains*. Biometrika **57** (1970), 129—139.
264. McCabe, B.J. *The asymptotic behaviour of a certain Markov chain*. Ann. Math Statist., **40** (1969), 665—666.
265. Mešalkin, L.D. *Limit theorems for finite state Markov chains*. Teor. Verojatnost. i Primenen. **3** (1958), 361—385 (Russian).
266. Metev, B. *On a nonhomogeneous Markov chain with a finite number of states*. In *Trans. Sixth Prague Conf. Information Theory, Statist. Decision Functions, Random Processes*. Prague, Academia, 1973, pp. 649—655 (Russian).
267. Mettler, L.E.; Gregg, T.G. *Population genetics and evolution*. Englewood Cliffs (N.J.), Prentice Hall, 1969.
268. Miehle, W. *Calculation of higher transitions in a Markov process*. Operations Res. **6** (1958), 693—698.
269. Mihoc, G. *On general properties of dependent statistical variables*. Bull. Math. Soc. Roumaine Sci **37** (1935) no. 1, 37—82; no. 2, 13—78 (Romanian).
270. Mihoc, G. *Sur les lois limites des variables liées en chaîne*. Bul. Fac. Stiinţe din Cernăuţi **10** (1936), 1—26.
271. Mihoc, G. *Sur le problème des itérations dans une suite d'épreuves*. Bull. Math. Soc. Roumaine Sci. **45** (1943), 81—95.
272. Mihoc, G. *The law of rare events in Markov chains*. Bul. Şti. Secţ. Şti. Mat. Fiz. Acad. R.P.R. **4** (1952), 783—790 (Romanian).
273. Mihoc, G. *Extending the Poisson law to homogeneous multiple Markov chains*. Bul. Şti. Sect. Şti. Mat. Fiz. Acad. R.P.R. **6** (1954), 5—15 (Romanian).
274. Mihoc, G. *Über verschiedene Ausdehnungen des Poissonschen Gesetzes auf endliche konstante Markoffsche Ketten*. In *Bericht über die Tagung Wahrscheinlichkeitsrechnung und mathematische Statistik in Berlin (Oktober, 1954)*. Berlin, Deutscher Verlag der Wissenschaften, 1956, pp. 43—49.
275. Mihoc, G. *Sur les lois limites des variables vectorielles enchaînées au sens de Markoff*. Teor. Verojatnost. i Primenen. **1** (1956), 103—112.
276. Mihoc, G.; Craiu, Mariana. *Statistical inference for dependent variables*. Bucharest, Editura Acad. R.S.R., 1972 (Romanian).
277. von Mises, R. *Wahrscheinlichkeitsrechnung und ihre Anwendung in der Statistik und theoretischen Physik*. Leipzig-Wien, Fr. Deuticke, 1931.
278. Monin, J.P.; Benayoun, R.; Sert, B. *Initiation to the mathematics of the processes of diffusion, contagion and propagation*. The Hague-Paris, Mouton, 1976.
279. Moran, P.A.P. *Random processes in genetics*. Proc. Cambridge Philos. Soc. **54** (1968), 60—71.
280. Moran, P.A.P. *The theory of storage*. London, Methuen, 1959.

281. Mott, J.L. *Conditions for the ergodicity of nonhomogeneous finite Markov chains.* Proc. Roy. Soc. Edinburgh Sect. A **64** (1957), 369—380.
282. Mott, J.L. *The central limit theorem for a convergent nonhomogeneous finite Markov chain.* Proc. Roy. Soc. Edinburgh Sect. A **65** (1959), 109—120.
283. Mott, J.L. *The latent roots of certain stochastic matrices.* Biometrika **49** (1962), 264—265.
284. Mott, J.L.; Schneider, H. *Matrix norms applied to weakly ergodic Markov chains.* Arch. Math. **8** (1957), 331—333.
285. Mullat, I. *On a class of absorbing Markov chains.* Izv. Akad. Nauk Estonskoĭ SSR Ser. Fiz.-Mat. **21** (1972), 294—296 (Russian).
286. Muthsam, H. *Teilprozesse Markoffscher Ketten.* Anz. Österreich. Akad. Wiss. math.-naturwiss. Kl. No. 4 (1972), 91—95.
287. Nakamura, M. *Some limit theorems for a class of network problems as related to finite Markov chains.* J. Appl. Probability **11** (1974), 94—101.
288. Neuts, M.F. *Absorption probabilities for a random walk between a reflecting and an absorbing barrier.* Bull. Soc. Math. Belg. **15** (1963), 253—258.
289. Neveu, J. *Sur le comportement asymptotique des chaînes de Markov.* In *Le calcul des probabilités et ses applications* (*Paris,* 1958). Colloques Internationaux du CNRS, LXXXVII. CNRS, Paris, 1959, pp. 185—193.
290. Newbould, M. *A classification of a random walk defined on a finite Markov chain.* Z. Wahrscheinlichkeitstheorie verw. Gebiete **26** (1973), 95—104.
291. Neyman, J. *Assessing the chain: energy crisis, pollution and health.* Internat. Statist. Rev. **43** (1975), 253—267.
292. Norman, M.F. *Markov processes and learning models.* New York, Academic Press, 1972.
293. Nuszkowski, H. *Analyse von Markoffketten mit Signalflussgraphen.* Wiss. Z. Techn. Univ. Dresden **24** (1975), 411—414.
294. Onicescu, O.; Mihoc, G. *Sopra le leggi-limite delle probabilità.* Giorn. Ist. Ital. Attuari **7** (1936), 54—69.
295. Onicescu, O.; Mihoc, G. *La dépendance statistique. Chaînes et familles de chaînes discontinues.* Act. Sci. Ind., 503. Paris, Hermann, 1937.
296. Onicescu, O.; Mihoc, G. *Propriétés asymptotiques des chaînes de Markoff étudiées à l'aide de la fonction caractéristique.* Mathematica (Cluj) **16** (1940), 13—43.
297. Onicescu, O.; Mihoc, G. *Les chaînes de variables aléatoires. Problèmes asymptotiques.* Bucarest, Études et Recherches 14, Acad. Roumaine, 1943.
298. Onicescu, O.; Mihoc, G. *Le coefficient de dispersion et la dépendence des épreuves.* Bull. Math. Soc. Roumaine Sci. **46** (1944), 77—80.
299. Onicescu, O.; Mihoc, G.; Ionescu Tulcea, C. T. *Calculus of probability and applications.* Bucharest, Editura Acad. R.P.R., 1956 (Romanian).
300. Oprişan, G.; Pipa, Gabriela *Sur l'utilisation des chaînes de Markov dans la théorie de la fiabilité.* In *Proc. Fourth Conf. Probability Theory (Braşov,* 1971*).* Bucharest, Editura Acad. R.S.R., 1973, pp. 571—576.
301. Ostenc, É. *Sur le principe ergodique dans les chaînes de Markov à éléments variables.* C. R. Acad. Sci. Paris **199** (1934), 175—176.
302. Parzen, E. *Stochastic processes.* San Francisco, Holden-Day, 1962.
303. Paz, A. *Definite and quasidefinite sets of stochastic matrices.* Proc. Amer. Math. Soc. **16** (1965), 634—641.
304. Pedler, P. J. *Occupation times for two state Markov chains.* J. Appl. Probability **8** (1971), 381—390.
305. Pegels, C. C.; Jelmert, A. E. *An evaluation of blood-inventory policies: a Markov chain application.* Operations Res. **18** (1970), 1087—1098.
306. Perkins, P. *A theorem on regular matrices.* Pacific J. Math. **11** (1961), 1529—1533.
307. Pitman, J. W. *Uniform rates of convergence for Markov chains transition probabilities.* Z. Wahrscheinlichkeitstheorie verw. Gebiete **29** (1974), 193—227.
308. Pitman, J. W. *Occupation measures for Markov chains.* Advances in Appl. Probability **9** (1977), 69—86.
309. Plucińska, A. *On the joint limiting distribution of times spent in particular states by a Markov process.* Colloq. Math. **9** (1962), 347—360.
310. Poincaré, H. *Calcul des probabilités, 2e édition.* Paris, Gauthier-Villars, 1912.

311. Pospišil, B. *Sur un problème de MM S. Bernstein et A. Kolmogoroff.* Časopis Pěst. Mat. Fys. **65** (1935/36), 64–76.
312. Prais, J. S. *Measuring social mobility.* J. Roy. Statist. Soc. Ser. A **118** (1955), 56–66.
313. Pruscha, H.; Theodorescu, R. *Recurrence and transience for discrete parameter stochastic processes.* In *Trans. Eighth Prague Conf. Information Theory, Statist. Decision Functions, Random Processes, Vol. B.* Prague, Academia, 1978, pp. 97–105.
314. Pullman, N. *The geometry of finite Markov chains.* Canad. Math. Bull. **8** (1965), 345–358.
315. Pullman, N. J. *Infinite products of substochastic matrices.* Pacific J. Math. **16** (1966), 537–544.
316. Pullman, N. J.; Styan, G. P.H. *The convergence of Markov chains with nonstationary transition probabilities and constant causative matrix.* Stochastic Processes Appl. **1** (1973), 279–285.
317. Pye, G. *A Markov model of the term structure.* Quart. J. Economics **80** (1966), 60–72.
318. Pyh, Ju. A. *An application of the ergodicity coefficient to the estimation of the spectral radius of real matrices.* Mat. Zametki **23** (1978), 137–142 (Russian).
319. Råde, L. *The two-state Markov process and additional events.* Amer. Math. Monthly **83** (1976), 354–356.
320. Rajarshi, M. B. *Success runs in a two-state Markov chain.* J. Appl. Probability **11** (1974), 190–192. Correction Ibid **14** (1977), 661.
321. Ray, W. D.; Margo, F. *The inverse problem in reducible Markov chains.* J. Appl. Probability **13** (1976), 49–56.
322. Rényi, A. *Representations for real numbers and their ergodic properties.* Acta Math.. Acad. Sci. Hungar. **8** (1957), 477–493.
323. Rényi, A. *Probability theory.* Budapest, Akadémiai Kiadó, 1970.
324. Romanovski, V. I. *Discrete Markov chains.* Moscow-Leningrad, G. I. T.-T.L., 1949 (Russian).
325. Rosenblatt, M. *Functions of a Markov process that are Markovian.* J. Math. Mech. **8** (1959), 585–596.
326. Rosenblatt, M. *Stationary Markov chains and independent random variables.* J. Math. Mech. **9** (1960), 945–950.
327. Rosenblatt, M. *Functions of a Markov process.* Z. Wahrscheinlichkeitstheorie verw. Gebiete **5** (1966), 232–243.
328. Rosenblatt, M. *Markov processes: structure and asymptotic behaviour.* Berlin, Springer, 1971.
329. Rosenblatt, M. *Note correcting a remark in a paper of Karl Bosch.* Z. Wahrscheinlichkeitstheorie verw. Gebiete **33** (1975), 219.
330. Rouanet, H. *Les modèles stochastiques d'apprentissage.* The Hague-Paris, Gauthier-Villars & Mouton, 1967.
331. Runnenburg, J. T. *On Elfving's problem of imbedding a time-discrete Markov chain in a time-continuous one for finitely many states. I.* Indag. Math. **24** (1962), 536–541.
332. Sapogov, N. A. *On singular Markov chains.* Dokl. Akad. Nauk SSSR **58** (1947), 193–196 (Russian).
333. Sapogov, N. A. *The Laplace-Ljapunov limit theorem for singular Markov chains.* Dokl. Akad. Nauk SSSR **58** (1947), 1905–1908 (Russian).
334. Sarymsakov, T. A. *On the ergodic principle for nonhomogeneous Markov chains.* Dokl. Akad. Nauk SSSR **90** (1953), 25–28 (Russian).
335. Sarymsakov, T. A. *The foundations of the theory of Markov processes.* Moscow, G.I.T.-T.L., 1954 (Russian).
336. Sarymsakov, T. A. *On the theory of nonhomogeneous Markov chains.* Dokl. Akad. Nauk UzSSR **8** (1956), 3–7 (Russian).
337. Sarymsakov, T. A. *On nonhomogeneous Markov chains.* Dokl. Akad. Nauk SSSR **120** (1958), 465–467 (Russian).
338. Sarymsakov, T. A. *Nonhomogeneous Markov chains.* Teor. Verojatnost. i Primenen. **6** (1961), 194–201 (Russian).
339. Sarymsakov, T. A.; Mustafin, H. A. *On an ergodic theorem for nonhomogeneous Markov chains.* Trudy Sredneaziatskogo Gosuniversiteta, N. S., Vyp. 74, 1957, Fiz.–Mat. Nauki, Kniga 15, pp. 1–38 (Russian).

340. Sarmanov, O. V. *Necessary and sufficient conditions of existence of a discrete limit law for two state Markov chains.* Dokl. Akad. Nauk SSSR **110** (1956), 735–738 (Russian).
341. Sarmanov, O. V.; Zaharov, V. K. *The combinatorial problem of N. V. Smirnov.* Dokl. Akad. Nauk SSSR **176** (1967), 530–532 (Russian).
342. Scarf, H. *Stationary operating characteristics of an inventory model with time lag.* In [5], pp. 298–318.
343. Scheffer, C. L. *On Elfving's problem of imbedding a time-discrete Markov chain in a time-continuous one for finitely many states. II.* Indag. Math. **24** (1962), 542–548.
344. Schultz, G. *Uber die Häufigkeit der Iterationen in einer Beobachtungsfolge.* Deutsche Math. **7** (1942), 23–38.
345. Schürger, K.; Tăutu, P. *Markov configuration processes on a lattice.* Rev. Roumaine Math. Pures Appl. **21** (1976), 233–244.
346. Schweitzer, P. J. *Perturbation theory and finite Markov chains.* J. Appl. Probability **5** (1968), 401–413.
347. Schweitzer, P. J. *Some invalid properties of Markov chains.* Operations. Res. **14** (1966), 1153–1154.
348. Scott, D. *Gaur II: a computational procedure to analyze a Markov chain.* Report No. 16 (Statistical numerical analysis and data processing series), Statistical Laboratory, Iowa State Univ., 1978.
349. Seim, G. *Die Klassification der Zustände bei inhomogenen Markoffschen Ketten.* Math. Z. **123** (1971), 21–31.
350. Seleacu, V. *A problem of statistical inference for absorbing Markov chains.* Stud. Cerc. Mat. **29** (1977), 419–422 (Romanian).
351. Senčenko, D. V. *Unique determination of Markov processes with a finite number of states.* Mat. Sb. (N. S.) **71** **(113)** (1966), 30–42 (Russian).
352. Senčenko, D. V. *The characteristics of inhomogeneous Markov processes with a finite number of states.* Teor. Verojatnost. i Primenen. **13** (1968), 548–555 (Russian).
353. Senčenko, D. V. *The tail σ-algebra of a nonhomogeneous finite Markov chain.* Mat. Zametki **12** (1972), 295–302 (Russian).
354. Seneta, E. *Quasi-stationary distributions and time-reversion in genetics (With discussion).* J. Roy, Statist. Soc. Ser. B **28** (1966), 253–277.
355. Seneta, E. *The random walk and bacterial growth.* Zastos. Mat. **9** (1967), 135–147.
356. Seneta, E. *On imbedding discrete chains in continuous time.* Austral. J. Statist. **9** (1967), 1–7.
357. Seneta, E. *On strong ergodicity of inhomogeneous products of finite stochastic matrices.* Studia Math. **46** (1973), 241–247.
358. Seneta, E. *On the historical development of the theory of finite inhomogeneous Markov chains.* Proc. Cambridge Philos. Soc. **74** (1973), 507–513.
359. Seneta, E. *Non-negative matrices.* London, George Allen & Unwin, 1973.
360. Severo, N. C. *A note on the Ehrenfest multiurn model.* J. Appl. Probability **7** (1970), 444–445.
361. Singer, A. *A random walk in which absorption takes place at the second visit to a point.* Bull. Soc. Math. Belg. **17** (1965), 27–36.
362. Siraždinov, S. H. *The ergodic principle for nonhomogeneous Markov chains.* Dokl. Akad. Nauk SSSR **71** (1950), 829–830 (Russian).
363. Siraždinov, S. H. *Limit theorems for homogeneous Markov chains.* Taškent, Izd. Akad. Nauk UzSSR, 1955 (Russian).
364. Smirnov, N. V. *On the statistical estimation of transition probabilities in Markov chains.* Vestnik Leningrad. Univ. **10** (1955) no. 11, 47–48 (Russian).
365. Smirnov, N. V.; Sarmanov, O. V.; Zaharov, V. K. *A local limit theorem for the number of transitions in a Markov chain and applications.* Dokl. Akad. Nauk SSSR **167** (1966), 1238–1241 (Russian).
366. Smith, P. E. *Markov chains, exchange matrices and regional development.* J. Regional Sci. **3** (1961) no. 1, 27–36.
367. Statulevicius, V. *Local limit theorems and asymptotic expansions for nonhomogeneous Markov chains.* Litovsk. Mat. Sb. **1** (1961), 231–314 (Russian).
368. Statulevicius, V. *Limit theorems for sums of random variables connected in a Markov chain. I–III.* Litovsk. Mat. Sb. **9** (1969), 345–362, 635–672; **10** (1970), 161–169 (Russian).

369. Stone, R. *A Markovian education model and other examples linking social behaviour to the economy.* J. Roy. Statist. Soc. Ser. A **135** (1972), 511—543.

370. Stormer, H. *Semi-Markoff-Prozesse mit endlich vielen Zuständen: Theorie und Anwendungen.* Lecture Notes in Operations Res. and Math. Systems, Vol. 34. Berlin. Springer, 1970.

371. Syski, R. *Markovian queues (With discussion).* In Smith, W. L.; Wilkinson, W. E. (Eds.) *Proc. sympos. congestion theory.* Chapel Hill, Univ. North Carolina Press, 1965, pp. 170—222.

372. Székely, G. *On an inhomogeneous Markov chain in patrolling machines.* I. Period. Math. Hungar. **1** (1971), 121—134.

373. Takács, L. *On certain sojourn time problems in the theory of stochastic processes.* Acta Math. Acad. Sci. Hungar. **8** (1957), 169—191.

374. Takács, L. *On a sojourn time problem in the of theory stochastic processes.* Trans. Amer. Math. Soc. **93** (1959), 531—540.

375. Takács, L. *A combinatorial method in the theory of Markov chains.* J. Math. Anal. Appl. **9** (1964), 153—161.

376. Takács, L. *On the classical ruin problems.* J. Amer. Statist. Assoc. **64** (1969), 889—906.

377. Takahashi, Y. *On the effects of small deviations in the transition matrix of a finite Markov chain.* J. Operations Res. Soc. Japan **16** (1973), 104—129.

378. Tan, W. Y. *On the absorption probabilities and absorption times of finite homogeneous birth-death processes.* Biometrics **32** (1976), 745—752.

379. Tankó, J. *A simple algorithm to determine the ergodic classes of a Markov chain.* In *Progress in statistics (Europeans Meeting of Statisticians, Budapest, 1972), Vol. II.* Colloq. Math. Soc. János Bolyai, Vol. 9. Amsterdam, North-Holland, 1974, pp. 803—812.

380. Teugels, J. L. *An example on geometric ergodicity of a finite Markov chain.* J. Appl. Probability **9** (1972), 466—469.

381. Teugels, J. L. *A bibliography on semi-Markov processes.* J. Comput. Appl. Math. **2** (1976), 125—144.

382. Thionet, P. *Contribution à l'étude de "Jouez le gagnant" (Play the Winner) — procédure séquentielle de sélection.* Bull. Inst. Internat. Statist. **46** (1975), book 4, 392—395.

383. Thomas, M. U.; Barr, D. R. *An approximate test of Markov chain lumpability.* J. Amer. Statist. Assoc. **72** (1977), 175—179.

384. Tong, H. *Determination of the order of a Markov chain by Akaike's information criterion.* J. Appl. Probability **12** (1975), 488—497.

385. Tortrat, A. *Etude d'une méthode d'iteration propre à certaines matrices, applications aux processus de Markoff correspondants, cas des processus continus homogènes par rapport à l'espace, cas non homogène.* Publ. Sci. Univ. Alger Sér. A **4** (1957), 145—189.

386. Uebe, G. *Classifying the states of a finite Markov chain.* In *Proc. Operations Res. (Papers of the DGU Annual Meeting 1971).* Würzburg-Vienna, Physica-Verlag, 1972, pp. 587—603.

387. Uppuluri, V. R. R.; Carpenter, J. A. *A generalization of the classical occupancy problem.* J. Math. Anal. Appl. **34** (1971), 316—324.

388. Vere-Jones, D. *Ergodic properties of non-negative matrices. I—II.* Pacific J. Math. **22** (1967), 361—386; **26** (1968), 601—620.

389. Vere-Jones, D.; Kendall, D. G. *A commutativity problem in the theory of Markov chains.* Teor. Verojatnost. i Primenen. **4** (1959), 97—100.

390. Vincze, I. *Über das Ehrenfestsche Modell der Wärmeübertragung.* Arch. Mat. **15** (1964), 394—400.

391. Vladimirescu, I. *Formulae for the moments of the number of occurrences of a transient state and the moments of the time to absorption in a homogeneous finite Markov chain.* Stud. Cerc. Mat. **32** (1980) (Romanian).

392. Volkov, I. S. *On the distribution of sums of random variables defined on homogeneous finite Markov chains.* Teor. Verojatnost. i Primenen. **3** (1958), 413—429 (Russian).

393. Vrânceanu, G. G. *Interprétation géométrique des processus probabilistiques continus.* Paris, Gauthier-Villars, 1969.

394. Watson, D. S. *Molecular biology of the gene, 2nd Ed.* New York, W. A. Benjamin, Inc., 1970.
395. Watterson, G. A. *Markov chains with absorbing states: a genetic example.* Ann. Math. Statist. **32** (1961), 716–729.
396. Waugh, W.A. O'N. *Conditioned Markov processes.* Biometrika **45** (1958), 241–249.
397. Weesakul, B. *The random walk between a reflecting and an absorbing barrier.* Ann. Math. Statist. **32** (1961), 765–769.
398. Williams, D. *The Q-matrix problem.* In *Séminaire de Probabilités, X.* Lecture Notes in Math., Vol. 511. Berlin, Springer, 1976, pp. 216–234.
399. Whittle, P. *Some distributions and moment formulae for the Markov chain.* J. Roy. Statist. Soc. Ser. B **17** (1955), 235–242.
400. Whittle, P. *Reversibility and acyclicity.* In Gani, J. (Ed.). *Perspectives in probability and statistics: papers in honour of M. S. Bartlett.* London, Academic Press, 1975, pp. 217–224.
401. Wielandt, H. *Unzerlcgbare nichtnegative Matrizen.* Math. Z. **52** (1950), 642–648.
402. Wilson, E. B.; Burgess, A. R. *Multiple sampling plans viewed as finite Markov chains.* Technometrics **13** (1971), 371–383.
403. Wolfowitz, J. *Products of indecomposable, aperiodic, stochastic matrices.* Proc. Amer. Math. Soc. **14** (1963), 733–737.
404. Wright, S. *Evolution in Mendelian populations.* Genetics **16** (1931), 97–159.
405. Wynn, H. P. *Limiting second moments for transient states of Markov chains.* J. Appl. Probability **10** (1973), 891–894.
406. Yeh, R. Z. *A geometric proof of Markov ergodic theorem.* Proc. Amer. Math. Soc. **26** (1970), 335–340.
407. Zaharov, V. K.; Sarmanov, O. V. *On the distribution law of the number of runs in a homogeneous Markov chain.* Dokl. Akad. Nauk SSSR **179** (1968), 526–528 (Russian).
408. Zitek, F. *On cstimating transition probabilities.* Apl. Mat. **2** (1957), 251–257.

List of Symbols

Subject Index

Mathematics

FUNCTIONAL ANALYSIS (Second Corrected Edition), George Bachman and Lawrence Narici. Excellent treatment of subject geared toward students with background in linear algebra, advanced calculus, physics and engineering. Text covers introduction to inner-product spaces, normed, metric spaces, and topological spaces; complete orthonormal sets, the Hahn-Banach Theorem and its consequences, and many other related subjects. 1966 ed. 544pp. 6⅛ x 9¼. 0-486-40251-7

ASYMPTOTIC EXPANSIONS OF INTEGRALS, Norman Bleistein & Richard A. Handelsman. Best introduction to important field with applications in a variety of scientific disciplines. New preface. Problems. Diagrams. Tables. Bibliography. Index. 448pp. 5⅜ x 8½. 0-486-65082-0

VECTOR AND TENSOR ANALYSIS WITH APPLICATIONS, A. I. Borisenko and I. E. Tarapov. Concise introduction. Worked-out problems, solutions, exercises. 257pp. 5⅝ x 8¼. 0-486-63833-2

AN INTRODUCTION TO ORDINARY DIFFERENTIAL EQUATIONS, Earl A. Coddington. A thorough and systematic first course in elementary differential equations for undergraduates in mathematics and science, with many exercises and problems (with answers). Index. 304pp. 5⅜ x 8½. 0-486-65942-9

FOURIER SERIES AND ORTHOGONAL FUNCTIONS, Harry F. Davis. An incisive text combining theory and practical example to introduce Fourier series, orthogonal functions and applications of the Fourier method to boundary-value problems. 570 exercises. Answers and notes. 416pp. 5⅜ x 8½. 0-486-65973-9

COMPUTABILITY AND UNSOLVABILITY, Martin Davis. Classic graduate-level introduction to theory of computability, usually referred to as theory of recurrent functions. New preface and appendix. 288pp. 5⅜ x 8½. 0-486-61471-9

ASYMPTOTIC METHODS IN ANALYSIS, N. G. de Bruijn. An inexpensive, comprehensive guide to asymptotic methods–the pioneering work that teaches by explaining worked examples in detail. Index. 224pp. 5⅜ x 8½ 0-486-64221-6

APPLIED COMPLEX VARIABLES, John W. Dettman. Step-by-step coverage of fundamentals of analytic function theory–plus lucid exposition of five important applications: Potential Theory; Ordinary Differential Equations; Fourier Transforms; Laplace Transforms; Asymptotic Expansions. 66 figures. Exercises at chapter ends. 512pp. 5⅜ x 8½. 0-486-64670-X

INTRODUCTION TO LINEAR ALGEBRA AND DIFFERENTIAL EQUATIONS, John W. Dettman. Excellent text covers complex numbers, determinants, orthonormal bases, Laplace transforms, much more. Exercises with solutions. Undergraduate level. 416pp. 5⅜ x 8½. 0-486-65191-6

RIEMANN'S ZETA FUNCTION, H. M. Edwards. Superb, high-level study of landmark 1859 publication entitled "On the Number of Primes Less Than a Given Magnitude" traces developments in mathematical theory that it inspired. xiv+315pp. 5⅜ x 8½. 0-486-41740-9

CALCULUS OF VARIATIONS WITH APPLICATIONS, George M. Ewing. Applications-oriented introduction to variational theory develops insight and promotes understanding of specialized books, research papers. Suitable for advanced undergraduate/graduate students as primary, supplementary text. 352pp. 5⅜ x 8½.
0-486-64856-7

COMPLEX VARIABLES, Francis J. Flanigan. Unusual approach, delaying complex algebra till harmonic functions have been analyzed from real variable viewpoint. Includes problems with answers. 364pp. 5⅜ x 8½. 0-486-61388-7

AN INTRODUCTION TO THE CALCULUS OF VARIATIONS, Charles Fox. Graduate-level text covers variations of an integral, isoperimetrical problems, least action, special relativity, approximations, more. References. 279pp. 5⅜ x 8½.
0-486-65499-0

COUNTEREXAMPLES IN ANALYSIS, Bernard R. Gelbaum and John M. H. Olmsted. These counterexamples deal mostly with the part of analysis known as "real variables." The first half covers the real number system, and the second half encompasses higher dimensions. 1962 edition. xxiv+198pp. 5⅜ x 8½. 0-486-42875-3

CATASTROPHE THEORY FOR SCIENTISTS AND ENGINEERS, Robert Gilmore. Advanced-level treatment describes mathematics of theory grounded in the work of Poincaré, R. Thom, other mathematicians. Also important applications to problems in mathematics, physics, chemistry and engineering. 1981 edition. References. 28 tables. 397 black-and-white illustrations. xvii + 666pp. 6⅛ x 9¼.
0-486-67539-4

INTRODUCTION TO DIFFERENCE EQUATIONS, Samuel Goldberg. Exceptionally clear exposition of important discipline with applications to sociology, psychology, economics. Many illustrative examples; over 250 problems. 260pp. 5⅜ x 8½.
0-486-65084-7

NUMERICAL METHODS FOR SCIENTISTS AND ENGINEERS, Richard Hamming. Classic text stresses frequency approach in coverage of algorithms, polynomial approximation, Fourier approximation, exponential approximation, other topics. Revised and enlarged 2nd edition. 721pp. 5⅜ x 8½. 0-486-65241-6

INTRODUCTION TO NUMERICAL ANALYSIS (2nd Edition), F. B. Hildebrand. Classic, fundamental treatment covers computation, approximation, interpolation, numerical differentiation and integration, other topics. 150 new problems. 669pp. 5⅜ x 8½. 0-486-65363-3

THREE PEARLS OF NUMBER THEORY, A. Y. Khinchin. Three compelling puzzles require proof of a basic law governing the world of numbers. Challenges concern van der Waerden's theorem, the Landau-Schnirelmann hypothesis and Mann's theorem, and a solution to Waring's problem. Solutions included. 64pp. 5⅜ x 8½.
0-486-40026-3

THE PHILOSOPHY OF MATHEMATICS: AN INTRODUCTORY ESSAY, Stephan Körner. Surveys the views of Plato, Aristotle, Leibniz & Kant concerning propositions and theories of applied and pure mathematics. Introduction. Two appendices. Index. 198pp. 5⅜ x 8½. 0-486-25048-2